TAIWAN ARMY WEAPONS and EQUIPMENT (Including Marine Corps)

WENDELL MINNICK

EDITOR

ISBN: 9798410709873

NOTE TO READER

This material was gathered at Taiwan defense industry exhibitions and military exercises in Taiwan from 2000-2016. All materials provided are for research and educational purposes. It is raw and unvarnished with no analysis. It may very well be the only source on this subject from original sources provided by the Taiwan government and military.

CONTENTS

Collection Notes/Recommended Reading 1
Weapons Brochures 2
Sketches (MND 2021 Report) 354
Media Brochures 364
US Military/Defense Contractors 447
Taiwan Domestic Militia Training 462
Recommended Periodicals 485
About the Editor 508
INDEX 543

COLLECTION NOTES

The editor of this book collected these brochures over a 20-year career covering Taiwan defense/military affairs (2000-2020) for *Jane's Defence Weekly*, *Defense News*, and *Shephard Media*.

RECOMMENDED READING

Easton, Ian. *The Chinese Invasion Threat: Taiwan's Defense and American Strategy in Asia*. Project 2049 Institute, 2017.

Fun Patch: The Republic of China Military Emblem Book (English/Traditional Chinese). January 2022. ISBN: 978-957-43-9589-7. This book is the only military patch book in Taiwan. It only covers Air Force/Navy old and new patches. There are no Army aviation patches.

Illustrated Guide for Weapons and Tactics (Traditional Chinese). Issue 112. August 10, 2020.

Military Weapon Systems of R.O.C. Armed Forces (Traditional Chinese). December 19, 2007 (ROC Year 96). ISBN: 978-986-83872-0-1. ycbook@popularworld.com.

Tsang, Steve, editor. *If China Attacks Taiwan: Military Strategy, Politics and Economics*. Routledge, 2006.

Bryen, Stephen D., and (Ret.) Lt. Gen. Hailston, Earl, editors. *Stopping A Taiwan Invasion: Findings & Recommendations from The Center for Security Policy Panel of Experts*. Self-Published: April 26, 2022. ASIN: B09YV9PB3; ISBN-13: 979-8806319709.

Salary for Military Personnel. 2018 brochure from the Hackers-In-Taiwan Conference (HITCON). Note: US $1 = Taiwan NT $30 (appox.). 1/3.

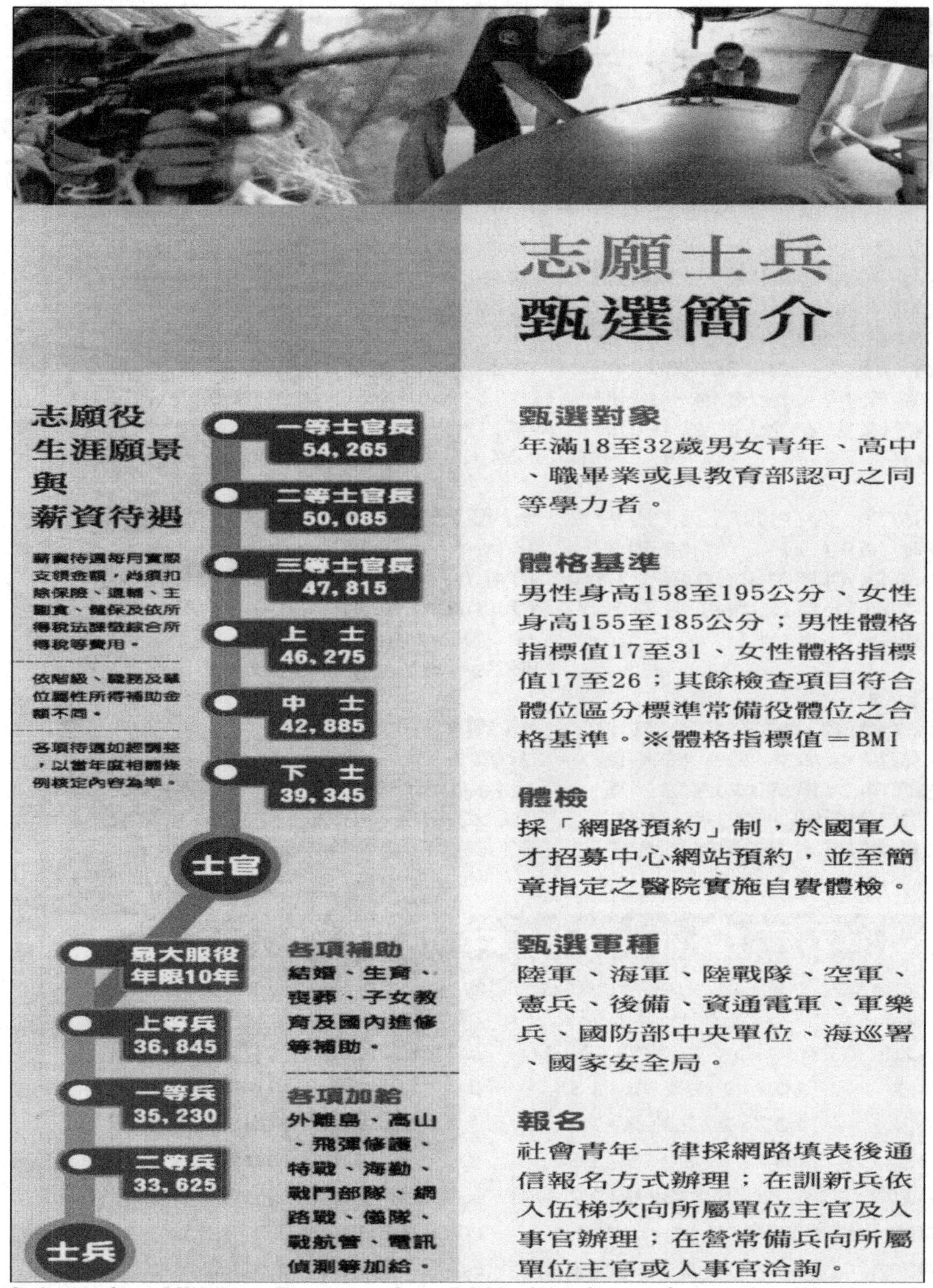

志願士兵
甄選簡介

志願役
生涯願景
與
薪資待遇

薪資待遇每月實際支領金額，尚須扣除保險、退輔、主副食、健保及依所得稅法課徵綜合所得稅等費用。

依階級、職務及單位屬性所得補助金額不同。

各項待遇如經調整，以當年度相關條例核定內容為準。

士官
一等士官長 54,265
二等士官長 50,085
三等士官長 47,815
上士 46,275
中士 42,885
下士 39,345

士兵
最大服役年限10年
上等兵 36,845
一等兵 35,230
二等兵 33,625

各項補助
結婚、生育、喪葬、子女教育及國內進修等補助。

各項加給
外離島、高山、飛彈修護、特戰、海勤、戰鬥部隊、網路戰、儀隊、戰航管、電訊偵測等加給。

甄選對象
年滿18至32歲男女青年、高中、職畢業或具教育部認可之同等學力者。

體格基準
男性身高158至195公分、女性身高155至185公分；男性體格指標值17至31、女性體格指標值17至26；其餘檢查項目符合體位區分標準常備役體位之合格基準。※體格指標值＝BMI

體檢
採「網路預約」制，於國軍人才招募中心網站預約，並至簡章指定之醫院實施自費體檢。

甄選軍種
陸軍、海軍、陸戰隊、空軍、憲兵、後備、資通電軍、軍樂兵、國防部中央單位、海巡署、國家安全局。

報名
社會青年一律採網路填表後通信報名方式辦理；在訓新兵依入伍梯次向所屬單位主官及人事官辦理；在營常備兵向所屬單位主官或人事官洽詢。

Salary for Military Personnel. 2018 brochure from the Hackers-In-Taiwan Conference (HITCON). Note: US $1 = Taiwan NT $30 (appox.). 2/3.

測驗項目

社會青年：智力測驗、口試及體能鑑測（徒手跑步：男性10分鐘30秒內完成1600公尺、女性5分鐘30秒內完成800公尺）。報考軍樂兵人員加測專長測驗；報考國家安全局人員加測專長口試。

在營常備兵(含在訓新兵)智力測驗成績須90分以上；體測依國軍體能鑑測基準均須達全項合格；在訓新兵比照社會青年體測基準。

遴選分發

未役社會青年、補充兵列管或替代役備役錄取人員，於基礎訓練完訓前14日完成遴選（配合民專軍用原則，依用人單位需求條件辦理）及分發（配合民專軍用，依各單位戶籍地區劃分、外島及戰鬥部隊補充需求辦理）作業。

遴選或分發至外島地區服役者，服役期滿2年後（東引地區1年）得依個人意願申請調整至(鄰近)戶籍地區服役。

服役規定

基礎訓練成績合格者，自核定日起服志願士兵現役４年。服役期滿得依國防軍事需要及志願，按規定申請自願繼續留營服役，最大年限10年。

薪資待遇

志願士兵生效日起，支領二等兵待遇33,625元、一等兵待遇35,230元、上等兵待遇36,845元，另有各項補助與加給。

107年度志願士兵甄選期程

梯次	報名日期	考試日期	入營報到
第1梯次	106/12/6-12/20	107/2/3	107/3/7
第2梯次	106/12/27-107/1/10	107/3/10	107/4/11
第3梯次	107/1/17-2/13	107/3/24	107/4/25
第4梯次	107/2/22-3/29	107/5/26	107/7/11
第5梯次	107/4/10-5/16	107/7/14	107/8/15
第6梯次	107/5/23-6/27	107/8/25	107/9/26
第7梯次	107/7/4-8/22	107/10/20	107/11/14
第8梯次	107/8/29-10/3	107/11/24	107/12/19

Salary for Military Personnel. 2018 brochure from the Hackers-In-Taiwan Conference (HITCON). Note: US $1 = Taiwan NT $30 (appox.). 3/3.

ARMAMENTS BUREAU

The Armaments Bureau logo is expected to change in the near future as there have been political calls to remove the image of mainland China from military patches.

Under the Armaments Bureau is the Materiel Production Center (MPC) with numerous arsenals and factories.

MPC
No. 165, Kunyang Street
Nangang District
Taipei City, 11551
(886) 2-2785-4610
mpc.mnd@msa.hinet.net

202nd Arsenal

The 202nd Arsenal: artillery, munitions, bullet proof vests/helmets.

205th Arsenal
Kaohsiung City
Develops and produces handguns, rifles, machine guns, bullets, boots, tents. Also involved in unmanned quadcopter development and production.

209th Arsenal
Responsible for development of CM-32 Armored Vehicle. Jili, Nantou County.

401/402 Factory
Night Vision.

Materiel Production Center (MPC). Armaments Bureau. Ministry of National Defense. 2013 Taipei Aerospace and Defense Technology Exhibition (TADTE). Brochure 1/12.

Materiel Production Center (MPC). Armaments Bureau. Ministry of National Defense. 2013 Taipei Aerospace and Defense Technology Exhibition (TADTE). Brochure 2/12.

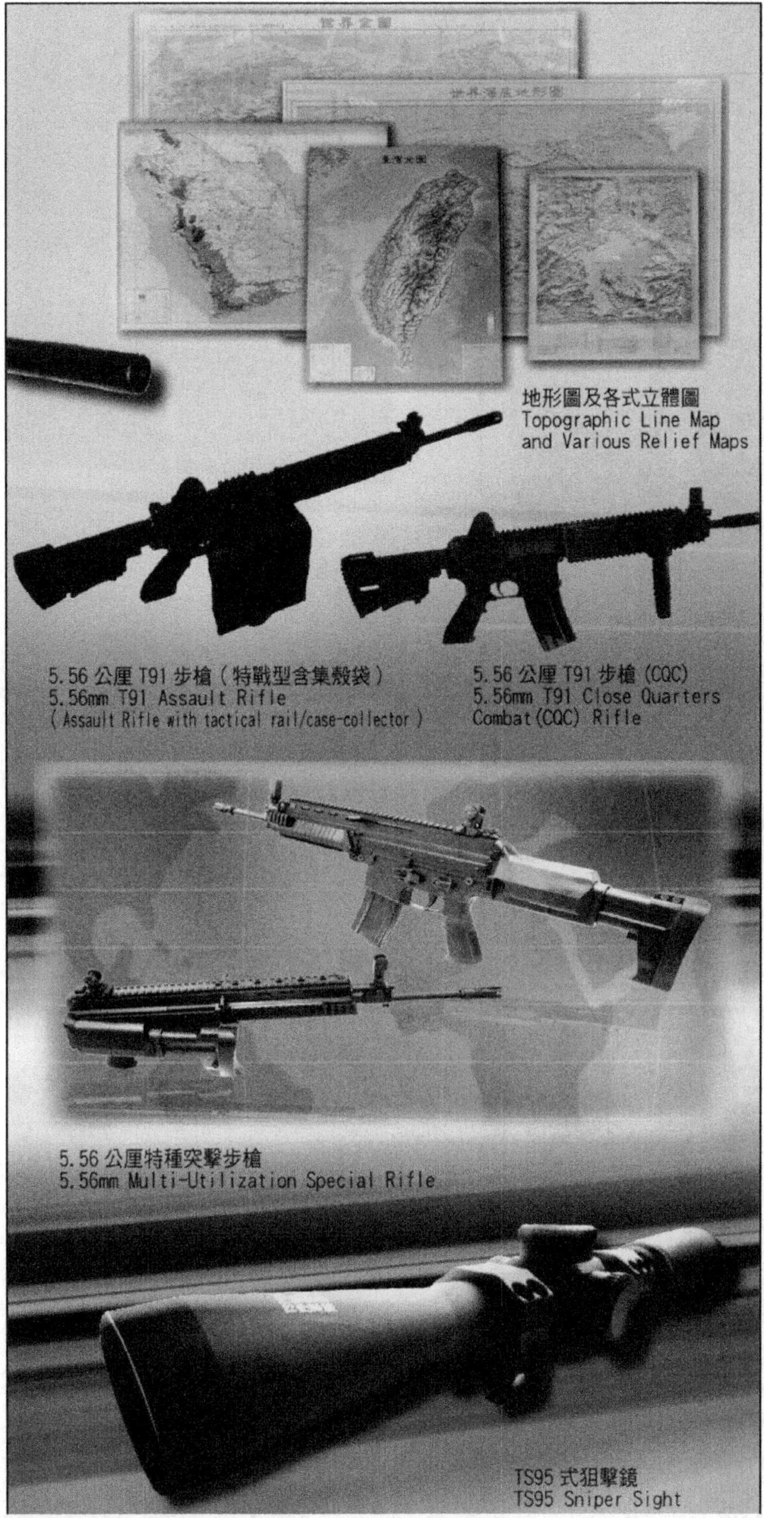

Materiel Production Center (MPC). Armaments Bureau. Ministry of National Defense. 2013 Taipei Aerospace and Defense Technology Exhibition (TADTE). Brochure 3/12.

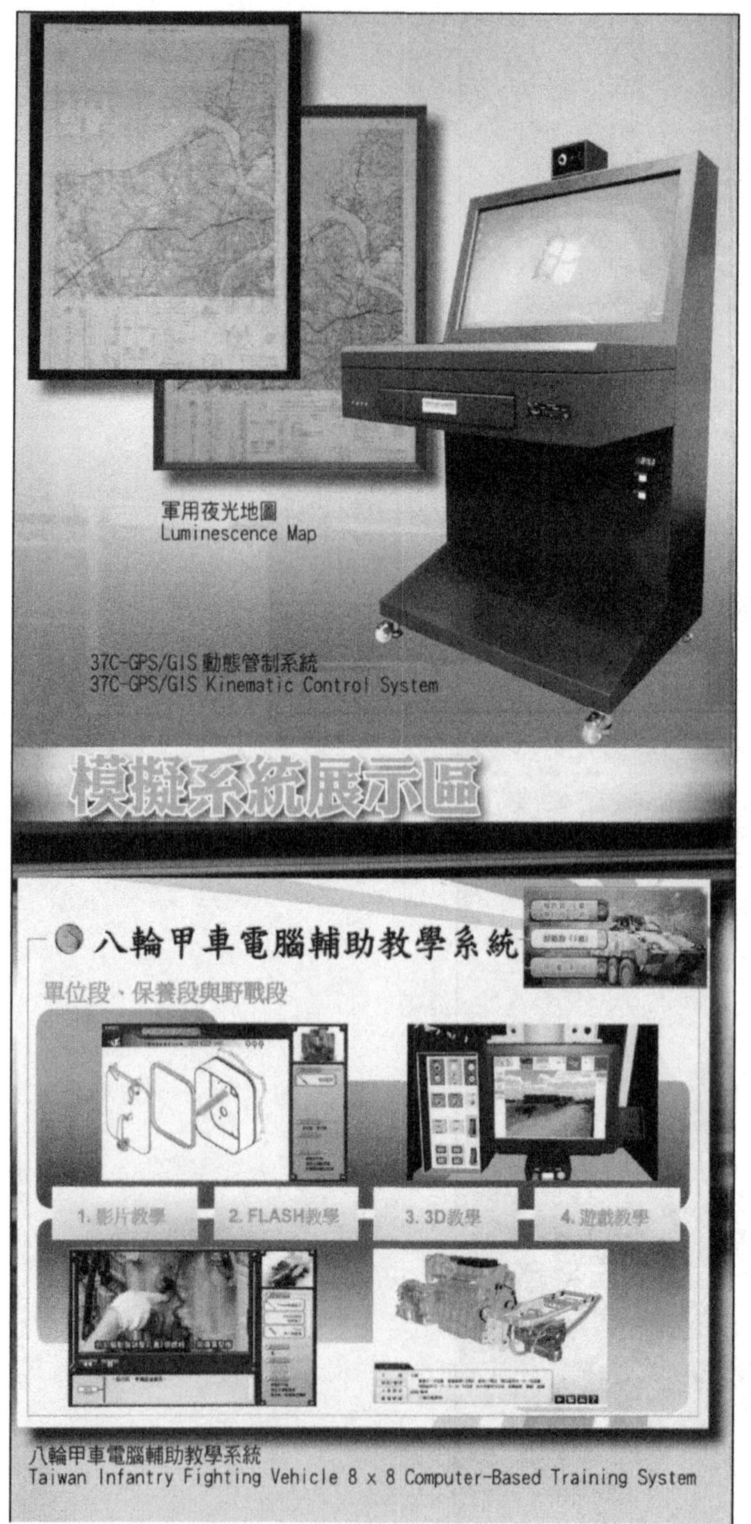

Materiel Production Center (MPC). Armaments Bureau. Ministry of National Defense. 2013 Taipei Aerospace and Defense Technology Exhibition (TADTE). Brochure 4/12.

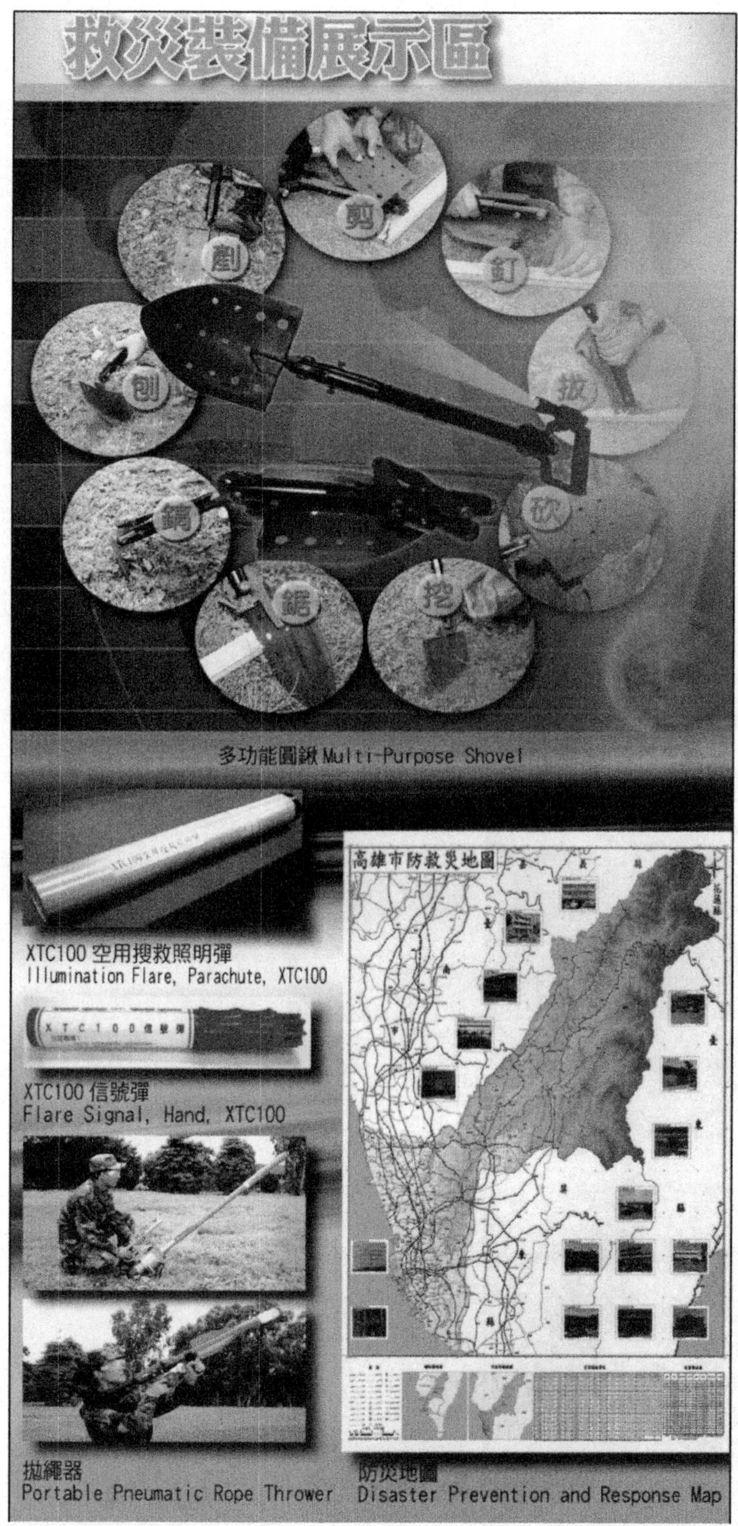

Materiel Production Center (MPC). Armaments Bureau. Ministry of National Defense. 2013 Taipei Aerospace and Defense Technology Exhibition (TADTE). Brochure 5/12.

Materiel Production Center (MPC). Armaments Bureau. Ministry of National Defense. 2013 Taipei Aerospace and Defense Technology Exhibition (TADTE). Brochure 6/12.

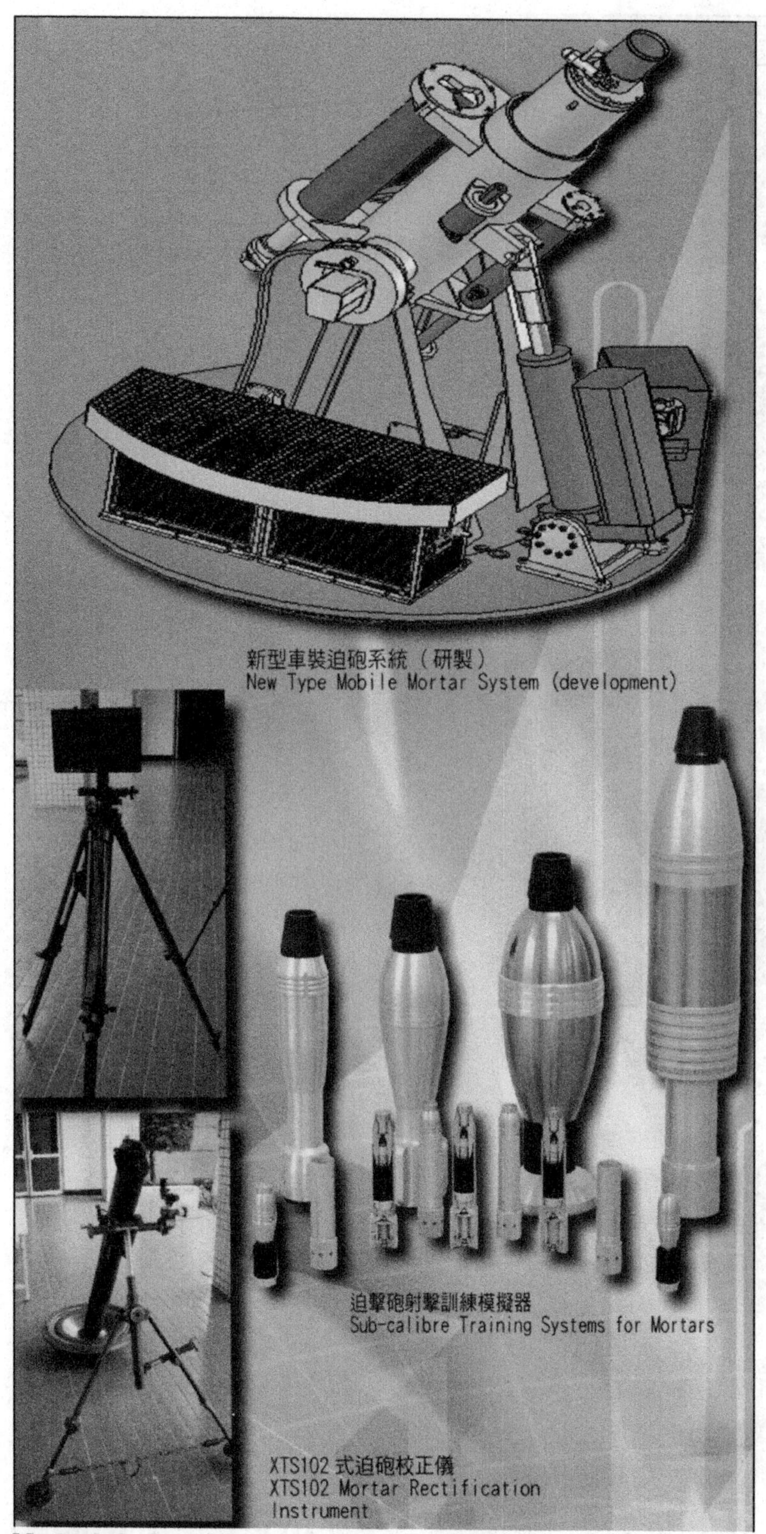

Materiel Production Center (MPC). Armaments Bureau. Ministry of National Defense. 2013 Taipei Aerospace and Defense Technology Exhibition (TADTE). Brochure 7/12.

Materiel Production Center (MPC). Armaments Bureau. Ministry of National Defense. 2013 Taipei Aerospace and Defense Technology Exhibition (TADTE). Brochure 8/12.

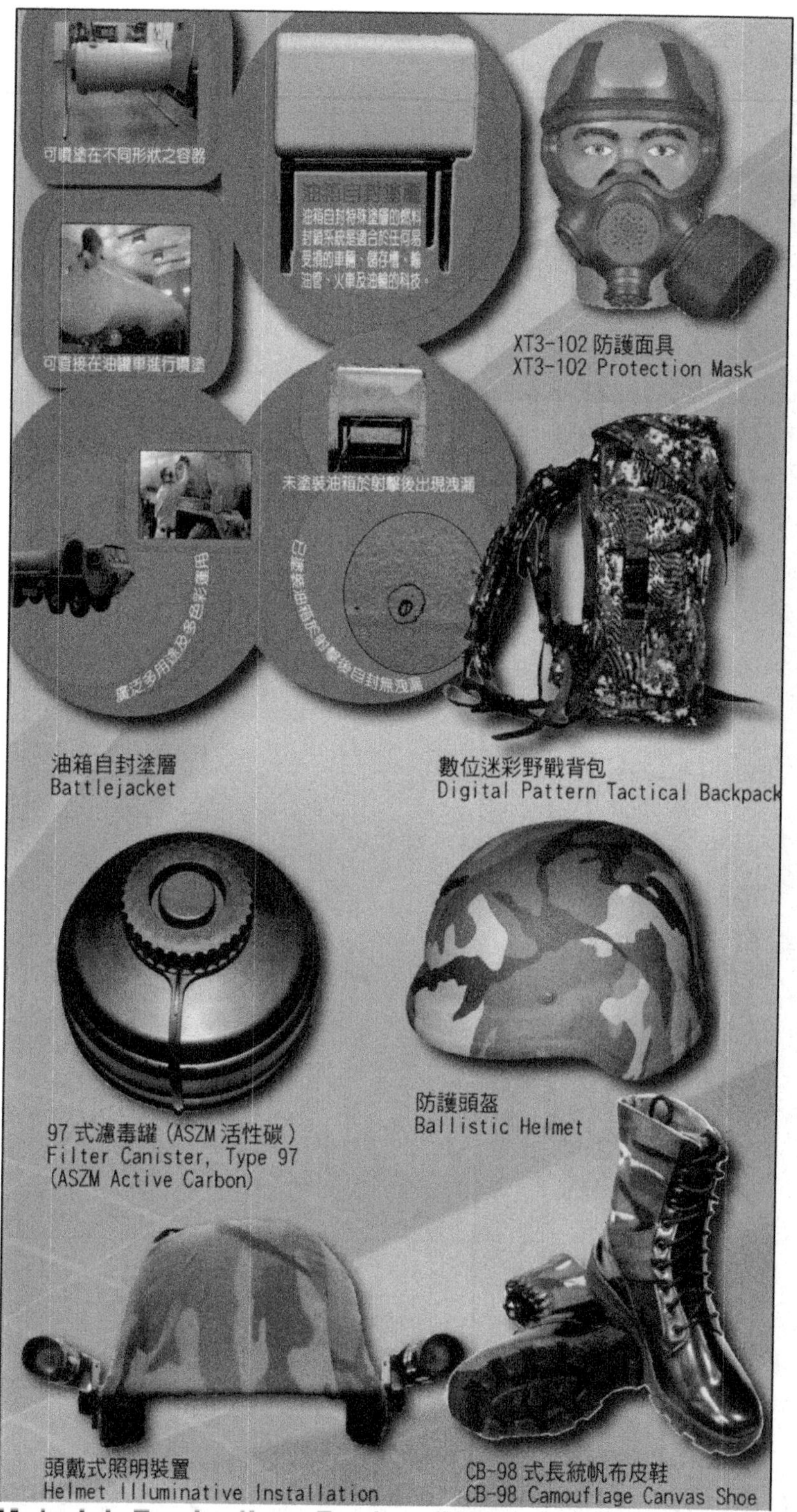

Materiel Production Center (MPC). Armaments Bureau. Ministry of National Defense. 2013 Taipei Aerospace and Defense Technology Exhibition (TADTE). Brochure 9/12.

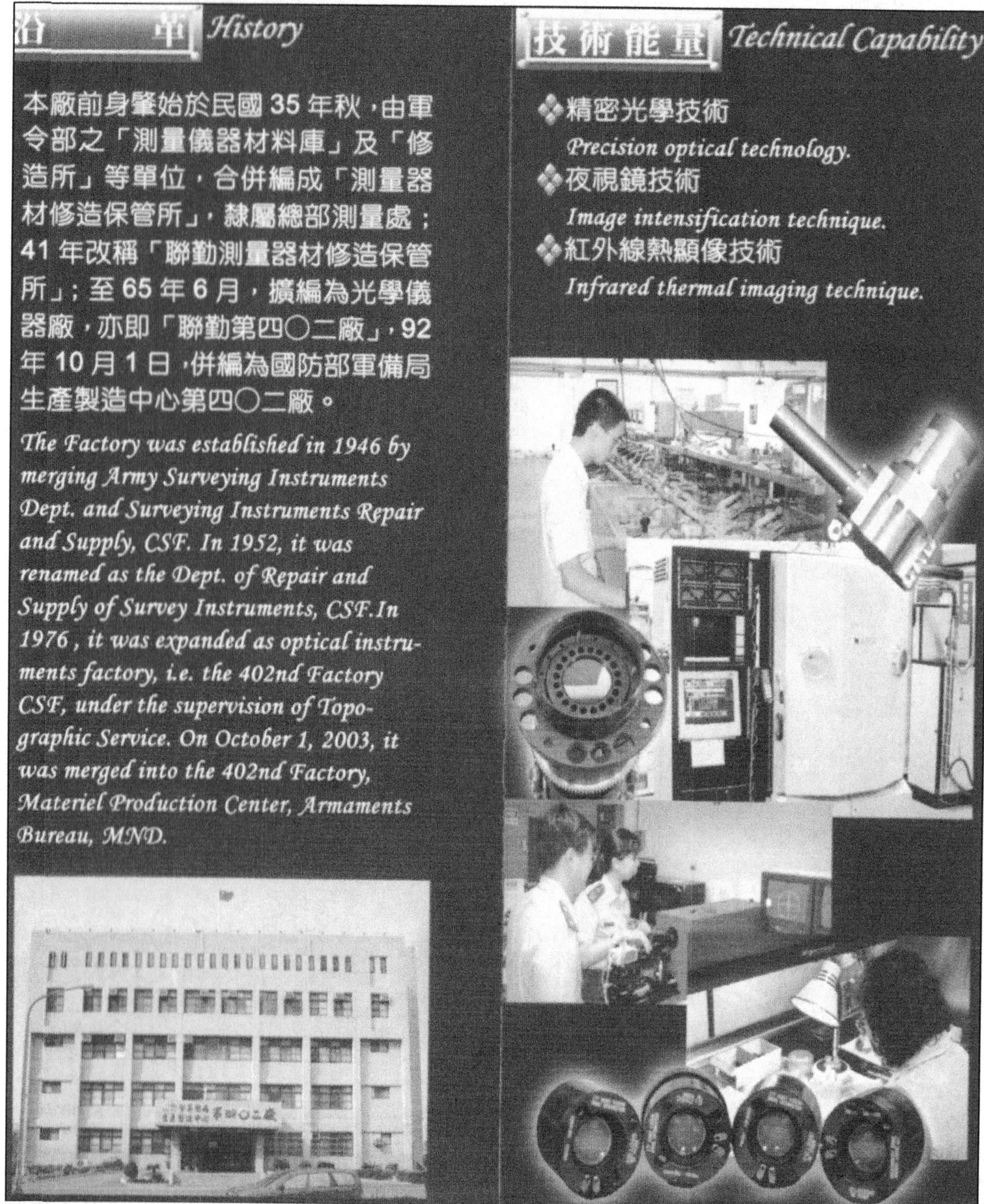

Materiel Production Center (MPC). 402nd Factory Brochure. Unidentified Taipei Aerospace and Defense Technology Exhibition (TADTE). Brochure 2/3.

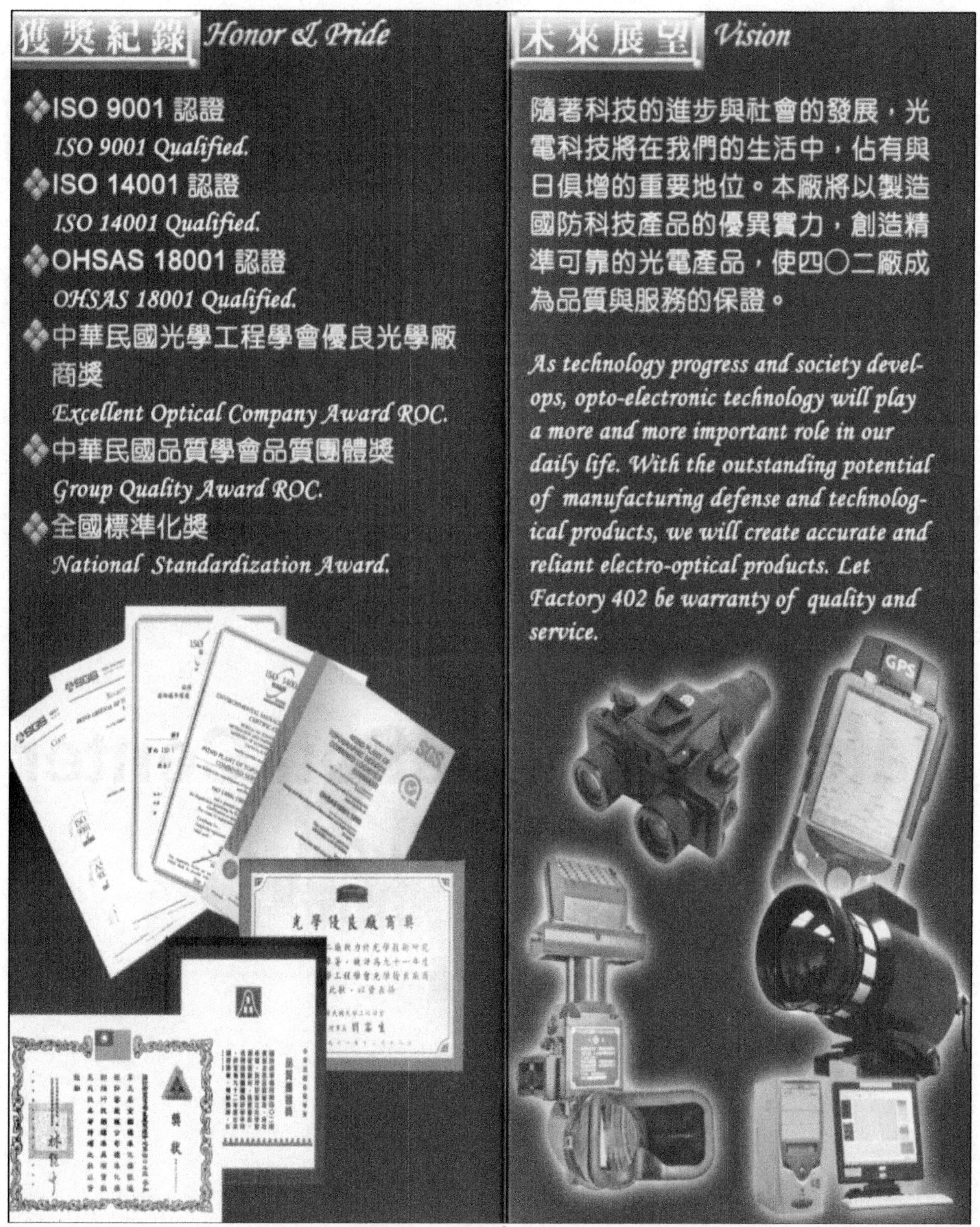

Materiel Production Center (MPC). 402nd Factory Brochure. Unidentified Taipei Aerospace and Defense Technology Exhibition (TADTE). Brochure 3/3.

VEHICLES

Materiel Production Center (MPC)

Materiel Production Center (MPC). 209th Arsenal. Clouded Leopard 8x8 Mortar Carrier. 2018 Defence Services Asia (Malaysia). Brochure 1/2.

Cloud-Leopard
8X8 Mortar Carrier

Description:

The prototype of Mortar Carrier is developed by Cloud-Leopard Infantry Fighting Vehicle in order to enhance the vehicle off-road and comfortable performance.

Suspension system is replaced by hydro-pneumatic and the four-wheel steering upgraded to six-wheel steering to reduce the radius of steering. The horizontal sliding hatch is easy to operate, and the omnidirectional surveillance system can let the crew control the outside situation of the whole vehicle. The streamlined shape can raise the survival rate at the battlefield.

Specification:

- **Combat weight:** 24 ton
- **Horse power:** 450hp
- **Max. range:** 600 Km
- **Max. speed:** 100 km/h
- **Slope:** 60%
- **Side slope:** 30%
- **Vertical obstacle:** 0.7 m
- **Trench:** 2 m
- **Fording:** 1 m
- **Radius of gyration:** 9 m

Materiel Production Center (MPC). 209th Arsenal. Clouded Leopard 8x8 Mortar Carrier. 2018 Defence Services Asia (Malaysia). Brochure 2/2.

國防部軍備局生產製造中心第二〇九廠

◆展品名稱：輪型迫砲車載具暨環車監控系統

輪型迫砲車載具暨環車監控裝置圖

◆技術功能與特色

以輪型迫砲車載具為展示主體，搭載以16個鏡頭構成之全周無縫視野，支援艙門關閉時的駕駛行為。另外車長可經由多重視窗顯示螢幕清楚掌握行軍環境狀況，快速下達戰術指令。

- 駕駛可清楚掌握行軍時環境狀況。
- 車輛周圍盲點消除，使車內人員可完整掌握車輛周邊資訊。
- 車長可經由多重視窗顯示螢幕快速下達戰術指令。
- 鏡頭保護蓋可防止鏡頭遭外力破壞或環境汙損。

◆展品名稱：八輪甲車電腦輔助教學系統

八輪甲車電腦輔助教學系統軟體示意圖

◆技術功能與特色

本 CBT 系統軟體以滿足教學者之教學與學習者人機介面需求為要。利用各式圖像及動畫互動技術，配合系統介紹旁白，達到吸引學員注意力、提升學習興趣，並加深學習者的記憶，強化學習效果。

- 系統充分運用動畫、影片、3D 互動及遊戲開發等技術建構互動式教學教具。
- 依照八輪甲車技術手冊建構操作、保養及維修等 3D 虛擬實境互動學習場景。
- 教學不受天候、場地及時間限制，且符合人因工程人機介面設計需求。

國防部軍備局生產製造中心第209廠
The 209th Plant, Materiel Production Center, Armament Bureau, M.N.D.

Materiel Production Center (MPC). 209th Arsenal. Clouded Leopard Armament and Equipment. Unidentified Taipei Aerospace and Defense Technology Exhibition (TADTE). 1/1 Brochure.

國防部軍備局生產製造中心第二0九廠

◆展品名稱：遙控式輕型戰鬥載具系統

戰鬥偵察機器人裝置圖

多功能爆裂物移除機器人裝置圖

◆技術功能與特色

遙控式輕型戰鬥載具系統，展示輕型戰鬥模組與爆裂物移除機械夾爪系統等兩組系統，以電力驅動通用履帶載具掛載應用模組，支援戰場攻堅、搜索、爆裂物(不明物體)排除等高危險性任務執行，有效降低人員損傷，提高任務執行效率駕駛可清楚掌握行軍時環境狀況。

● 戰鬥偵察機器人：
搭載各式口徑機槍與反裝甲武器，作為戰場火力平台，具備低姿勢、低噪音、低熱散等特性，可運用於搜索、偵查監視及城鎮或複雜地形目標獲取，協助士兵攻堅，減少戰鬥人員傷亡。

● 多功能爆裂物移除機器人：
配備多功能爆裂物移除機械手臂，使操作人員在安全環境下，移除爆裂物、危險化學物質、輻射物質及高危險汙染物體。

國防部軍備局生產製造中心第209廠
The 209th Plant, Materiel Production Center, Armament Bureau, M.N.D.

Materiel Production Center (MPC). 209th Arsenal. Robotic Vehicles. Unidentified Taipei Aerospace and Defense Technology Exhibition (TADTE). 1/1 Brochure.

GUNS

Materiel Production Center (MPC)

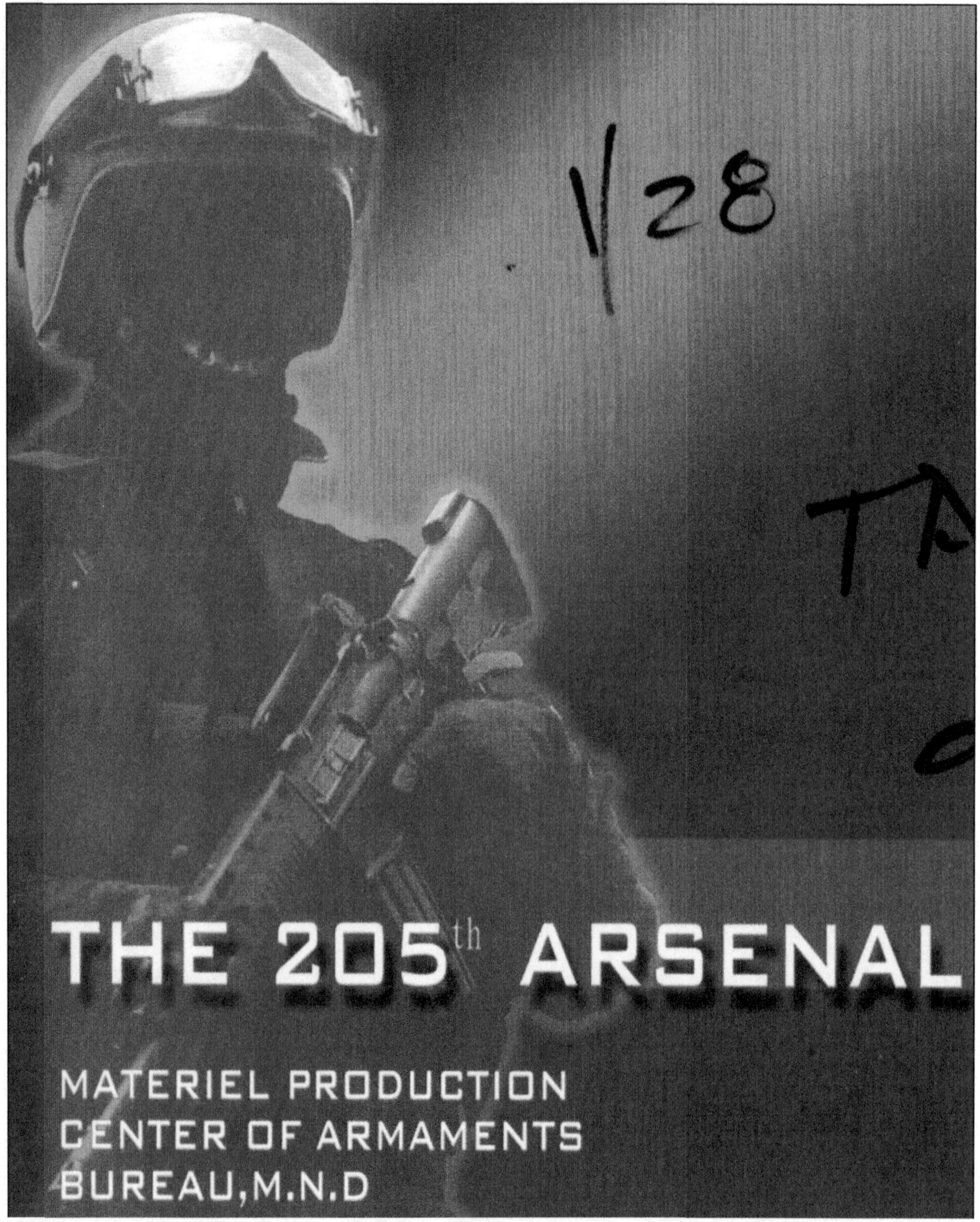

Materiel Production Center (MPC). 205th Arsenal. 2015 Taipei Aerospace and Defense Technology Exhibition (TADTE). Brochure 1/5.

Materiel Production Center (MPC). 205th Arsenal. Timeline. 2015 Taipei Aerospace and Defense Technology Exhibition (TADTE). Brochure 2/5.

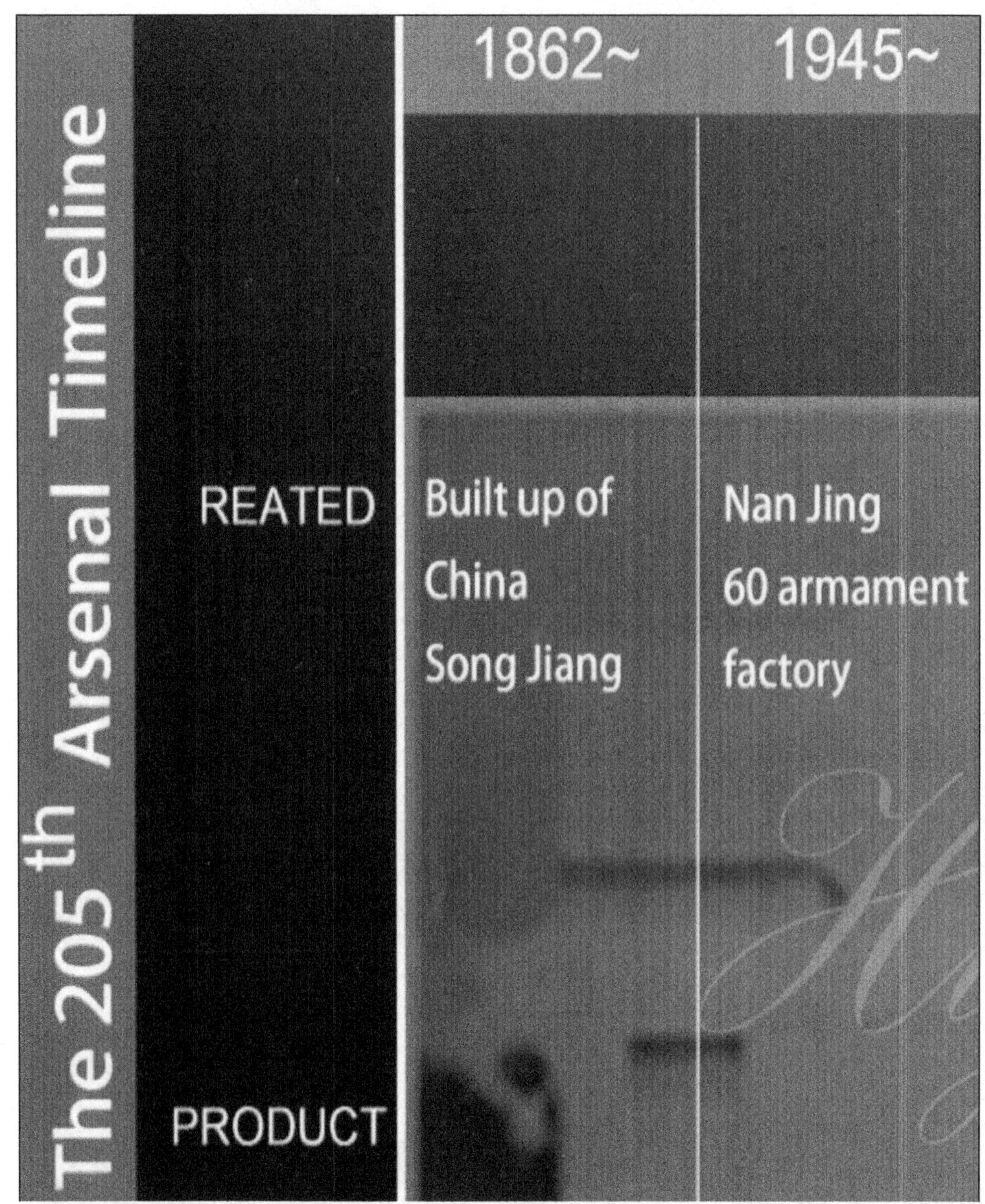

Materiel Production Center (MPC). 205th Arsenal. Timeline. 2015 Taipei Aerospace and Defense Technology Exhibition (TADTE). Brochure 3/5 (CLOSE-UP).

1948~	1976~
Moved to Taiwan Kaoshiung 60 armament factory	The 205th Arsenal Joint Logistics
0.45" T51 Pistol	9mm T75 Pistol
9mm T39 Submachine Gun	9mm T75K1 Pistol
7.62mm T57 Rifle	9mm T77 Submachine Gun
7.62mm T57 Machine Gun	5.56mm T65 Rifle
7.62mm T41 Machine Gun	5.56mm T65K2 Rifle
	5.56mm T86 Rifle
	5.56mm T75 Machine Gun
	7.62mm T74 Machine Gun
	40mm T85 Grenade Launcher
	20mm T75 Cannon

Materiel Production Center (MPC). 205th Arsenal. Timeline. 2015 Taipei Aerospace and Defense Technology Exhibition (TADTE). Brochure 4/5 (CLOSE-UP).

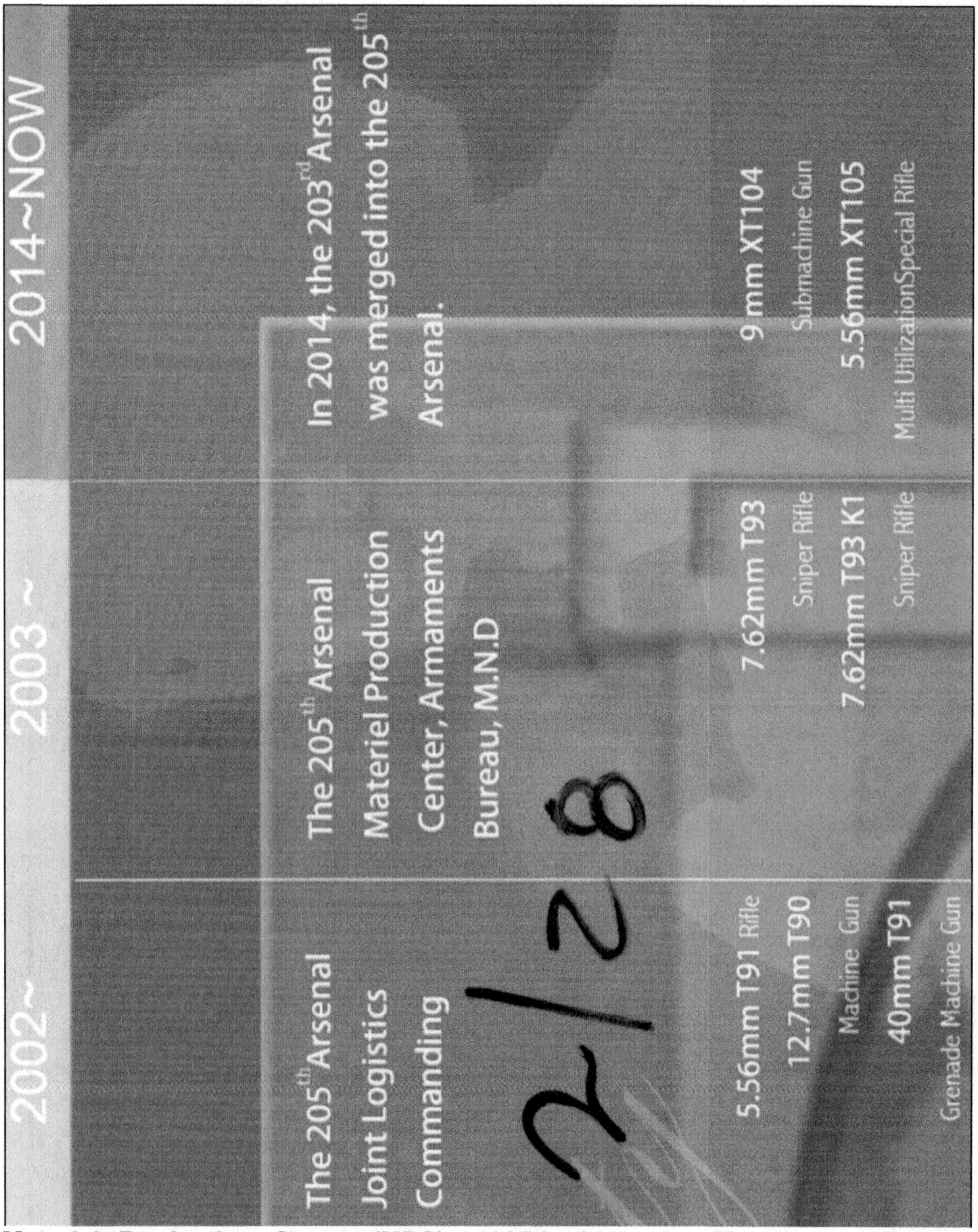

Materiel Production Center (MPC). 205th Arsenal. Timeline. 2015 Taipei Aerospace and Defense Technology Exhibition (TADTE). Brochure 5/5 (Close-Up).

Materiel Production Center (MPC). 205th Arsenal. 9mm T75 Pistol. 2015 Taipei Aerospace and Defense Technology Exhibition (TADTE). Note: Manufactured under licensed by Beretta for the M92FS Beretta/M9. Brochure 1/2.

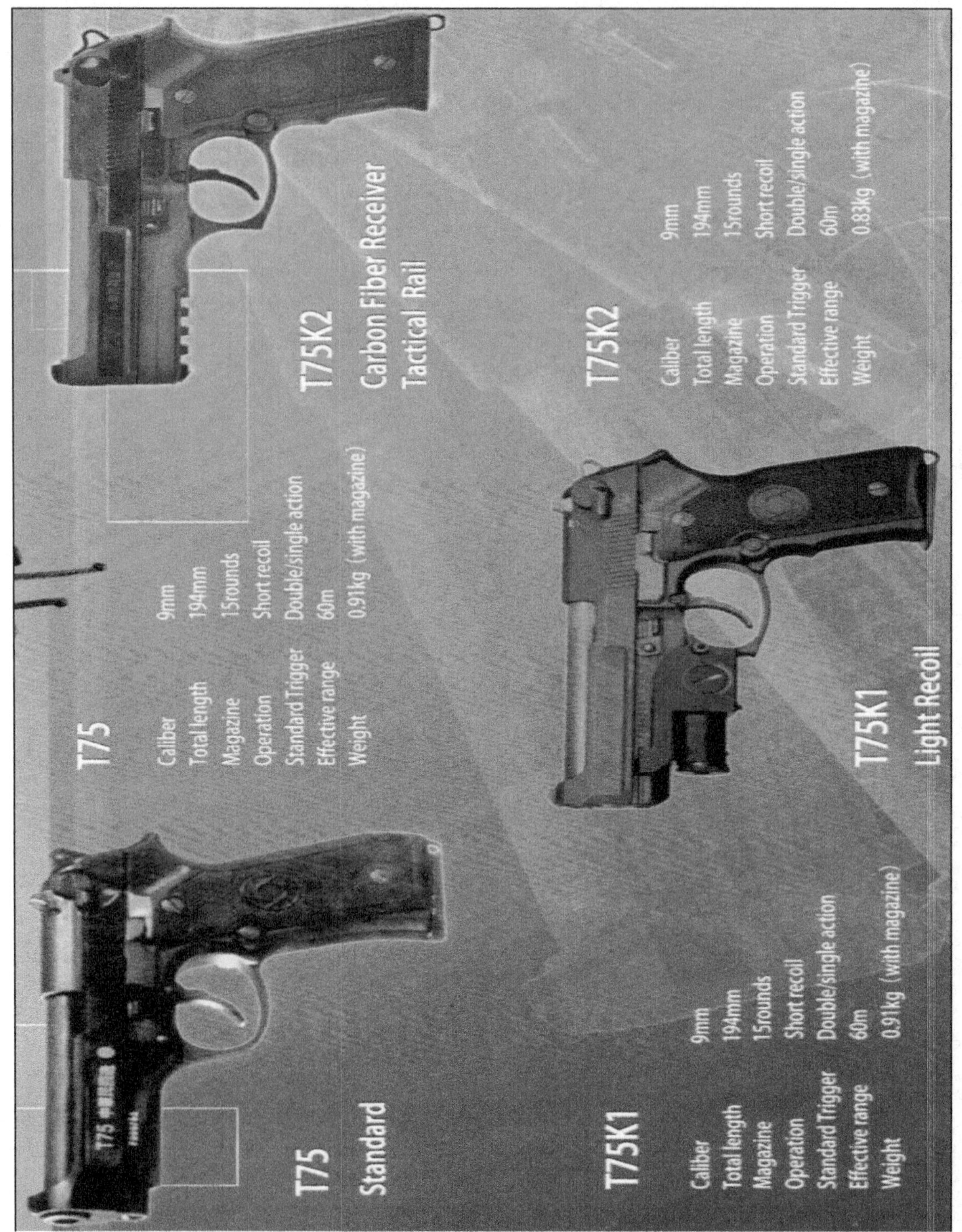

Materiel Production Center (MPC). 205th Arsenal. 9mm T75 Pistol. 2015 Taipei Aerospace and Defense Technology Exhibition (TADTE). Note: Manufactured under licensed by Beretta for the M92FS Beretta/M9. Brochure 2/2.

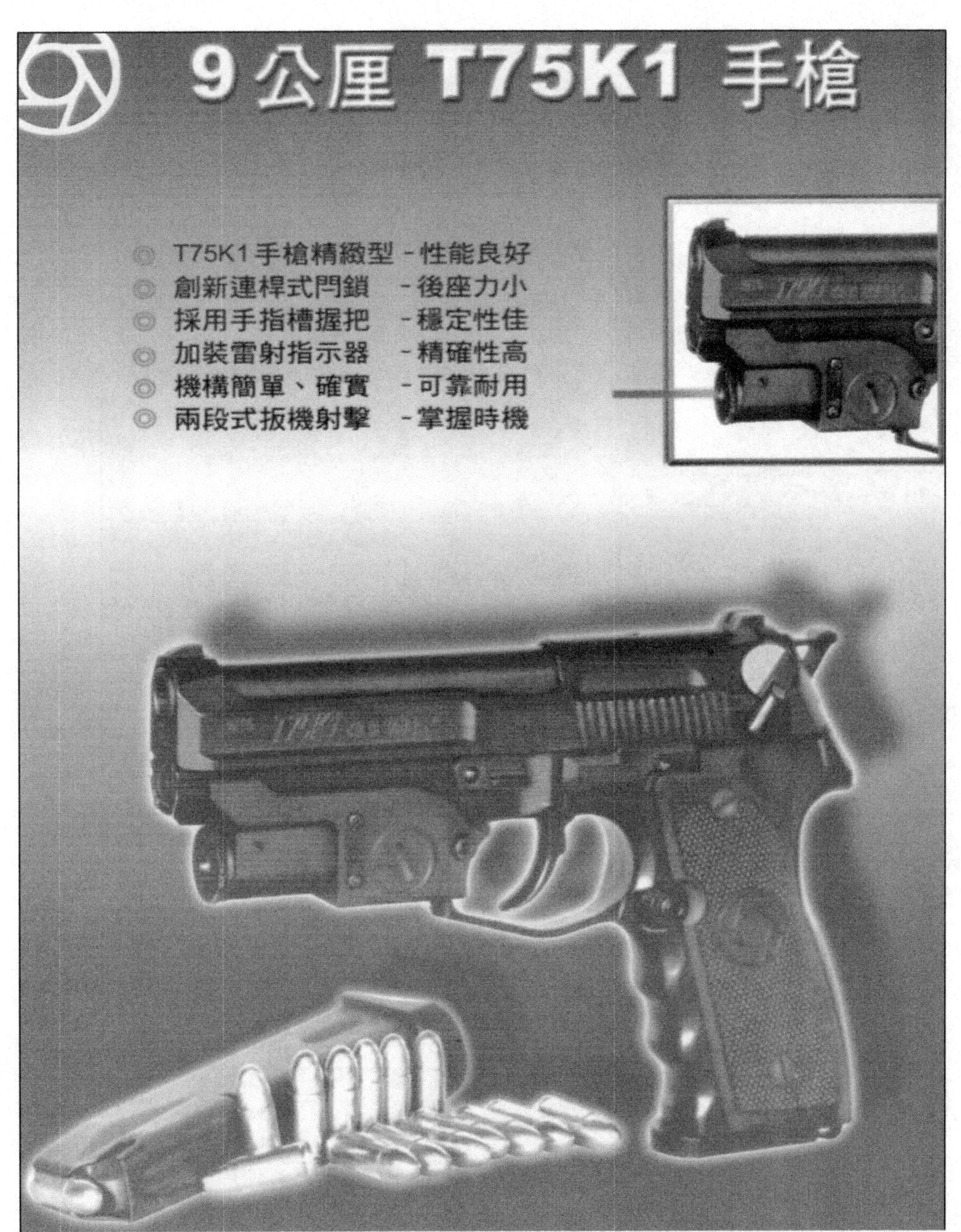

Materiel Production Center (MPC). 205th Arsenal. 9mm T75K1 Pistol. 2011 Taipei Aerospace and Defense Technology Exhibition (TADTE). Note: Manufactured under licensed by Beretta for the M92FS Beretta/M9. Brochure 1/3.

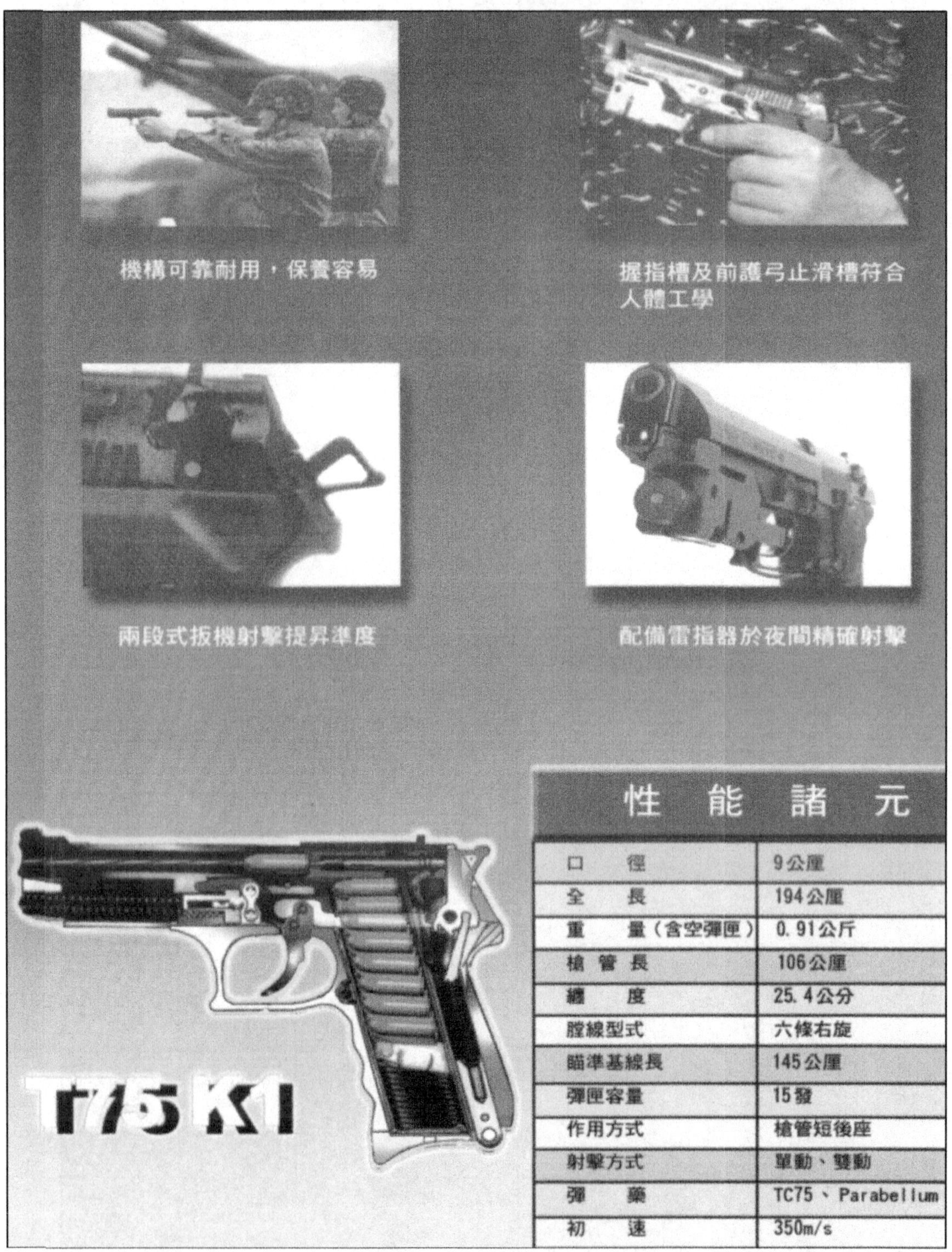

性能諸元	
口　徑	9公厘
全　長	194公厘
重　量（含空彈匣）	0.91公斤
槍管長	106公厘
纏　度	25.4公分
膛線型式	六條右旋
瞄準基線長	145公厘
彈匣容量	15發
作用方式	槍管短後座
射擊方式	單動、雙動
彈　藥	TC75、Parabellum
初　速	350m/s

Materiel Production Center (MPC). 205th Arsenal. 9mm T75K1 Pistol. 2011 Taipei Aerospace and Defense Technology Exhibition (TADTE). Note: Manufactured under licensed by Beretta for the M92FS Beretta/M9. Brochure 2/3.

性　能　諸　元	
口　徑	9公厘
全　長	194公厘
重　量（含空彈匣）	0.91公斤
槍管長	106公厘
纏　度	25.4公分
膛線型式	六條右旋
瞄準基線長	145公厘
彈匣容量	15發
作用方式	槍管短後座
射擊方式	單動、雙動
彈　藥	TC75、Parabellum
初　速	350m/s

Materiel Production Center (MPC). 205th Arsenal. 9mm T75K1 Pistol. 2011 Taipei Aerospace and Defense Technology Exhibition (TADTE). Note: Manufactured under licensed by Beretta for the M92FS Beretta/M9. Brochure 3/3.

Materiel Production Center (MPC). 205th Arsenal. 9mm T97 Handgun. 2011 Taipei Aerospace and Defense Technology Exhibition (TADTE). Note: Based on the Beretta 92. Brochure 1/3.

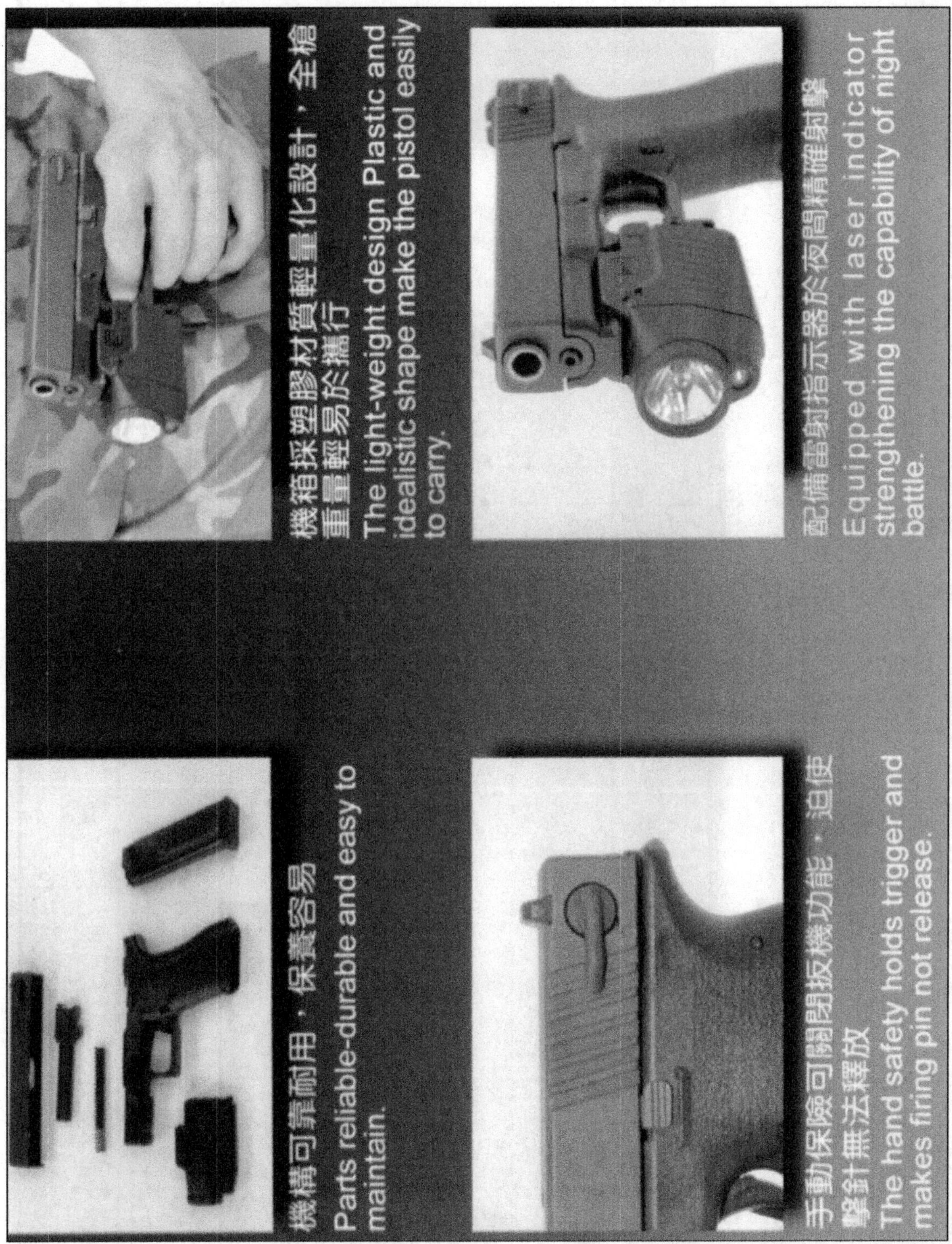

Materiel Production Center (MPC). 205th Arsenal. 9mm T97 Handgun. 2011 Taipei Aerospace and Defense Technology Exhibition (TADTE). Based on the Beretta 92. Brochure 2/3.

性能諸元 TECHNICAL DATA	
彈藥 Cartridge	9X19公厘 9X19 mm
全長 Length	175. 5公厘 175. 5 mm
重量 Weight (unloaded)	0. 676公斤 0. 676 kg
槍管長 Barrel Length	102公厘 102 mm
膛線 Rifling	六條右旋，254公厘/轉 6 grooves, rh, 1 turn in 254mm
瞄準基線長 Sight Radius	153公厘 153 mm
彈匣容量 Magazine Capacity	10發 10 rounds
作用方式 Operation	槍管短後座 Short recoil
射擊方式 Firing selection	擊針撞擊 Firing pin action

Materiel Production Center (MPC). 205th Arsenal. 9mm T97 Handgun. 2011 Taipei Aerospace and Defense Technology Exhibition (TADTE). Based on the Beretta 92. **Brochure 3/3.**

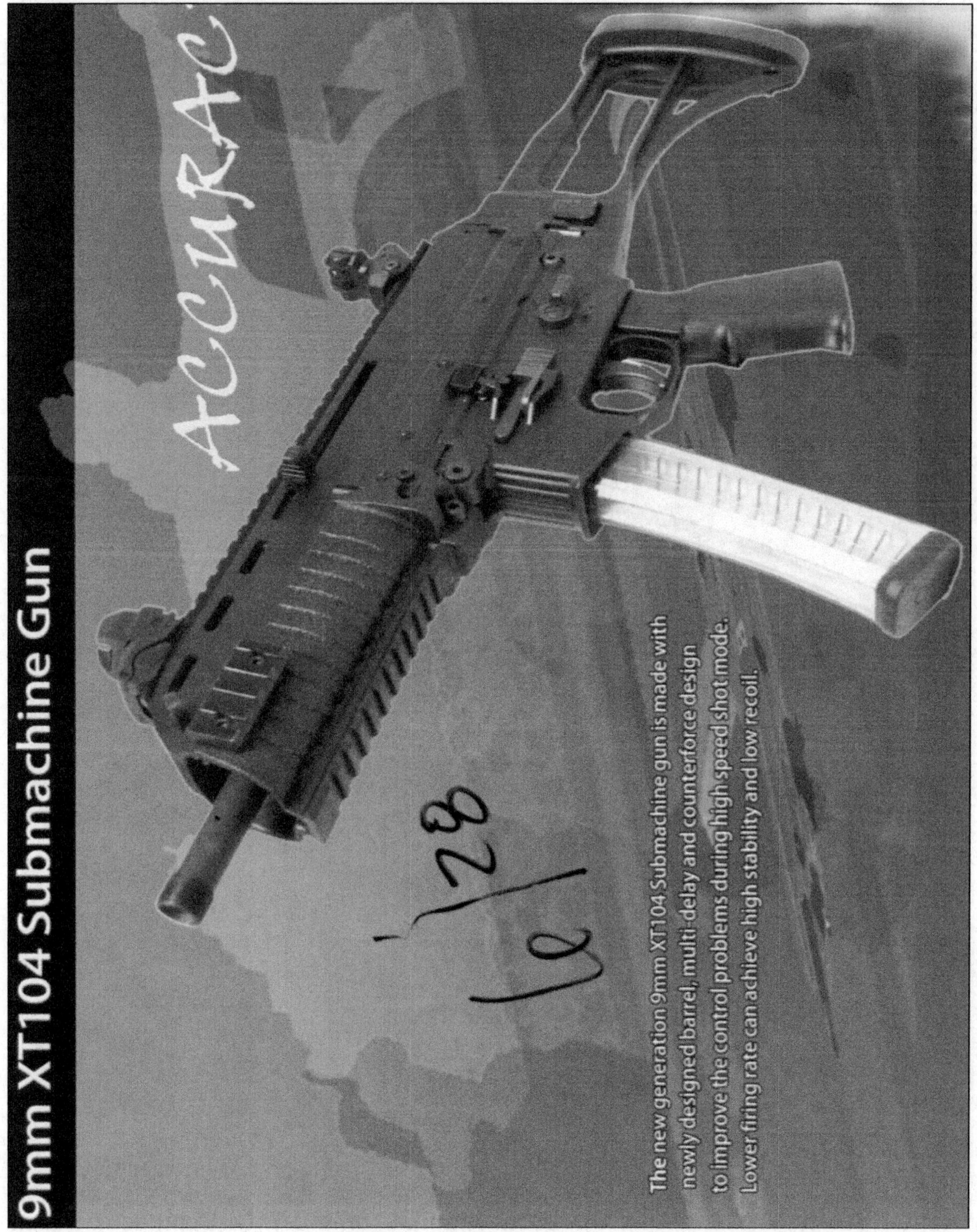

Materiel Production Center (MPC). 205th Arsenal. 9mm XT104 Submachine Gun. 2015 Taipei Aerospace and Defense Technology Exhibition (TADTE). Brochure 1/2.

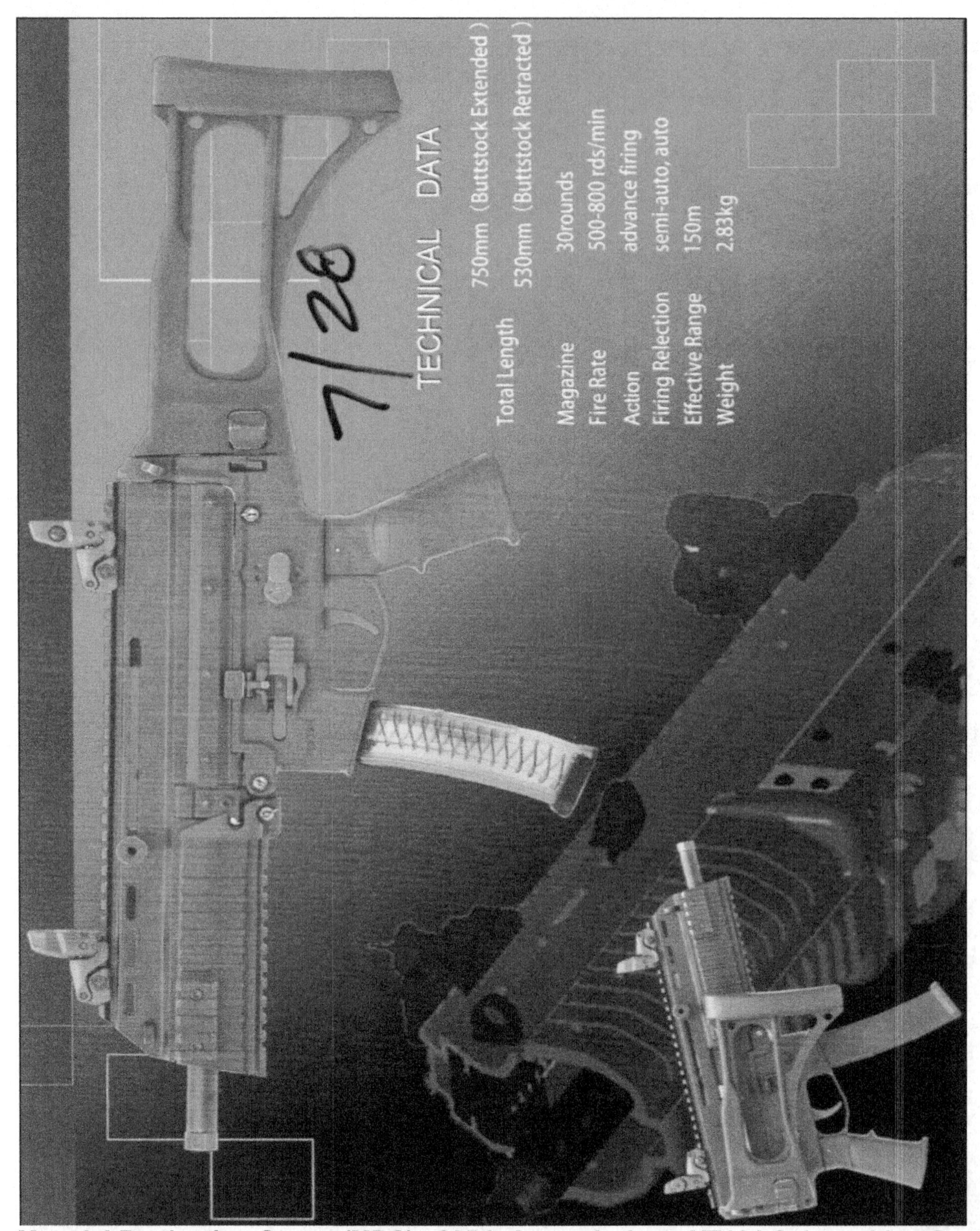

Materiel Production Center (MPC). 205th Arsenal. 9mm XT104 Submachine Gun. 2015 Taipei Aerospace and Defense Technology Exhibition (TADTE). Brochure 2/2.

Materiel Production Center (MPC). 205th Arsenal. 5.56mm XT105 Multi-Utilization Special Rifle. 2015 Taipei Aerospace and Defense Technology Exhibition (TADTE). Brochure 1/2.

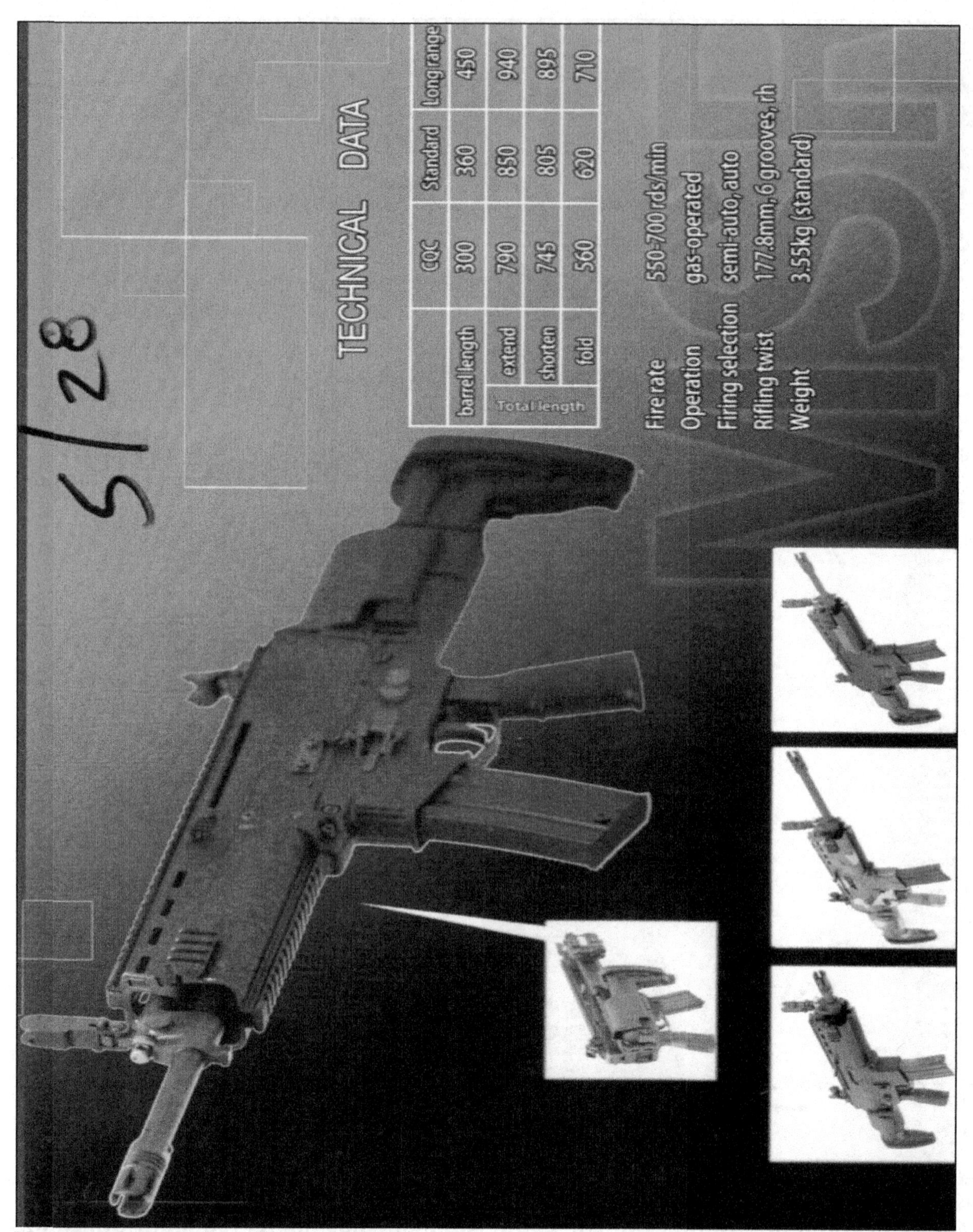

Materiel Production Center (MPC). 205th Arsenal. 5.56mm XT105 Multi-Utilization Special Rifle. 2015 Taipei Aerospace and Defense Technology Exhibition (TADTE). Brochure 2/2.

Materiel Production Center (MPC). 205th Arsenal. Rifle Shooting Simulator System. 2015 Taipei Aerospace and Defense Technology Exhibition (TADTE). Brochure 1/1.

5.56mm T65K2 Rifle
The T65K2 5.56mm assault rifle was developed to replace the US original 7.62mm M14 rifles in 1976. The T65K2 assault rifle is gas operated and selective fired weapon. Its barrel is an advanced design with 7-inch twist of rifling. With the new feature of three shots, it can add high explosive launcher and other aiming scopes to enhance combat capabilities.

Material Production Center. 205th Arsenal. 5.56mm T65K2 Rifle. 2015 Taipei Aerospace and Defense Technology Exhibition (TADTE). Brochure 1/2.

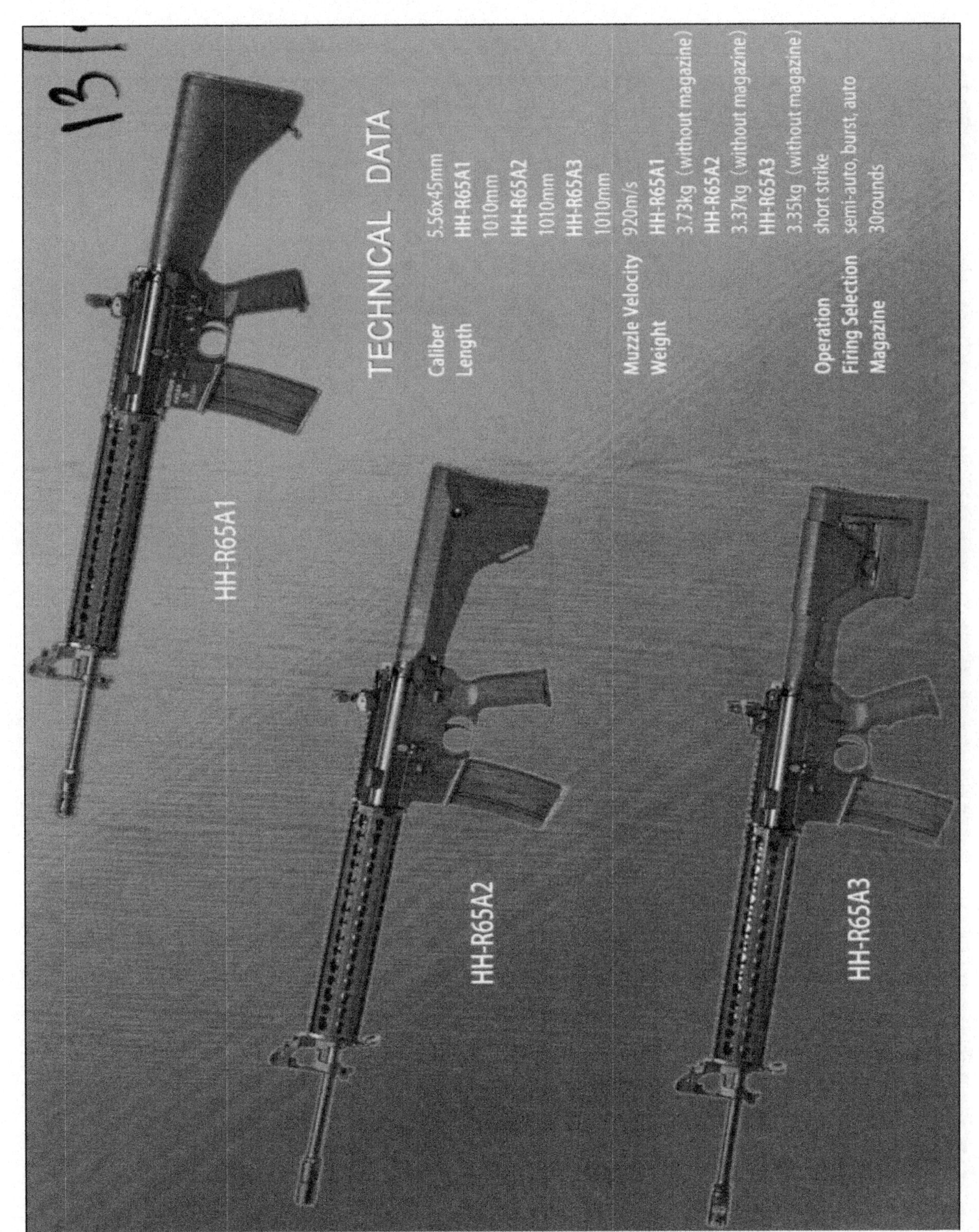

Materiel Production Center (MPC). 205th Arsenal. 5.56mm T65K2 Rifle. 2015 Taipei Aerospace and Defense Technology Exhibition (TADTE). Brochure 2/2.

Materiel Production Center (MPC). 205th Arsenal. 5.56mm T86 Rifle. 2011 Taipei Aerospace and Defense Technology Exhibition (TADTE). Brochure 1/2.

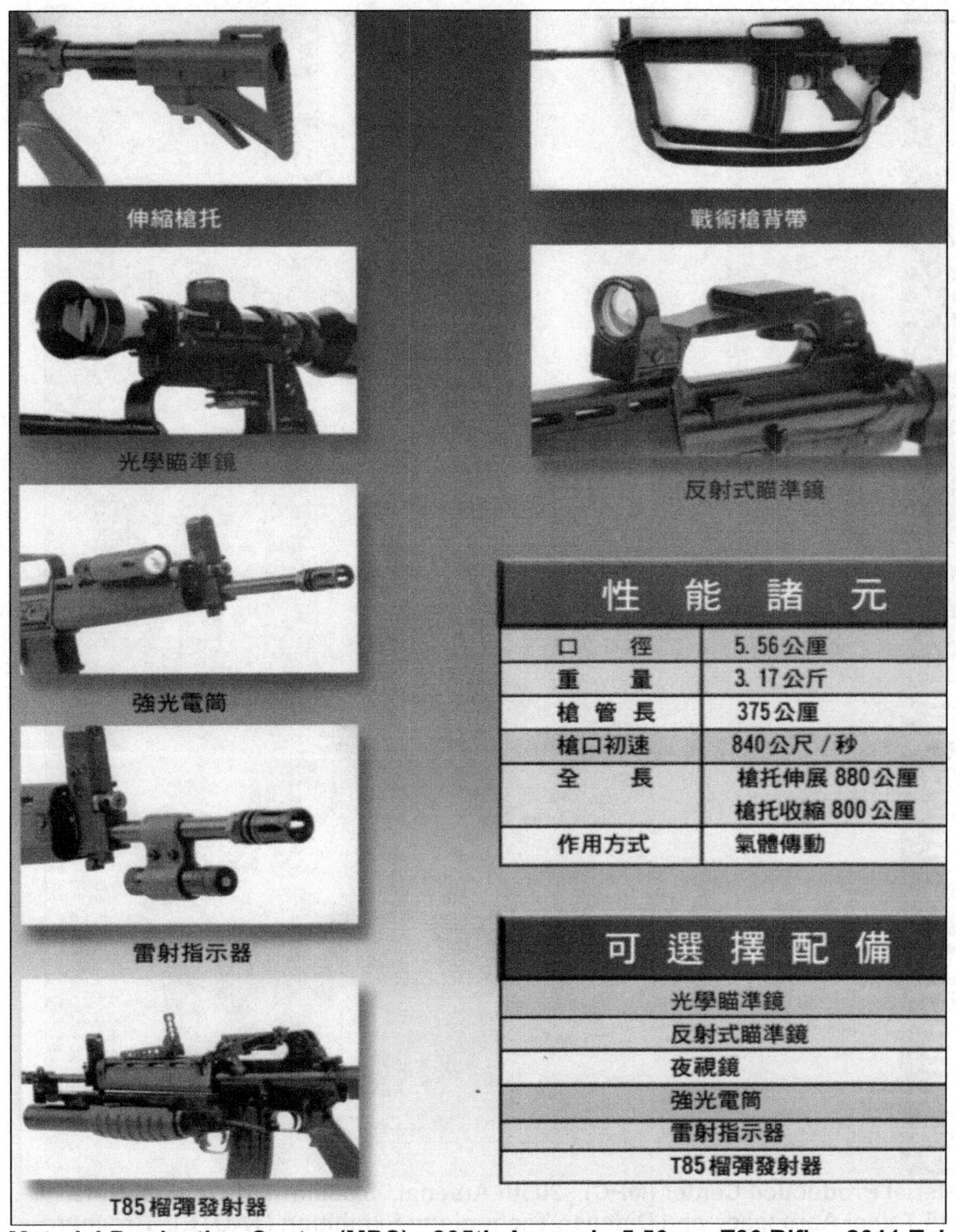

性能諸元	
口　　徑	5. 56公厘
重　　量	3. 17公斤
槍 管 長	375公厘
槍口初速	840公尺／秒
全　　長	槍托伸展 880 公厘 槍托收縮 800 公厘
作用方式	氣體傳動

可選擇配備
光學瞄準鏡
反射式瞄準鏡
夜視鏡
強光電筒
雷射指示器
T85榴彈發射器

Materiel Production Center (MPC). 205th Arsenal. 5.56mm T86 Rifle. 2011 Taipei Aerospace and Defense Technology Exhibition (TADTE). Brochure 2/2.

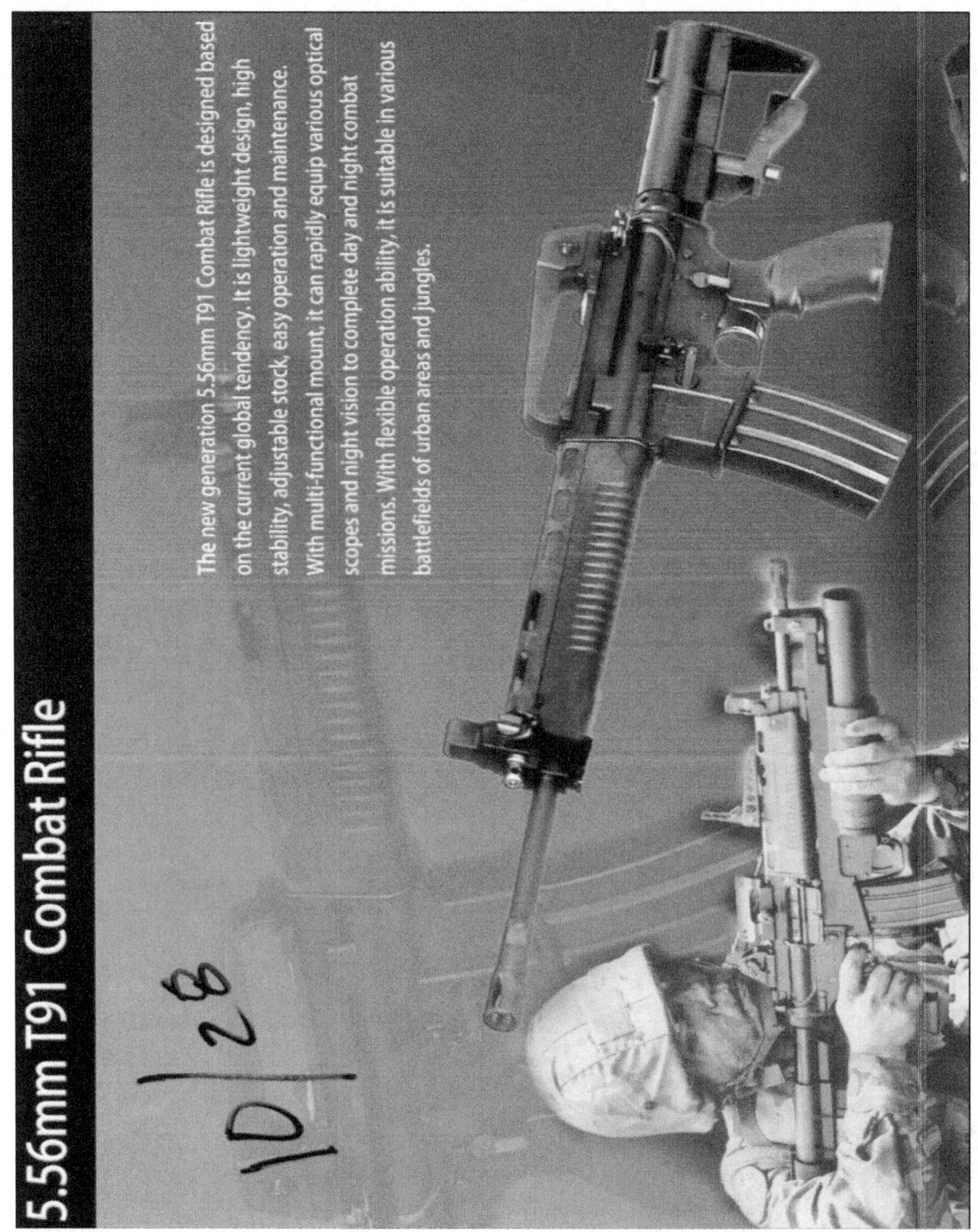

Materiel Production Center (MPC). 205th Arsenal. 5.56mm T91 Combat Rifle. 2015 Taipei Aerospace and Defense Technology Exhibition (TADTE). Brochure 1/2.

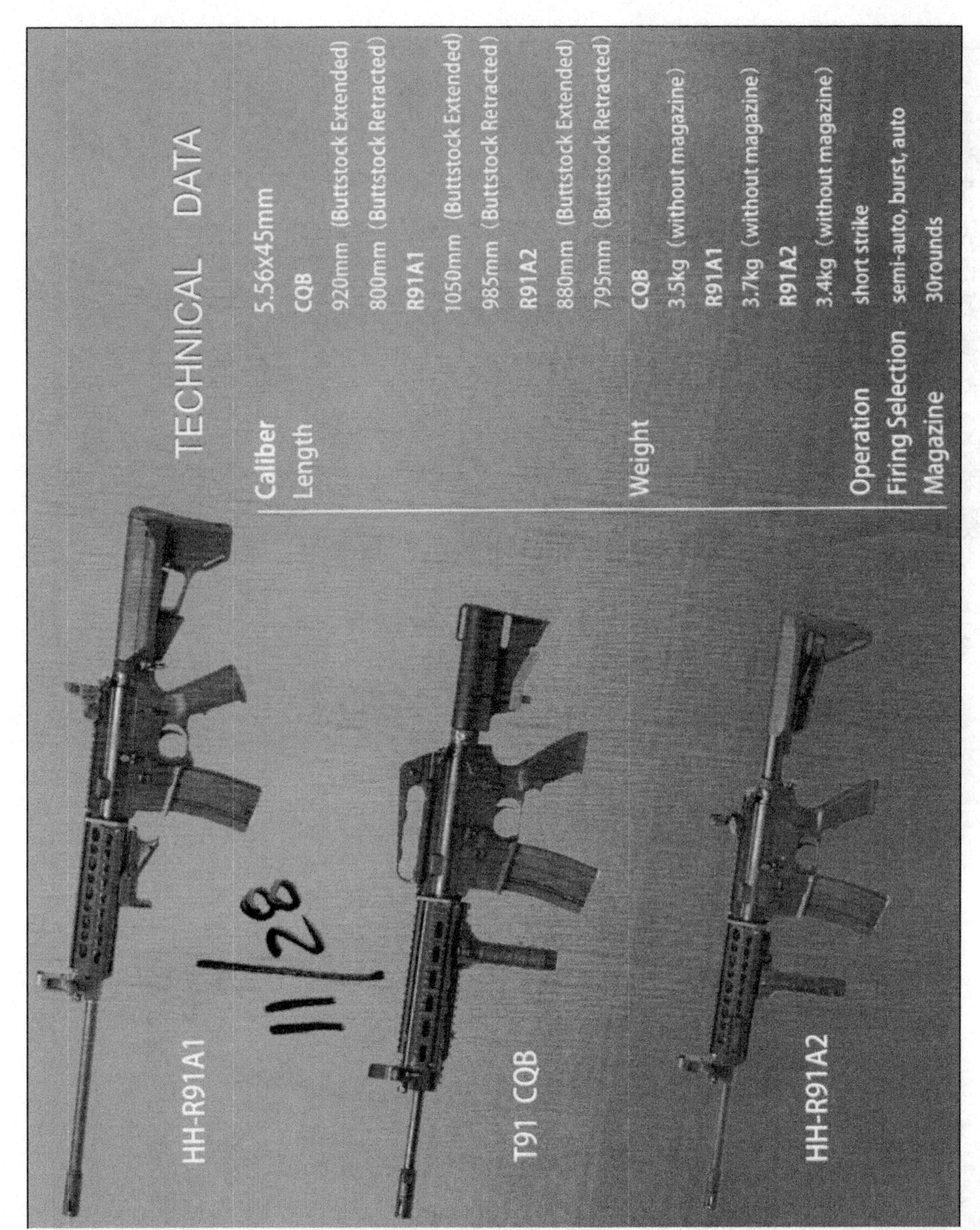

Materiel Production Center (MPC). 205th Arsenal. 5.56mm T91 Combat Rifle. 2015 Taipei Aerospace and Defense Technology Exhibition (TADTE). Brochure 2/2.

Materiel Production Center (MPC). 205th Arsenal. 5.56mm T91 Rifle. 2011 Taipei Aerospace and Defense Technology Exhibition (TADTE). Brochure 1/2.

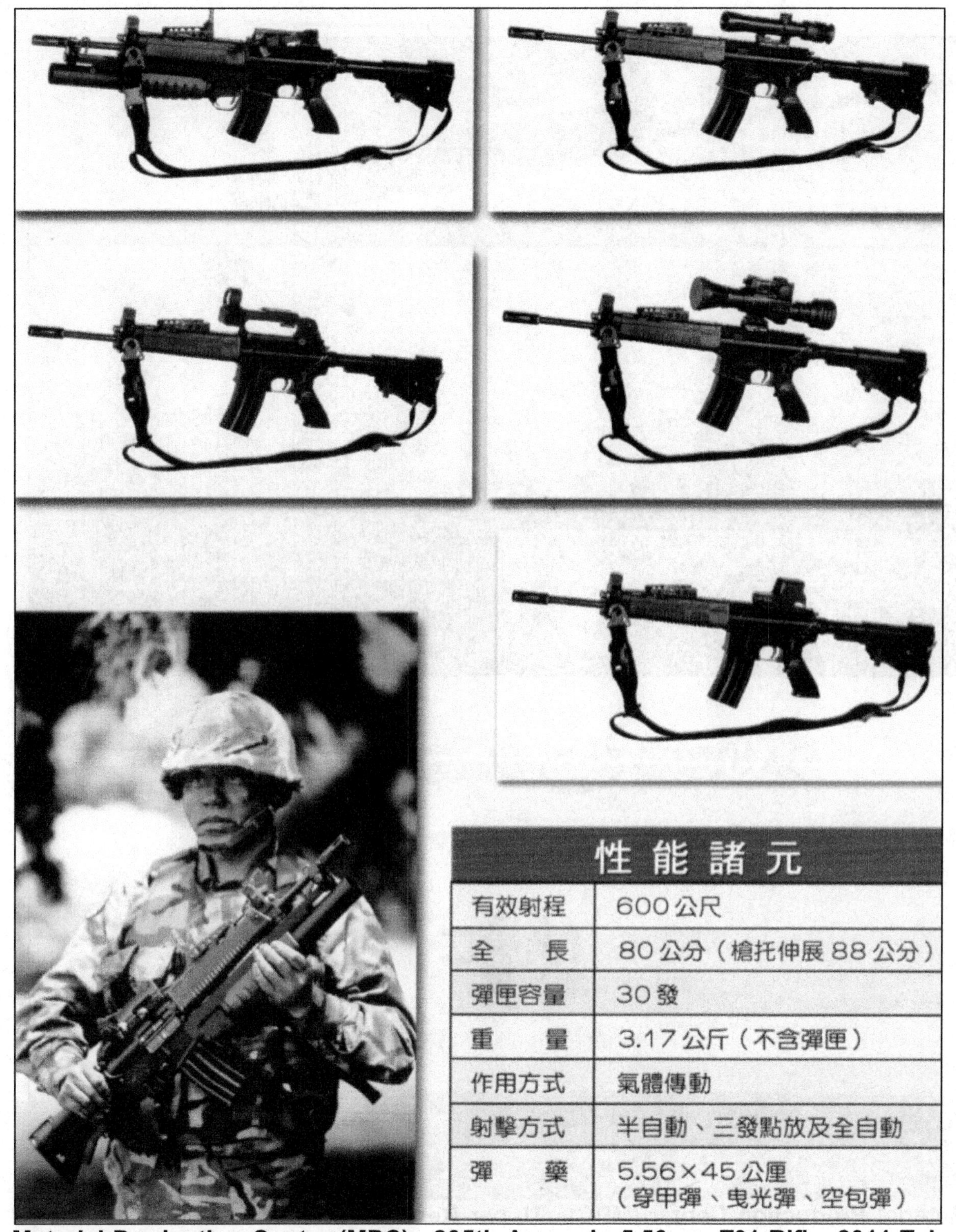

性能諸元	
有效射程	600公尺
全　長	80公分（槍托伸展88公分）
彈匣容量	30發
重　量	3.17公斤（不含彈匣）
作用方式	氣體傳動
射擊方式	半自動、三發點放及全自動
彈　藥	5.56×45公厘 （穿甲彈、曳光彈、空包彈）

Materiel Production Center (MPC). 205th Arsenal. 5.56mm T91 Rifle. 2011 Taipei Aerospace and Defense Technology Exhibition (TADTE). Brochure 2/2.

Materiel Production Center (MPC). Upper Receiver for T91 Rifle 5.56mm. 2018 Defence Services Asia (Malaysia). Brochure 1/2.

5.56mm Upper Receiver

Description:

This upper receiver is based on T91 assault rifles and used for commercial purpose. With cold hammer forged barrels, short stroke piston system and one-piece blot carrier to keep excellent durability and reliability.

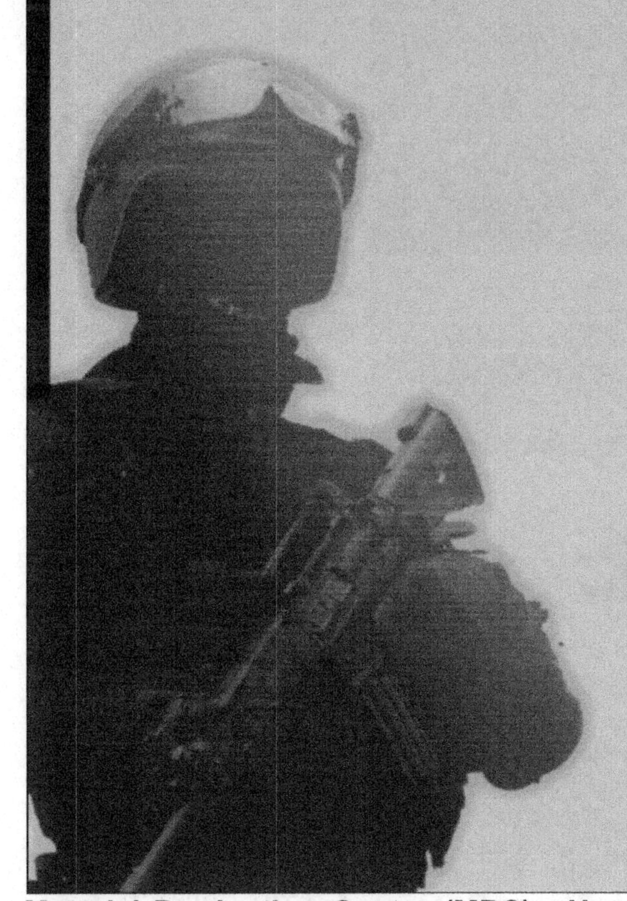

Specification:

- **Short Stroke Piston System.**
- **Cold Forged Barrel (5.56mm NATO).**
- **One-Piece Bolt Carrier.**
- **Suitable for AR15 and .223REM series Lower Receiver.**
- **Military Standard Technology for shooting sport purposes.**

Materiel Production Center (MPC). Upper Receiver for T91 Rifle 5.56mm. 2018 Defence Services Asia (Malaysia). Brochure 2/2.

Materiel Production Center (MPC). 5.56mm Upper Receiver for T91. 2018 Defence Services Asia (Malaysia). Brochure 1/2.

5.56mm Upper Receiver

Description:

This upper receiver is based on T91 assault rifles and used for commercial purpose. With cold hammer forged barrels, short stroke piston system and one-piece blot carrier to keep excellent durability and reliability.

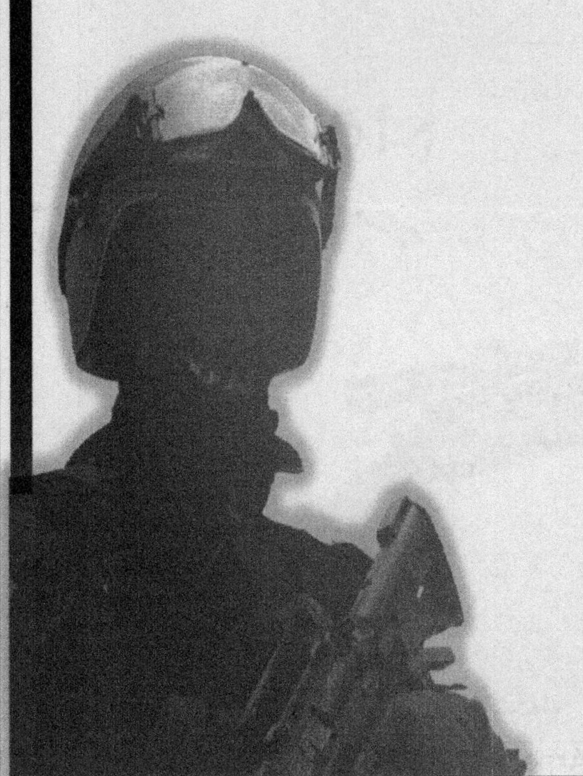

Specification:

- **Short Stroke Piston System.**
- **Cold Forged Barrel (5.56mm NATO).**
- **One-Piece Bolt Carrier.**
- **Suitable for AR15 and .223REM series Lower Receiver.**
- **Military Standard Technology for shooting sport purposes.**

Materiel Production Center (MPC). 5.56mm Upper Receiver for T91. 2018 Defence Services Asia (Malaysia). Brochure 2/2.

MPC

7.62公厘T93狙擊槍

7.62 mm T93 SNIPER RIFLE

狙擊槍係結合用兵造兵專業及技術，完美地將槍枝、彈藥、狙擊鏡及戰士四形體結合互動，以發揮槍枝整體效能。狙擊槍適於特勤、維安及狙擊作戰編制使用；在各種環境條件操作下，精於遠距精準射擊及追求一擊必殺的效果，為國軍狙敵致勝的槍枝極品。

This sniper gun is a product of the combination of strategic design and fabrication technology. The basic premise of this weapon is to integrate the weapon itself, ammunition, telescope and user into an ultimate realization of a product of sate of the art. This weapon is applicable to the purpose of special force, security surveillance, and sniper operation. Under various combat conditions, this weapon can still achieve high accuracy and lethality at long distance to target.

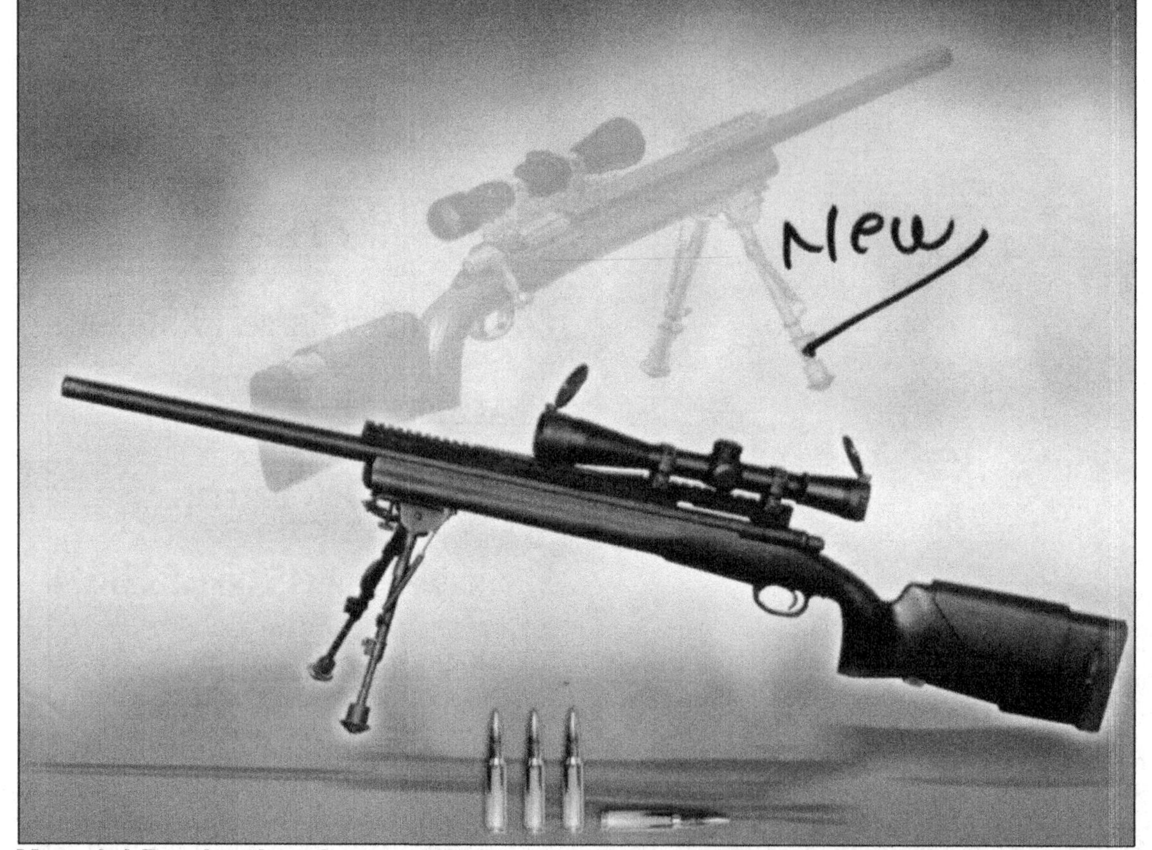

Materiel Production Center (MPC). 205th Arsenal. 7.62mm T93 Sniper Rifle. 2011 Taipei Aerospace and Defense Technology Exhibition (TADTE). Based on the Remington M24 Sniper Weapon System (SWS). Brochure 1/2.

性能諸元 TECHNICAL DATA	
口徑 bore size	7.62公厘 mm
槍管纏度 rifling twist	285公分/轉(cm/rev)
槍管膛線 Number of grooves	5條右旋(5, right-handed)
作用方式 cycle action	單發手動 single shot/manual
給彈方式 feed	5發裝填彈倉(5 rounds magazine)
全長 total length	約110公分(about 110 cm)
重量 weight	約6公斤(不含瞄準鏡、兩腳架)about 6 kgs, without telescope and bi-pod
有效射程 effective range	800公尺(m)
槍托 butt stock	複合材料、長度可調、貼腮設計 composite material, length adjustable
適用彈種 applicable ammunition	7.62公厘專用狙擊彈 exclusive 7.62mm sniper bullet
精度 accuracy	100公尺射距，以3發射擊，彈著散佈面1MOA【2.91公分圓形範圍內(含)】。1 MOA for three rounds distribution at 100 meters (within 2.91 cm diameter)

Materiel Production Center (MPC). 205th Arsenal. 7.62mm T93 Sniper Rifle. 2011 Taipei Aerospace and Defense Technology Exhibition (TADTE). Based on the Remington M24 Sniper Weapon System (SWS). Brochure 2/2.

Materiel Production Center (MPC). 7.62mm Sniper Rifle Barrel. 2018 Defence Services Asia (Malaysia). Brochure 1/2.

7.62mm Sniper Rifle Barrel

Description:

7.62mm Sniper Rifle Barrel applies with 7.62mm match cartridges. It offers high accuracy and lethal capability at long distances. It also provides outstanding quality for special-force, guarded and sniped missions.

Specification:

- Military Standard Processing and Quality.
- Accuracy and Durability.
- Material: Stainless Steel
- Barrel Length: 612mm (Noise suppressor not included)
- Barrel Weight: around 2.5kg
- Cold Forged and Chrome Treatment Barrel.

Materiel Production Center (MPC). 7.62mm Sniper Rifle Barrel. 2018 Defence Services Asia (Malaysia). Brochure 2/2.

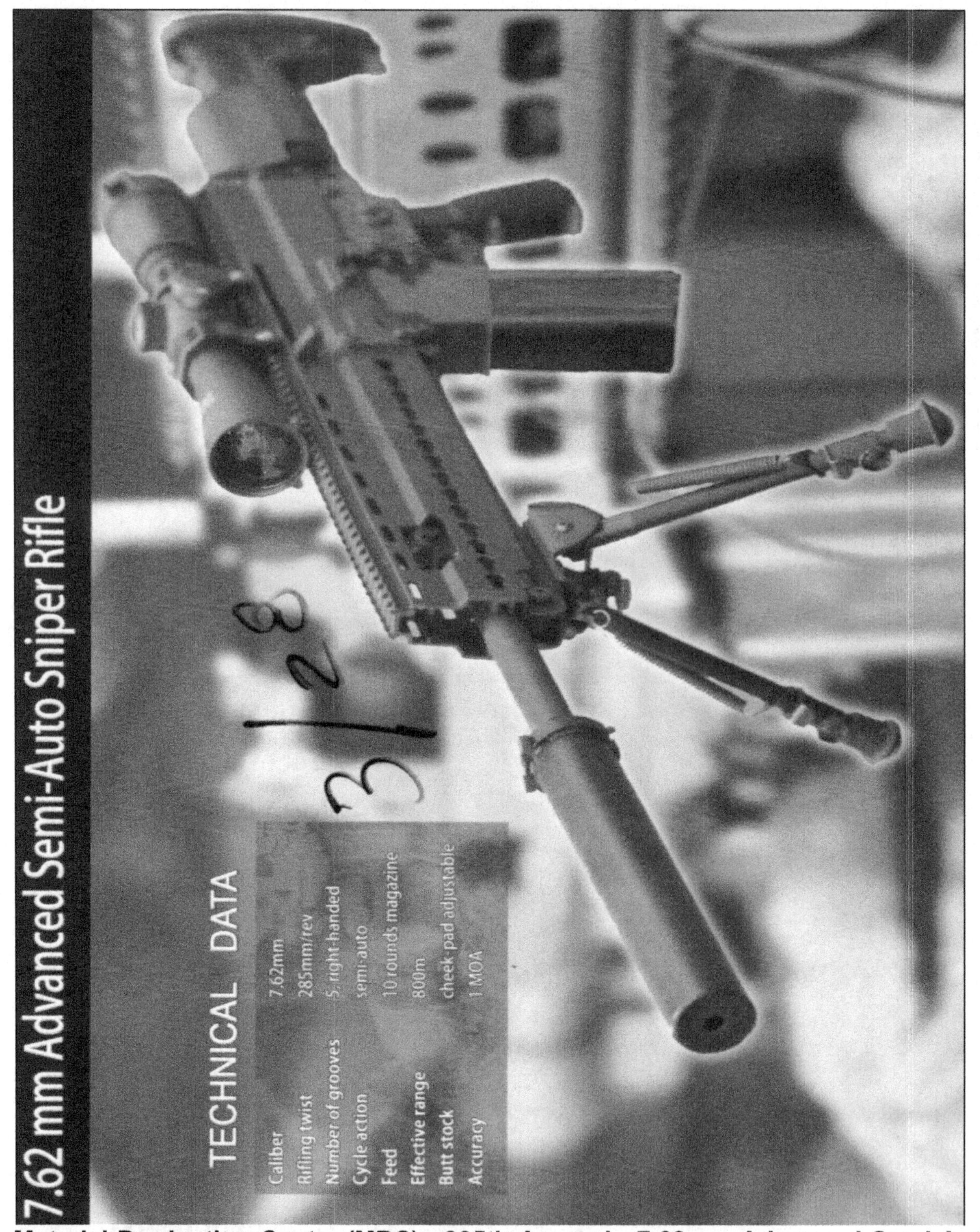

Materiel Production Center (MPC). 205th Arsenal. 7.62mm Advanced Semi-Auto Sniper Rifle. 2015 Taipei Aerospace and Defense Technology Exhibition (TADTE). Brochure 1/1.

Army Aviation and Special Forces Command (陸軍航空特戰指揮部). Army Sniper Company. 1/2.

Army Aviation and Special Forces Command (陸軍航空特戰指揮部). Army Sniper Company. 2/2.

Army Aviation and Special Forces Command (陸軍航空特戰指揮部). Sniper School; Army Infantry Training Command.

Previous sniper patch used from 2013-2016 for the 6th Army, Northern Command.

Fan Patch.

MUNITIONS
Bullets
Grenades
Large Caliber
Mortars

Materiel Production Center (MPC). 202nd Arsenal. Munitions. 2011 Taipei Aerospace and Defense Technology Exhibition (TADTE). Brochure 1/12.

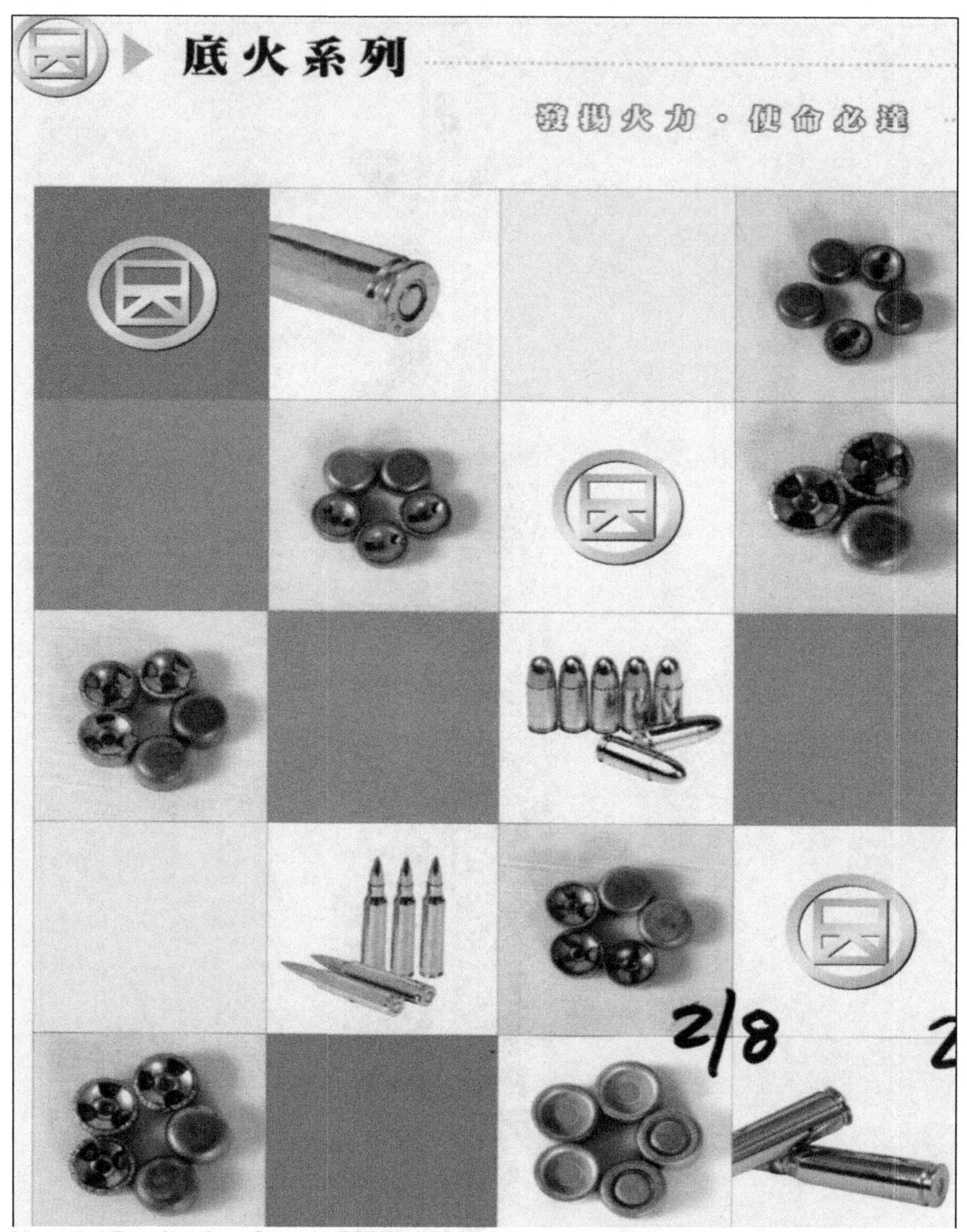

Materiel Production Center (MPC). 202nd Arsenal. Munitions. 2011 Taipei Aerospace and Defense Technology Exhibition (TADTE). Brochure 2/12.

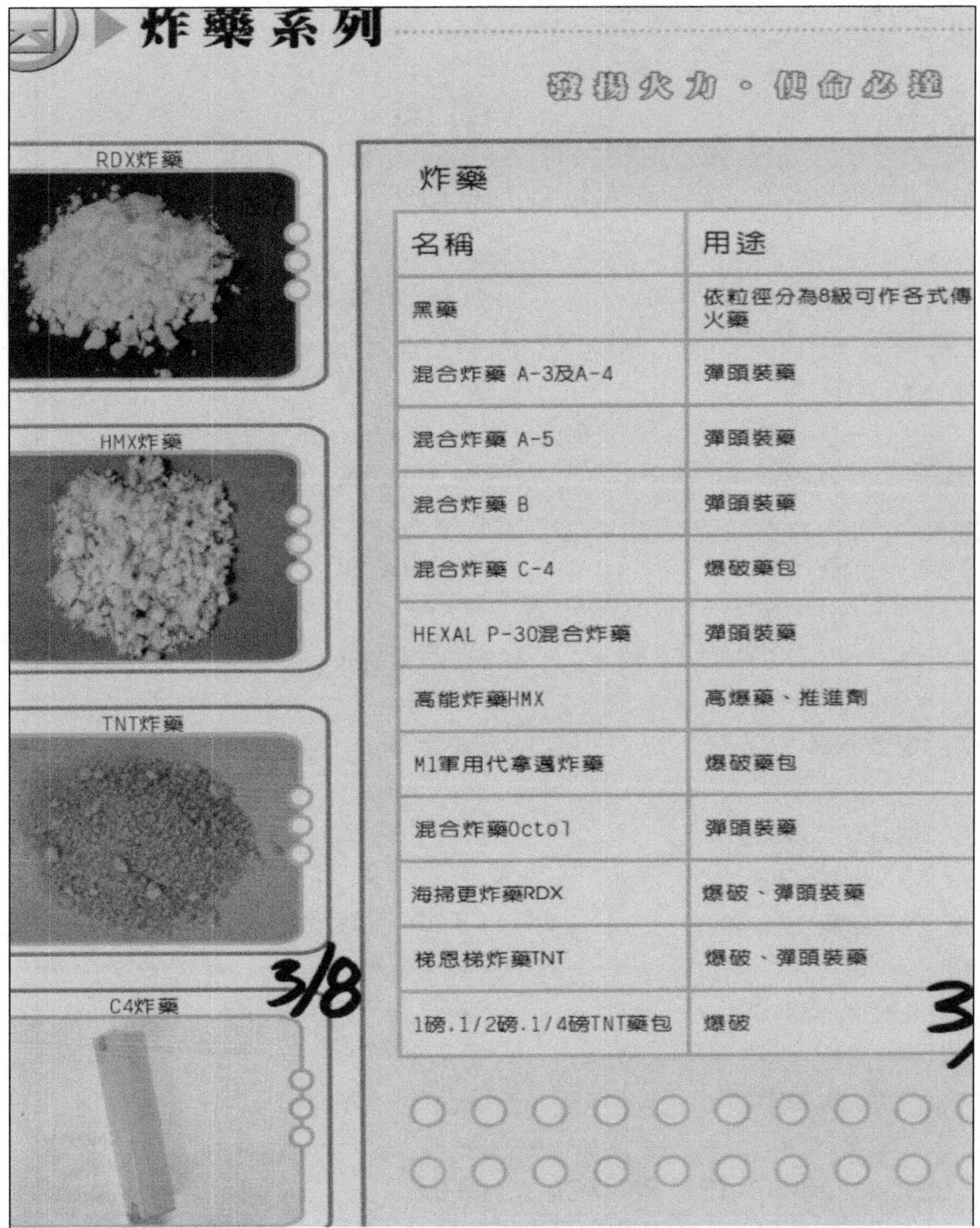

炸藥系列

發揚火力・使命必達

RDX炸藥

HMX炸藥

TNT炸藥

C4炸藥

炸藥

名稱	用途
黑藥	依粒徑分為8級可作各式傳火藥
混合炸藥 A-3及A-4	彈頭裝藥
混合炸藥 A-5	彈頭裝藥
混合炸藥 B	彈頭裝藥
混合炸藥 C-4	爆破藥包
HEXAL P-30混合炸藥	彈頭裝藥
高能炸藥HMX	高爆藥、推進劑
M1軍用代拿邁炸藥	爆破藥包
混合炸藥Octol	彈頭裝藥
海掃更炸藥RDX	爆破、彈頭裝藥
梯恩梯炸藥TNT	爆破、彈頭裝藥
1磅.1/2磅.1/4磅TNT藥包	爆破

Materiel Production Center (MPC). 202nd Arsenal. Munitions. 2011 Taipei Aerospace and Defense Technology Exhibition (TADTE). Brochure 3/12.

炸藥	
名稱	用途
黑藥	依粒徑分為8級可作各式傳火藥
混合炸藥 A-3及A-4	彈頭裝藥
混合炸藥 A-5	彈頭裝藥
混合炸藥 B	彈頭裝藥
混合炸藥 C-4	爆破藥包
HEXAL P-30混合炸藥	彈頭裝藥
高能炸藥HMX	高爆藥、推進劑
M1軍用代拿邁炸藥	爆破藥包
混合炸藥Octol	彈頭裝藥
海掃更炸藥RDX	爆破、彈頭裝藥
梯恩梯炸藥TNT	爆破、彈頭裝藥
1磅.1/2磅.1/4磅TNT藥包	爆破

3

Materiel Production Center (MPC). 202nd Arsenal. Munitions. 2011 Taipei Aerospace and Defense Technology Exhibition (TADTE). Brochure 4/12 (CLOSE-UP).

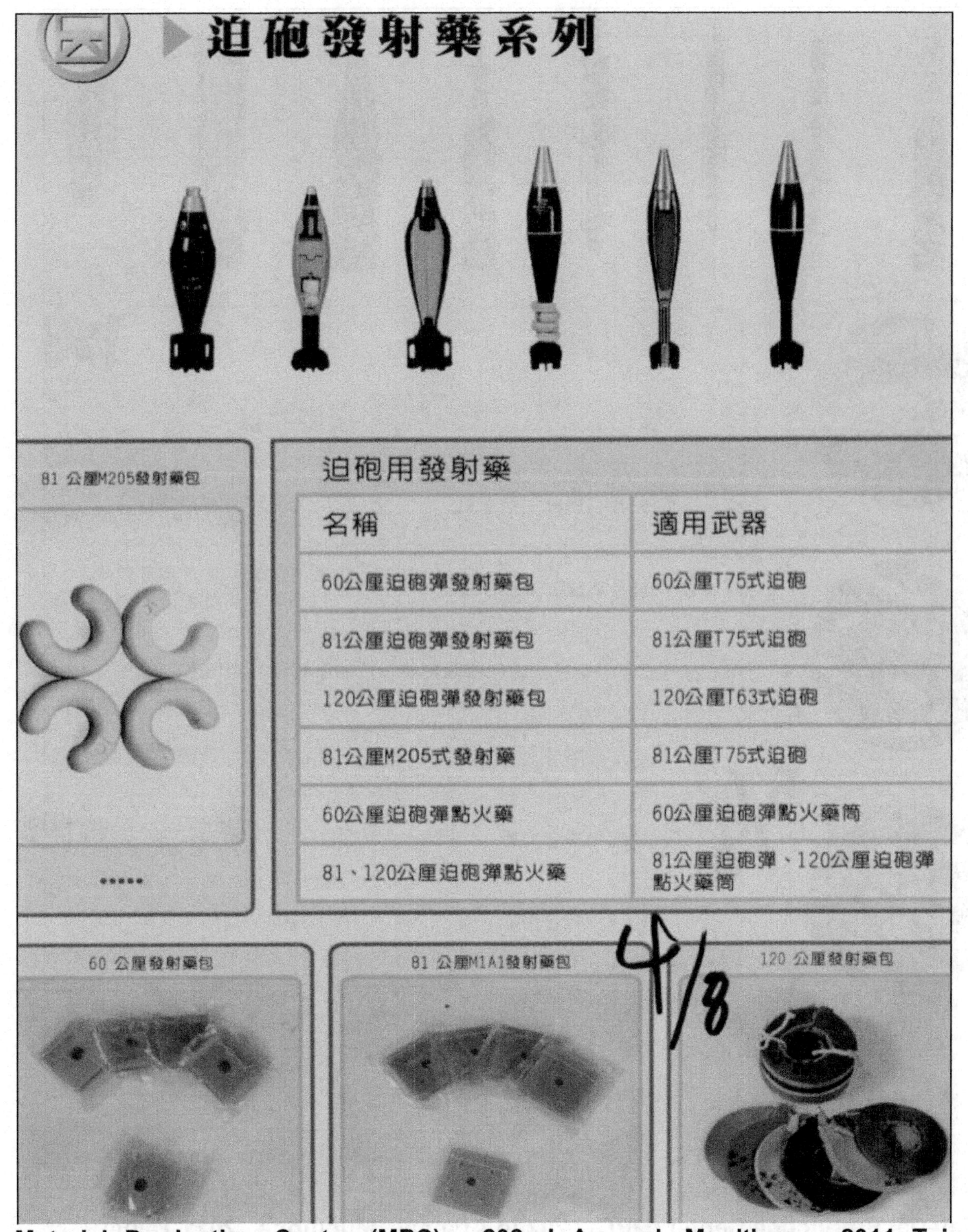

迫砲用發射藥

名稱	適用武器
60公厘迫砲彈發射藥包	60公厘T75式迫砲
81公厘迫砲彈發射藥包	81公厘T75式迫砲
120公厘迫砲彈發射藥包	120公厘T63式迫砲
81公厘M205式發射藥	81公厘T75式迫砲
60公厘迫砲彈點火藥	60公厘迫砲彈點火藥筒
81、120公厘迫砲彈點火藥	81公厘迫砲彈、120公厘迫砲彈點火藥筒

Materiel Production Center (MPC). 202nd Arsenal. Munitions. 2011 Taipei Aerospace and Defense Technology Exhibition (TADTE). Brochure 5/12.

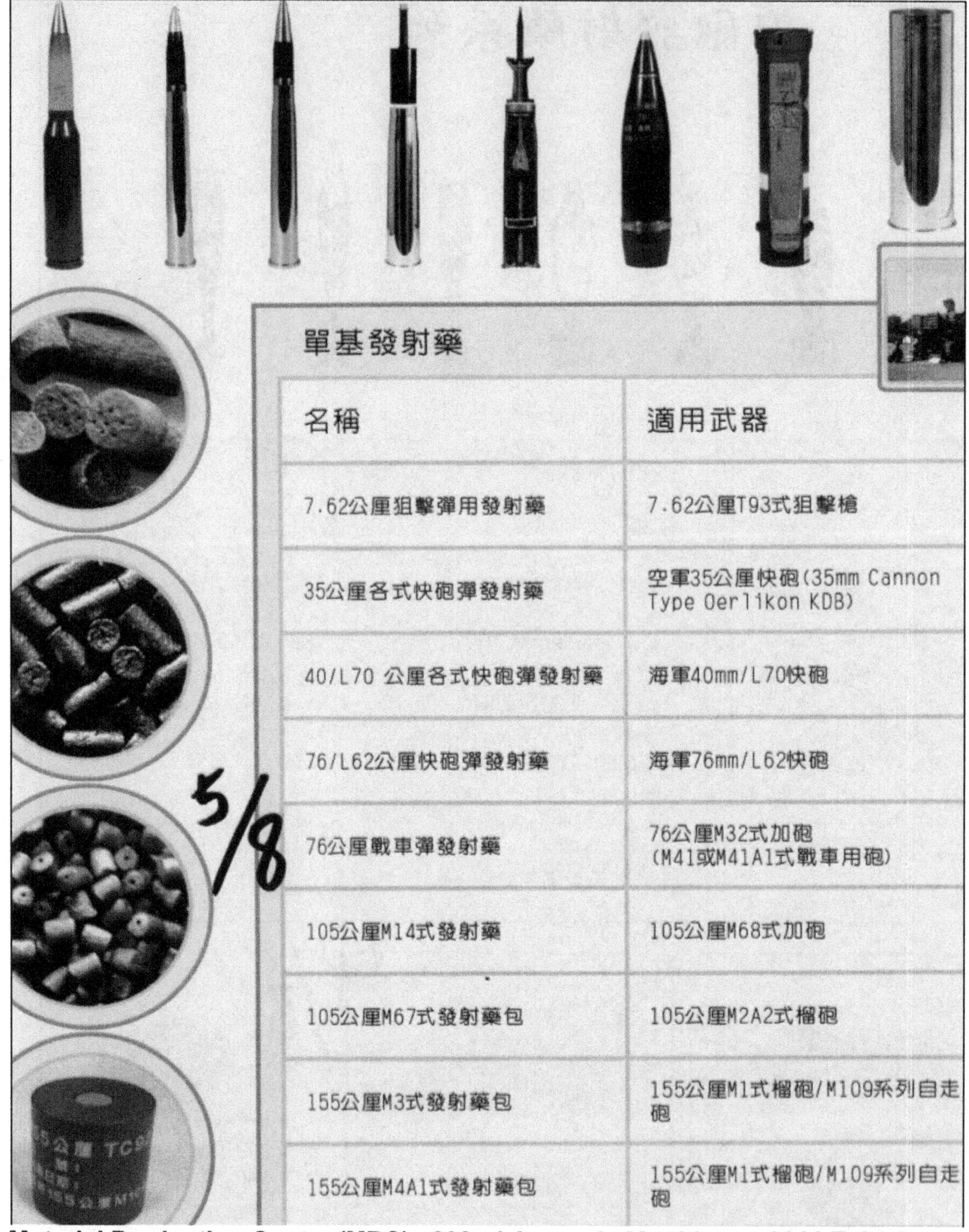

單基發射藥

名稱	適用武器
7.62公厘狙擊彈用發射藥	7.62公厘T93式狙擊槍
35公厘各式快砲彈發射藥	空軍35公厘快砲(35mm Cannon Type Oerlikon KDB)
40/L70 公厘各式快砲彈發射藥	海軍40mm/L70快砲
76/L62公厘快砲彈發射藥	海軍76mm/L62快砲
76公厘戰車彈發射藥	76公厘M32式加砲 (M41或M41A1式戰車用砲)
105公厘M14式發射藥	105公厘M68式加砲
105公厘M67式發射藥包	105公厘M2A2式榴砲
155公厘M3式發射藥包	155公厘M1式榴砲/M109系列自走砲
155公厘M4A1式發射藥包	155公厘M1式榴砲/M109系列自走砲

Materiel Production Center (MPC). 202nd Arsenal. Munitions. 2011 Taipei Aerospace and Defense Technology Exhibition (TADTE). Brochure 6/12.

單基發射藥

名稱	適用武器
7.62公厘狙擊彈用發射藥	7.62公厘T93式狙擊槍
35公厘各式快砲彈發射藥	空軍35公厘快砲(35mm Cannon Type Oerlikon KDB)
40/L70 公厘各式快砲彈發射藥	海軍40mm/L70快砲
76/L62公厘快砲彈發射藥	海軍76mm/L62快砲
76公厘戰車彈發射藥	76公厘M32式加砲 (M41或M41A1式戰車用砲)
105公厘M14式發射藥	105公厘M68式加砲
105公厘M67式發射藥包	105公厘M2A2式榴砲
155公厘M3式發射藥包	155公厘M1式榴砲/M109系列自走砲
155公厘M4A1式發射藥包	155公厘M1式榴砲/M109系列自走砲

Materiel Production Center. 202nd Arsenal. Munitions. 2011 Taipei Aerospace and Defense Technology Exhibition (TADTE). Brochure 7/12 (CLOSE-UP).

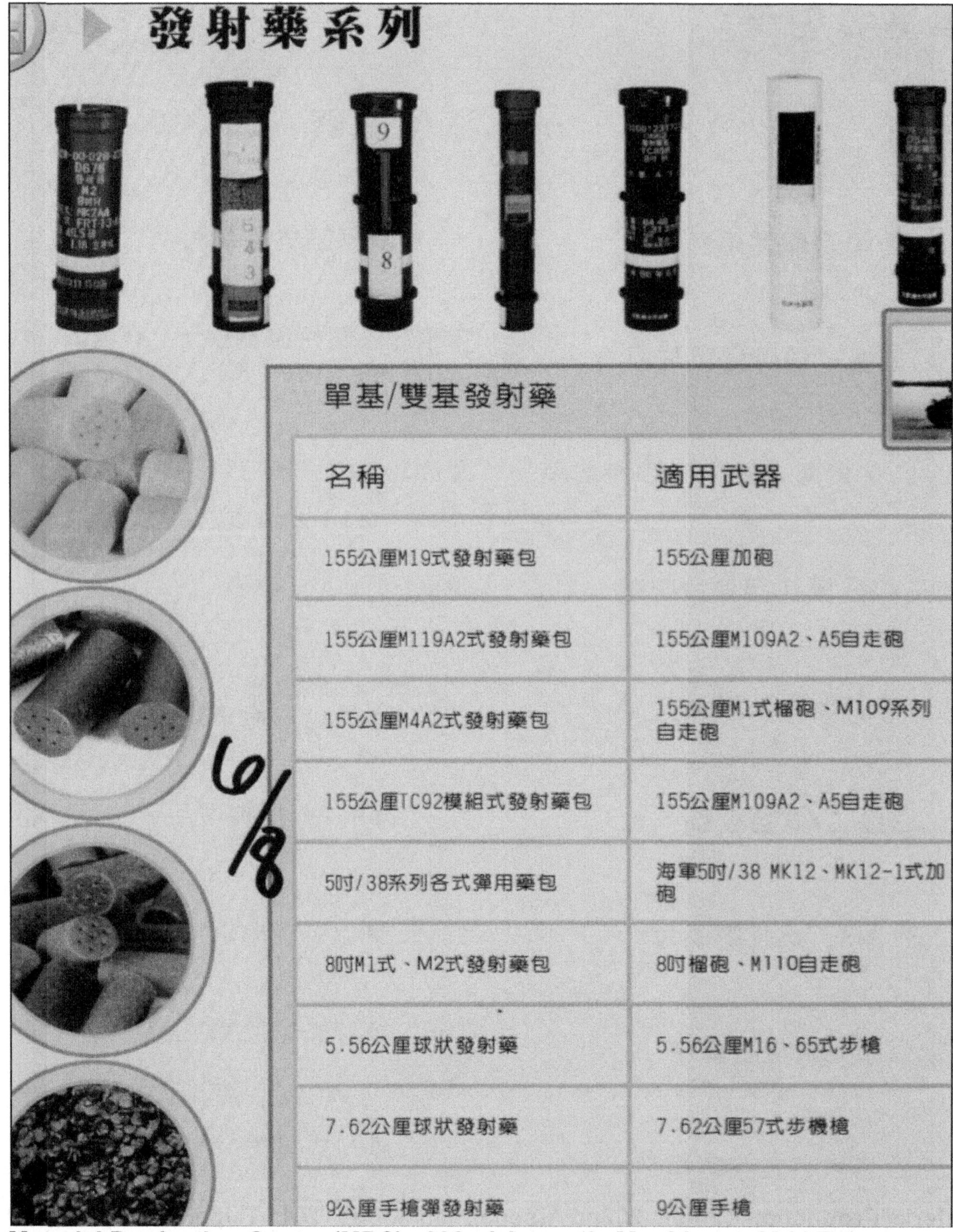

發射藥系列

單基/雙基發射藥

名稱	適用武器
155公厘M19式發射藥包	155公厘加砲
155公厘M119A2式發射藥包	155公厘M109A2、A5自走砲
155公厘M4A2式發射藥包	155公厘M1式榴砲、M109系列自走砲
155公厘TC92模組式發射藥包	155公厘M109A2、A5自走砲
5吋/38系列各式彈用藥包	海軍5吋/38 MK12、MK12-1式加砲
8吋M1式、M2式發射藥包	8吋榴砲、M110自走砲
5.56公厘球狀發射藥	5.56公厘M16、65式步槍
7.62公厘球狀發射藥	7.62公厘57式步機槍
9公厘手槍彈發射藥	9公厘手槍

Materiel Production Center (MPC). 202nd Arsenal. Munitions. 2011 Taipei Aerospace and Defense Technology Exhibition (TADTE). Brochure 8/12.

單基/雙基發射藥

名稱	適用武器
155公厘M19式發射藥包	155公厘加砲
155公厘M119A2式發射藥包	155公厘M109A2、A5自走砲
155公厘M4A2式發射藥包	155公厘M1式榴砲、M109系列自走砲
155公厘TC92模組式發射藥包	155公厘M109A2、A5自走砲
50吋/38系列各式彈用藥包	海軍50吋/38 MK12、MK12-1式加砲
80吋M1式、M2式發射藥包	80吋榴砲、M110自走砲
5.56公厘球狀發射藥	5.56公厘M16、65式步槍
7.62公厘球狀發射藥	7.62公厘57式步機槍
9公厘手槍彈發射藥	9公厘手槍

Materiel Production Center (MPC). 202nd Arsenal. Munitions. 2011 Taipei Aerospace and Defense Technology Exhibition (TADTE). Brochure 9/12 (CLOSE-UP).

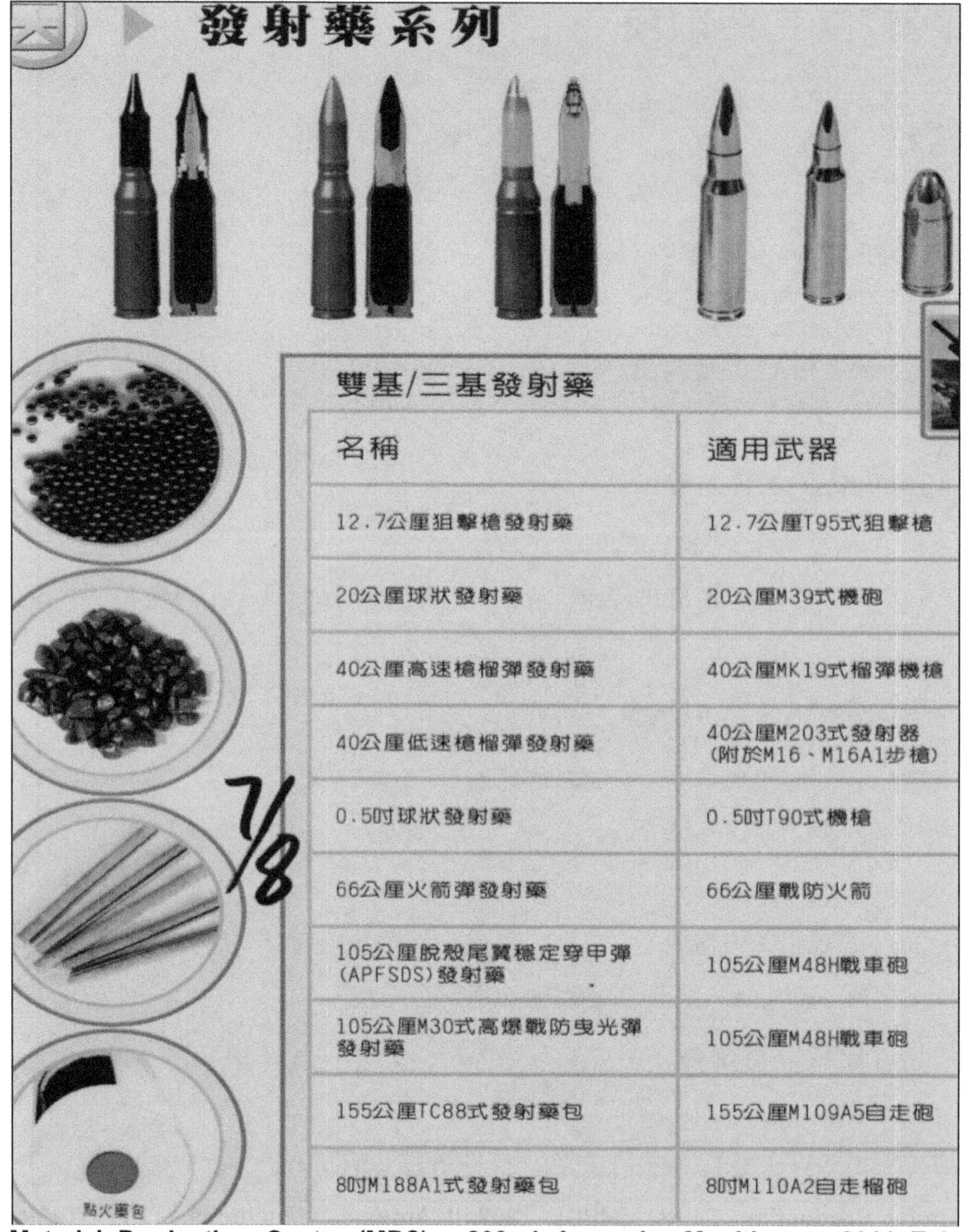

發射藥系列

雙基/三基發射藥

名稱	適用武器
12.7公厘狙擊槍發射藥	12.7公厘T95式狙擊槍
20公厘球狀發射藥	20公厘M39式機砲
40公厘高速槍榴彈發射藥	40公厘MK19式榴彈機槍
40公厘低速槍榴彈發射藥	40公厘M203式發射器 (附於M16、M16A1步槍)
0.50吋球狀發射藥	0.50吋T90式機槍
66公厘火箭彈發射藥	66公厘戰防火箭
105公厘脫殼尾翼穩定穿甲彈(APFSDS)發射藥	105公厘M48H戰車砲
105公厘M30式高爆戰防曳光彈發射藥	105公厘M48H戰車砲
155公厘TC88式發射藥包	155公厘M109A5自走砲
8吋M188A1式發射藥包	8吋M110A2自走榴砲

Materiel Production Center (MPC). 202nd Arsenal. Munitions. 2011 Taipei Aerospace and Defense Technology Exhibition (TADTE). Brochure 10/12.

雙基/三基發射藥

名稱	適用武器
12.7公厘狙擊槍發射藥	12.7公厘T95式狙擊槍
20公厘球狀發射藥	20公厘M39式機砲
40公厘高速槍榴彈發射藥	40公厘MK19式榴彈機槍
40公厘低速槍榴彈發射藥	40公厘M203式發射器 (附於M16、M16A1步槍)
0.50吋球狀發射藥	0.50吋T90式機槍
66公厘火箭彈發射藥	66公厘戰防火箭
105公厘脫殼尾翼穩定穿甲彈 (APFSDS)發射藥	105公厘M48H戰車砲
105公厘M30式高爆戰防曳光彈 發射藥	105公厘M48H戰車砲
155公厘TC88式發射藥包	155公厘M109A5自走砲
8吋M188A1式發射藥包	8吋M110A2自走榴砲

Materiel Production Center (MPC). 202nd Arsenal. Munitions. 2011 Taipei Aerospace and Defense Technology Exhibition (TADTE). Brochure 11/12 (CLOSE-UP).

Materiel Production Center (MPC). 202nd Arsenal. Munitions. 2011 Taipei Aerospace and Defense Technology Exhibition (TADTE). Brochure 12/12.

40 M430、M430A1 公厘 高速高爆穿甲彈

高爆穿甲彈具有預裂彈體及錐孔裝藥，彈頭作用後同時產生金屬噴流與彈體破片，可貫穿輕裝甲、破壞裝備及殺傷人員。M430A1為新型彈種，其穿甲力較M430提昇百分之三十。

—性 能 諸 元—

型　式	M430、M430A1 四〇公厘高速高爆穿甲彈
口　徑	40公厘
長　度	112.3公厘
全彈重	280公克，鐵皮箱包裝\每箱32顆
最大射程	2200公尺
適用武器	美造MK19 MODIII四〇公厘榴彈機槍

聯勤第二〇二廠

台北南港郵政 90582 信箱
TEL:02-27850271

Materiel Production Center (MPC). 202 Arsenal. M430/M30A1 High-Explosive, Dual-Purpose (HEDP) 40mm Grenade. Weapon: MK19 Model III. Unidentified Taipei Aerospace and Defense Technology Exhibition (TADTE). 1/1 Brochure.

40 M918 公厘 高速閃光音效彈

閃光音效彈內裝有閃光劑，彈頭作用後可產生閃光、聲音等標示與震撼效果，可作為部隊演訓或嚇阻、驅離使用。

—性 能 諸 元—

型 式	M918 四〇公厘高速閃光音效彈
口 徑	40公厘
長 度	112.3公厘
全 彈 重	280公克，鐵皮箱包裝\每箱32顆
最大射程	2200公尺
適用武器	美造MK19 MODIII四〇公厘榴彈機槍

聯勤第二〇二廠

台北南港郵政 90582 信箱
TEL:02-27850271

Materiel Production Center. 202mm Arsenal. M918 Target Practice (TP) 40mm Grenade. Weapon: MK19 Model III. Unidentified Taipei Aerospace and Defense Technology Exhibition (TADTE). 1/1 Brochure.

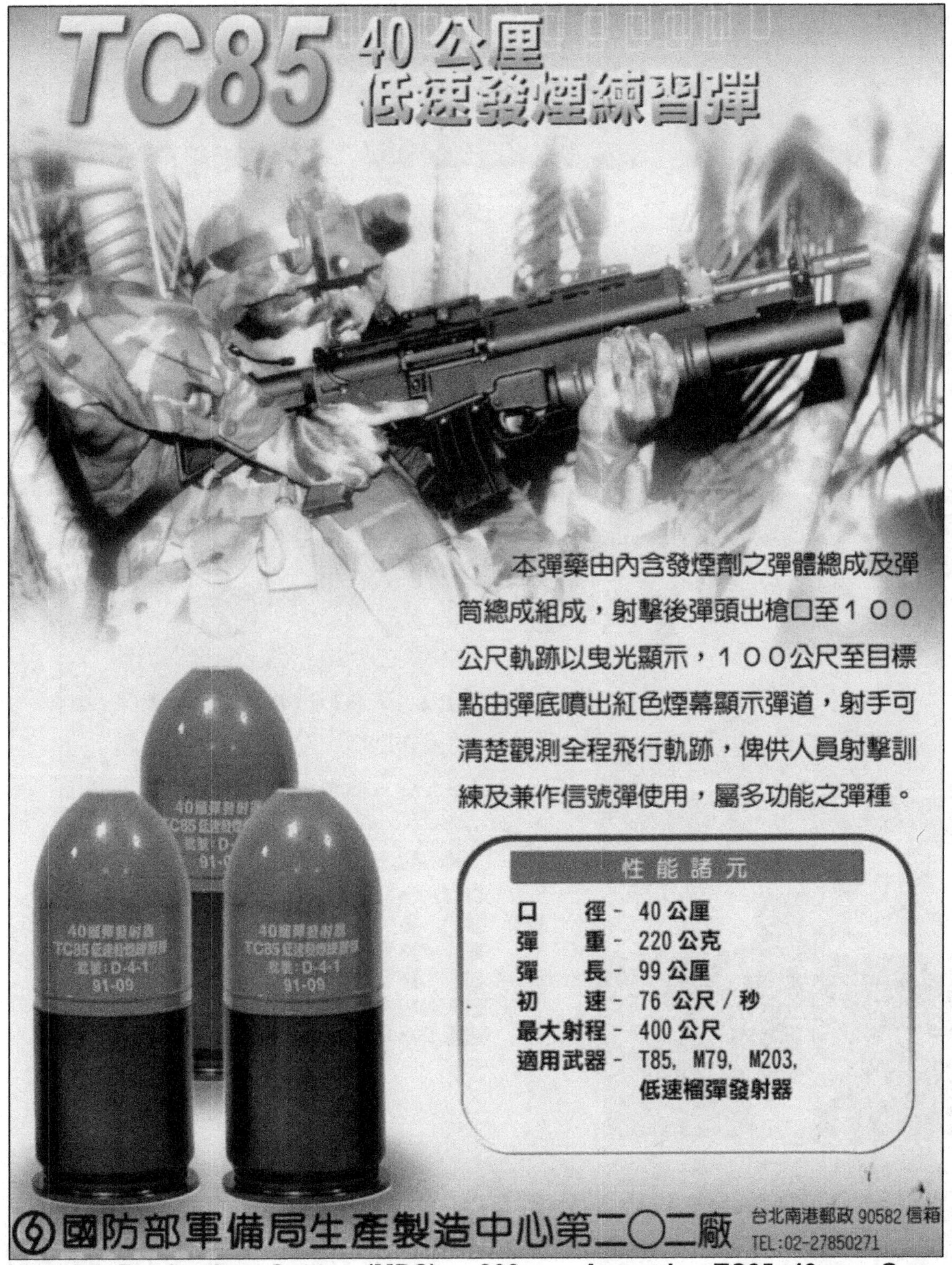

Materiel Production Center (MPC). 202mm Arsenal. TC85 40mm Grenade. Unidentified Taipei Aerospace and Defense Technology Exhibition (TADTE). 1/1

Brochure.

Materiel Production Center (MPC). 202mm Arsenal. TC86 High Explosive (HE) 40mm Grenade. Unidentified Taipei Aerospace and Defense Technology Exhibition (TADTE). 1/1 Brochure.

40 TC88 公厘 低速高爆穿甲彈

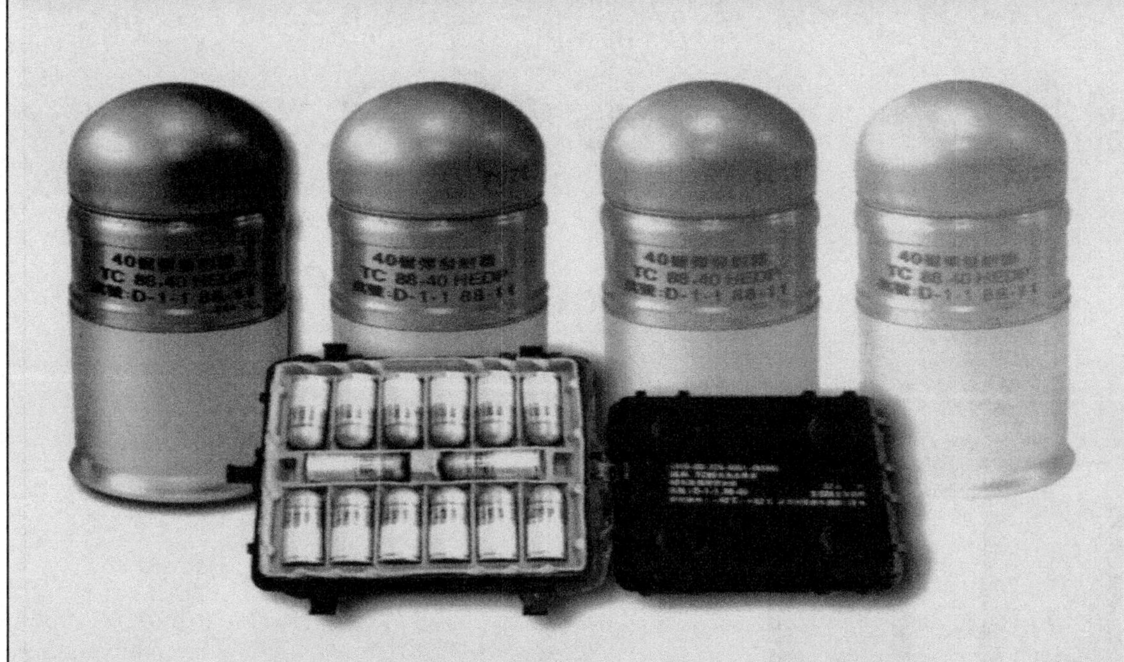

高爆穿甲彈具有預裂彈體及錐孔裝藥，彈頭作用後同時產生破片與金屬噴流，可貫穿輕裝甲、破壞裝備及殺傷人員之雙效功能。

—性 能 諸 元—

型 式	TC88 四〇公厘低速高爆穿甲彈
口 徑	40公厘
長 度	103公厘
全 彈 重	300公克，塑膠防潮包裝箱\每箱56顆
最大射程	400公尺
適用武器	T85、M79、M203 四〇公厘低速榴彈發射器

聯勤第二〇二廠

台北南港郵政 90582 信箱
TEL：02-27850271

Materiel Production Center (MPC). 202nd Arsenal. TC88 High-Explosive, Dual-Purpose (HEDP) 40mm Grenade. Unidentified Taipei Aerospace and Defense Technology Exhibition (TADTE). 1/1 Brochure.

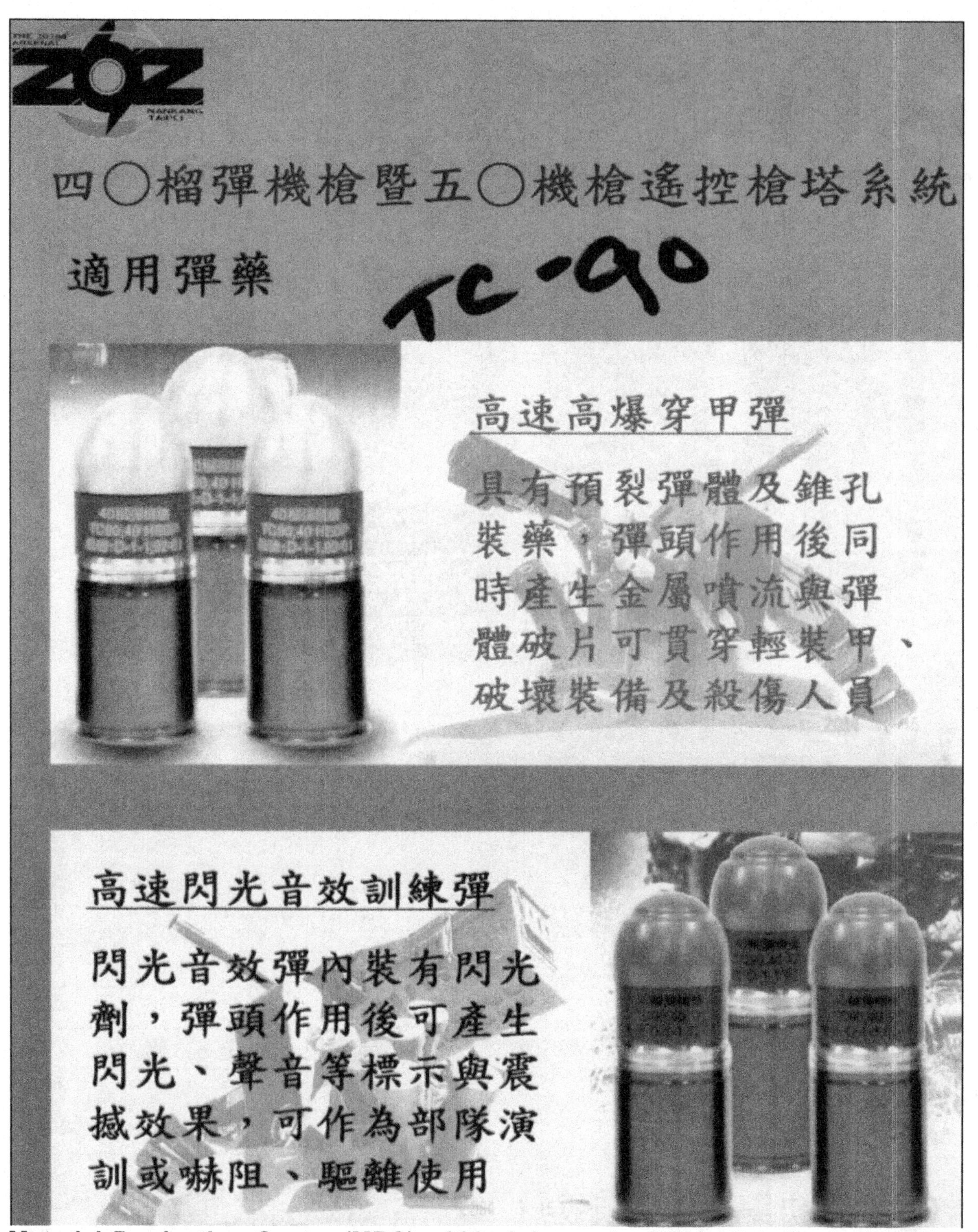

Materiel Production Center (MPC). 202nd Arsenal. TC90 High-Explosive, Dual-Purpose (HEDP) 40mm Grenade. Unidentified Taipei Aerospace and Defense Technology Exhibition (TADTE). 1/1 Brochure.

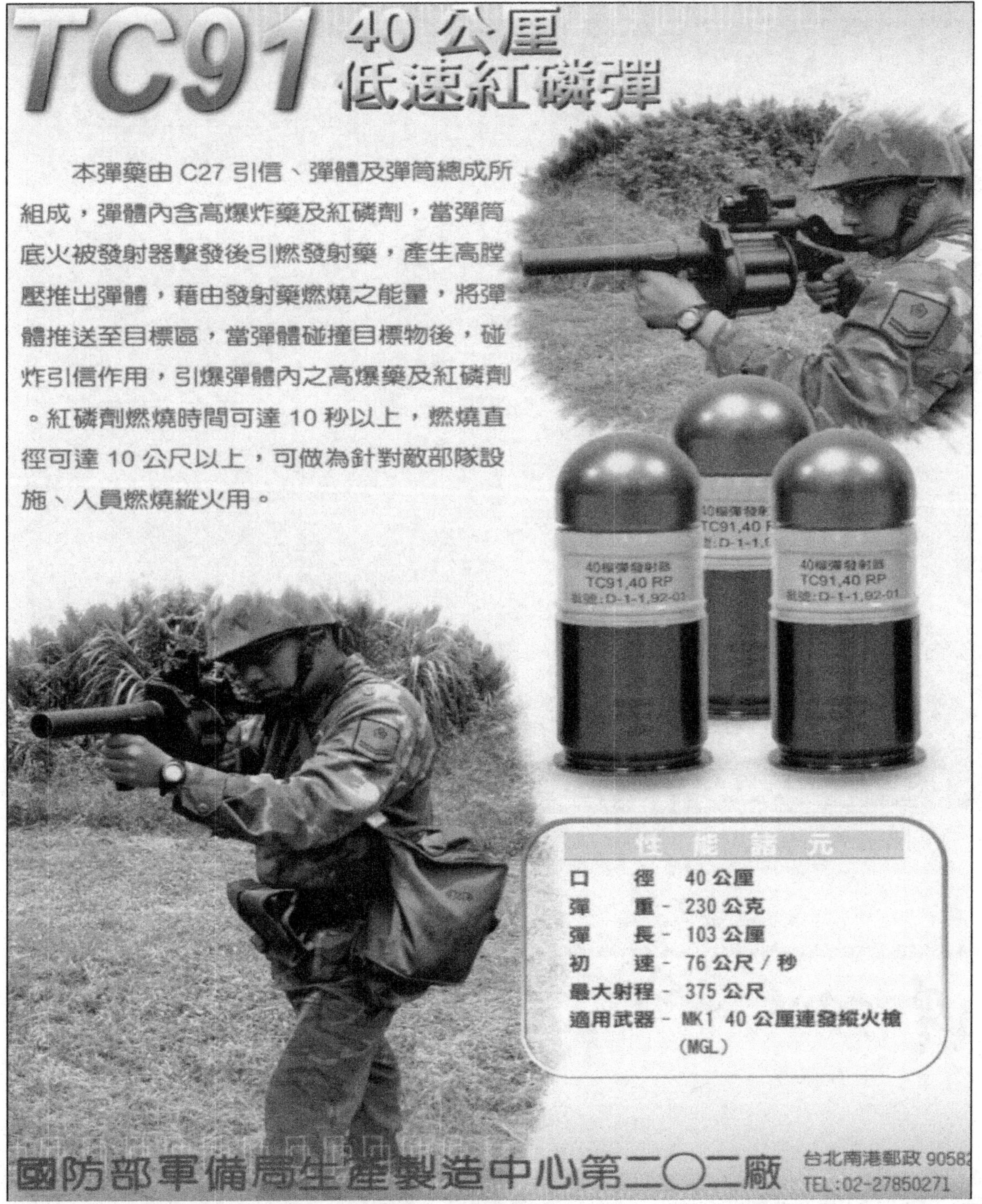

Materiel Production Center (MPC). 202nd Arsenal. TC91 Radio Frequency (RF) High-Explosive, Dual-Purpose (HEDP) 40mm Grenade. Weapon: MK-1 Multiple Grenade Launcher. Unidentified Taipei Aerospace and Defense Technology Exhibition (TADTE). 1/1 Brochure.

Materiel Production Center. 40mm Grenade Machine Gun Barrel for T91 Grenade Launcher. 2018 Defence Services Asia (Malaysia). Brochure 1/2.

40mm Grenade Machine Gun Barrel

Description:

40mm Grenade Machine Gun applies with this barrel, offers not only steady shooting ability but also powerful force. It can enhance lethal firepower for military troops.

Specification:

- Military Standard Processing and Quality.
- Accuracy and Durability.
- Material: Cr-Mo-V Steel
- Barrel Length: around 413mm
- Barrel Weight: around 1.7kg

Materiel Production Center. 40mm Grenade Machine Gun Barrel for T91 Grenade Launcher. 2018 Defence Services Asia (Malaysia). Brochure 2/2.

Materiel Production Center (MPC). TC103 81mm High-Explosive Mortar Round (ER). 2018 Defence Services Asia (Malaysia). Brochure 1/2.

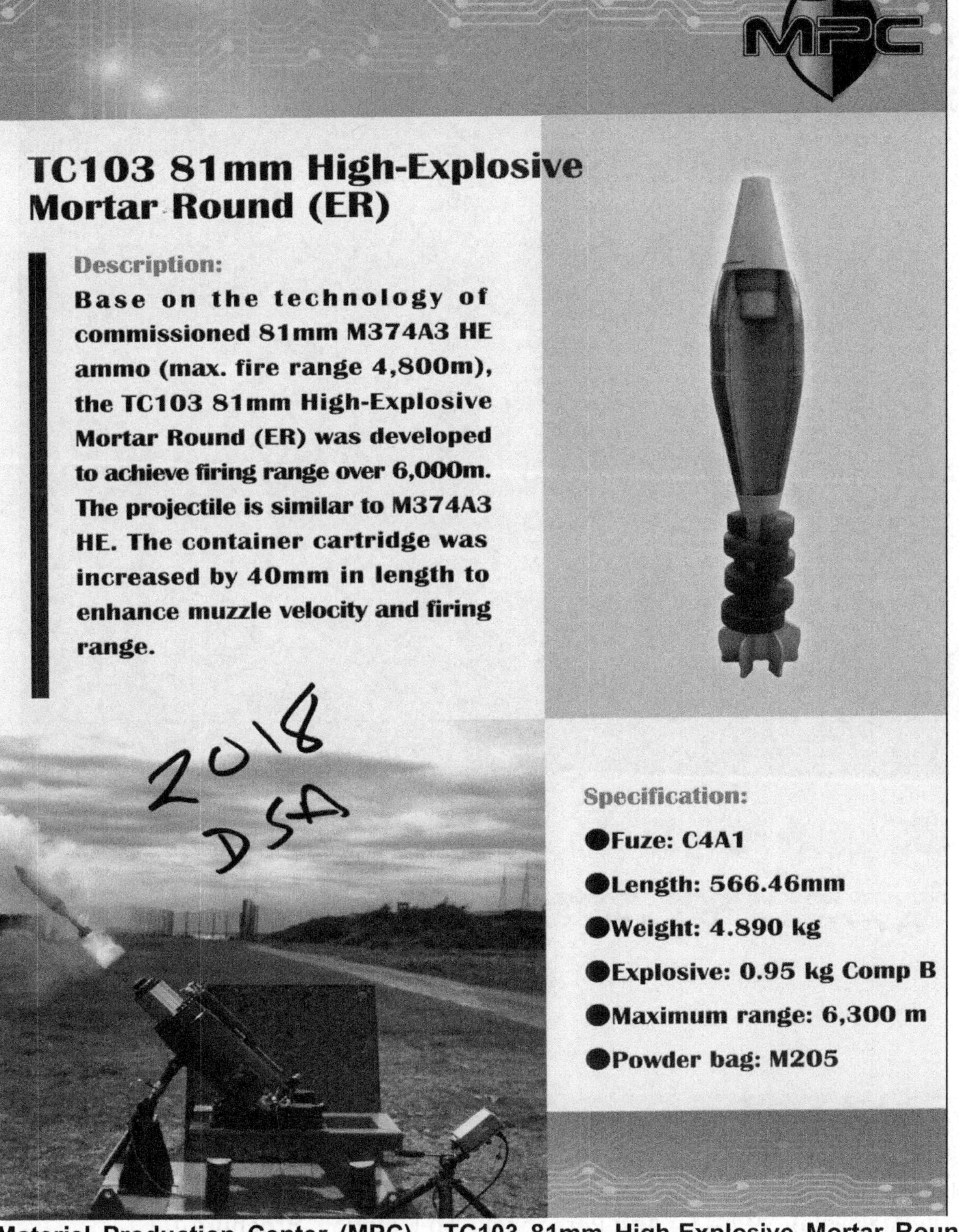

MPC
TC103 81mm High-Explosive Mortar Round (ER)
Description:
Base on the technology of commissioned 81mm M374A3 HE ammo (max. fire range 4,800m), the TC103 81mm High-Explosive Mortar Round (ER) was developed to achieve firing range over 6,000m. The projectile is similar to M374A3 HE. The container cartridge was increased by 40mm in length to enhance muzzle velocity and firing range.
2018 DSA
Specification:
●Fuze: C4A1
●Length: 566.46mm
●Weight: 4.890 kg
●Explosive: 0.95 kg Comp B
●Maximum range: 6,300 m
●Powder bag: M205

Materiel Production Center (MPC). TC103 81mm High-Explosive Mortar Round (ER). 2018 Defence Services Asia (Malaysia). Brochure 2/2.

Materiel Production Center. 81mm AP/AM Mortar Round. 2018 Defence Services Asia (Malaysia). AP/AM = Anti-Personnel/Anti-Materiel. Brochure 1/2.

81mm AP/AM Mortar Round

Description:

The 81mm AP/AM Mortar Round is generally similar to M374A3 HE projectile. The steel ball assembly is fitted in the new shell. There are more than 1,000 steel balls inside the projectile. It can create kinetic fragments (including shell fragments and preset steel balls) to increase the lethality.

2018 DSA

Specification:

- Fuze: C4A1
- Length: 566.46mm
- Explosive: >1,000 steel balls
- Maximum range: 6,300 m
- Powder bag: M205

Materiel Production Center. 81mm AP/AM Mortar Round. 2018 Defence Services Asia (Malaysia). AP/AM = Anti-Personnel/Anti-Materiel. Brochure 2/2.

TC63 120 公厘高爆彈

2011

砲彈自砲口前端裝填滑墜，當砲尾擊針撞擊底火，即引燃發射藥火藥鏈，將砲彈推送至目標區，再由引信發生作用，引爆彈裝炸藥，產生高爆功能，達到殺傷及破壞之目的。

【性能諸元】

- ❒ 彈重：12.61 公斤
- ❒ 彈長：580.6 公厘
- ❒ 裝填：TNT，2.3 公斤
- ❒ 引信：C4 碰炸引信
- ❒ 發射藥包：70 式
- ❒ 點火藥筒：120 式
- ❒ 底火：C2
- ❒ 初速：307 公尺/秒
- ❒ 最大射程：6,500 公尺
- ❒ 適用武器：T63，T86 迫砲

Materiel Production Center (MPC). 202nd Arsenal. T63 Mortar Shell (D-1-1) 120mm. 2011 Taipei Aerospace and Defense Technology Exhibition (TADTE). 1/1 Brochure.

Materiel Production Center. TC100 Sub-Caliber Mortar Training Simulation System. 2018 Defence Services Asia (Malaysia). Brochure 1/2.

TC100 Sub-Caliber Mortar Training Simulation System

DSA 2018

Description:

It is comprises of a full-caliber sabot and a sub-caliber cartridge. This system can be used in a small proving ground and to simulate the live-fire scenario to improve the teamwork and verify training result. The sub-caliber cartridge will be loaded in full-caliber sabot. The sub-caliber cartridge can produce the sound, light and smoke during firing.

Specification:

Main-Caliber Sabot

- **Calibre: 60mm, 81mm 120mm and 4.2inch**
- **Fire range: 2-20m (reusable)**

Sub-Caliber Cartridge

- **Calibre: 25mm**
- **Fire range: 70-500m**

Materiel Production Center. TC100 Sub-Caliber Mortar Training Simulation System. 2018 Defence Services Asia (Malaysia). Brochure 2/2.

海軍5吋38倍砲彈系列

2011

本系列彈藥包括 MK52 5 吋 38 倍對空普通彈及 TC97 5 吋清膛藥筒（38/54 倍徑通用）。射擊時，彈頭與發射藥筒分別上膛。擊發後，發射藥燃燒產生之高壓氣體，將彈頭往前推送出砲口，同時彈帶嵌入膛線使彈頭旋轉。彈頭持續發向預定目標區。另清膛藥筒底部之底火被火砲擊針撞擊通電發火後，點燃發射藥，瞬間產生高壓氣體，將留於砲管內部之彈頭推送出砲口，以達清膛之效果。

【性能諸元】

- ❒ 彈重：24.5 公斤
- ❒ 彈長：515 公厘
- ❒ 彈筒長：562 公厘
- ❒ 發射藥：SPDN 或 SPCF 發射藥
- ❒ 射程：16,500 公尺
- ❒ 底火：C9
- ❒ 適用武器：5 吋 38 倍加砲（對空普通彈）、5 吋 38/54 倍徑（通用）加砲（清膛藥筒）
- ❒ 彈筒重：9.984 公斤
- ❒ 初速：797 公尺/秒

Materiel Production Center (MPC). 202nd Arsenal. MK52. Unidentified Taipei Aerospace and Defense Technology Exhibition (TADTE). 1/2 Brochure.

Materiel Production Center (MPC). 202nd Arsenal. MK52. Unidentified Taipei Aerospace and Defense Technology Exhibition (TADTE). Brochure 2/2 (CLOSE-UP).

Materiel Production Center. 105mm Canister Cartridge. 2018 Defence Services Asia (Malaysia). Brochure 1/2.

105mm Canister Cartridge

Description:

This ammunition is composed of a cartridge of commissioned 105mm HEAT-T, new design projectile, and canister. The high density tungsten balls are filled in the canister to fulfill high lethality. The effective firing-range is over 350m and dispersion diameter is more than 100 meter which has proven by live-fire with M68 series barrel.

Specification:

- Length: 870 mm
- Weight: 24 kg
- Propellant: M30
- Maximum Range: 350 m
- Armament: M48H and M60A3

Materiel Production Center. 105mm Canister Cartridge. 2018 Defence Services Asia (Malaysia). Brochure 2/2.

AIR DEFENSE
GRENADE LAUNCHERS
MORTARS
TURRET GUNS

THE 202nd ARSENAL, MPC

This is a fully servo controlled weapon platform including the electric server, thermal imager, and fire control system. The main weapen of this system is a 40mm Automatic Grenade Launcher (AGL). The fire control system is equipped with ballistic computer and eye-safe laser range finder. Gyrostabilizers provide the functions of elevation and azimuth stabilization. This system is loaded on Cloud Leopard wheeled armor vehicle.

【Specifications】

- ❒ Weight：1300 kg
- ❒ Crew：Commander and Gunner
- ❒ Armament：40 mm automatic grenade launcher (AGL), 7.62 mm coaxial machine gun (MG)
- ❒ Ammunition：48 rds ready for use (40mm AGL), 400 rds ready for use (7.62mm MG)
- ❒ Azimuth：n × 360°
- ❒ Elevation：-5° to +60°
- ❒ Azimuth Drive：30 degrees per second
- ❒ Elevation Drive：30 degrees per second
- ❒ Sensor Unit：Day CCD TV Camera, Thermal Imager, Eye-safe Laser Range Finder

Materiel Production Center (MPC). 202nd Arsenal. 7.62mm/40MM Coaxial Turret Weapon for Clouded Leopard Eight-wheeled Armored Vehicle. Unidentified Taipei Aerospace and Defense Technology Exhibition (TADTE). 1/1 Brochure.

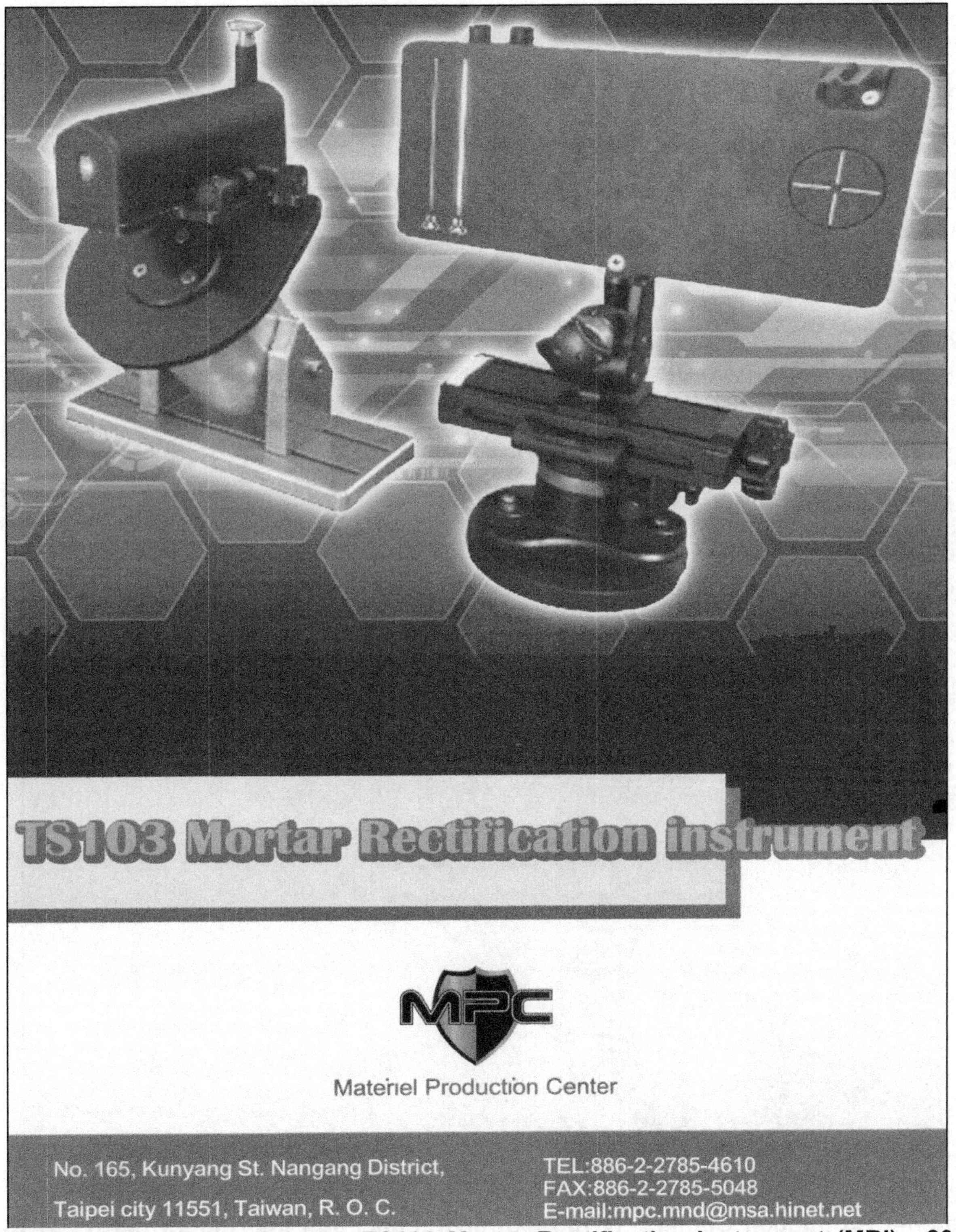

Materiel Production Center. TS103 Mortar Rectification Instrument (MRI). 2018 Defence Services Asia (Malaysia). Brochure 1/2.

TS103 Mortar Rectification instrument

Description:

Mortar Rectification Instrument is used for the alignment between the sighting-axis of a sight and the bore-axis of a mortar. It is suitable for various type of mortars, including mobile mortars. MRI simplifies the boresight procedure and is not limited by the weather and topography. It can easily achieve the alignment between a sight and a mortar and in turn improves the firing efficiency.

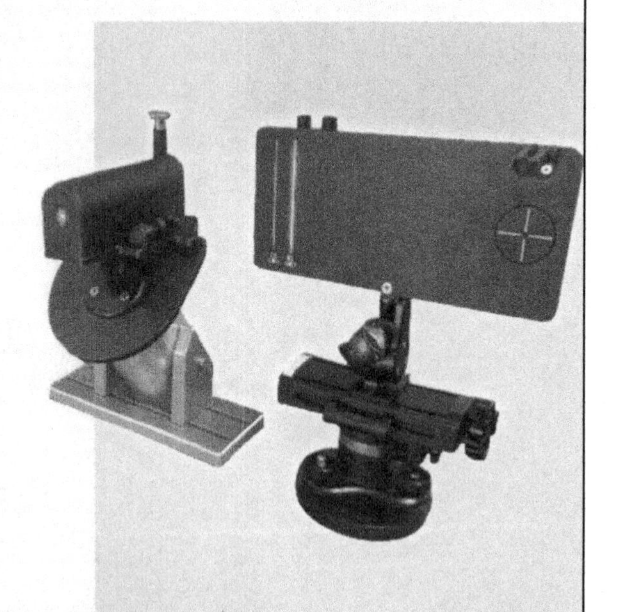

Specification:

- **Weight：**
  (1)Laser Indicator≦3100g
  (2)Adjustable Board≦3000g
- **Degrees：**
  900 mil ± 0.4 mil
- **Power Source：**
  Recharge Lithium Battery, Type 18650
- **Type of Laser Beam：**
  Green Laser≧20 mW
- **Type of Tripod：**
  1.8M height,5kg endurance

Materiel Production Center. TS103 Mortar Rectification Instrument (MRI). 2018 Defence Services Asia (Malaysia). Brochure 2/2.

Materiel Production Center. Advanced 81/120mm Mobile Mortar Weapon System. 2018 Defence Services Asia (Malaysia). Brochure 1/2.

Advanced 81/120mm Mobile Mortar Weapon System

Description:

The system is equipped with an electro-servo control system and advanced firing control modules. It can be integrated with traditional 81/120mm mortar and fire all different types of mortar rounds. The system features automatic fire control and fire direction functions. In the future, the system could be integrated with all kinds of combat vehicles to fulfill Army's tactical requirement.

Specification:

Hummer Type

- **Calibre: 81mm**
- **Length of tube: ≦1.5m**
- **Weight: ≦1,000 kg**
- **Fire range: ≧6,000 m**
- **Elevation: 45°-85°**
- **Angle of azimuth: 360°**

Materiel Production Center. Advanced 81/120mm Mobile Mortar Weapon System. 2018 Defence Services Asia (Malaysia). Brochure 2/2.

120 公厘 T63 迫擊砲

本砲為光膛、砲口裝填之曲射武器，由砲身、砲架、底盤及瞄準具等四大部分組成，其構造簡單、保養容易、火力強大，為支援第一線步兵之最佳武器。配合 120 公厘 T76 迫砲牽引架，更增加機動力，可迅速就戰鬥陣地。

【性能諸元】

- ❐ 口徑：120 公厘
- ❐ 全砲重：119.2 公斤
- ❐ 砲管長：1,620 公厘
- ❐ 底盤直徑：800 公厘
- ❐ 水平射角：13.5 度
- ❐ 最大射速：15 發/分
- ❐ 高低仰角：45 度～80 度
- ❐ 瞄準具：TS-67（M53A1）
- ❐ 最大射程：6,500 公尺
- ❐ 適用彈種：TC63 式高爆彈、TC68 式黃磷彈、TC69 式照明彈

Materiel Production Center (MPC). 202nd Arsenal. T63 Mortar 120mm. 2011 Taipei Aerospace and Defense Technology Exhibition (TADTE). 1/1 Brochure.

Materiel Production Center (MPC). 202nd Arsenal. 40mm/L70 T92 Air Defense Gun. Unidentified Taipei Aerospace and Defense Technology Exhibition (TADTE). 1/1 Brochure.

Materiel Production Center (MPC). 202nd Arsenal. 40mm/L70 T92 Air Defense Gun. Unidentified Taipei Aerospace and Defense Technology Exhibition (TADTE). 1/2 Brochure.

Materiel Production Center. 202nd Arsenal. 40mm/L70 T92 Air Defense Gun. Unidentified Taipei Aerospace and Defense Technology Exhibition (TADTE). 2/2 Brochure.

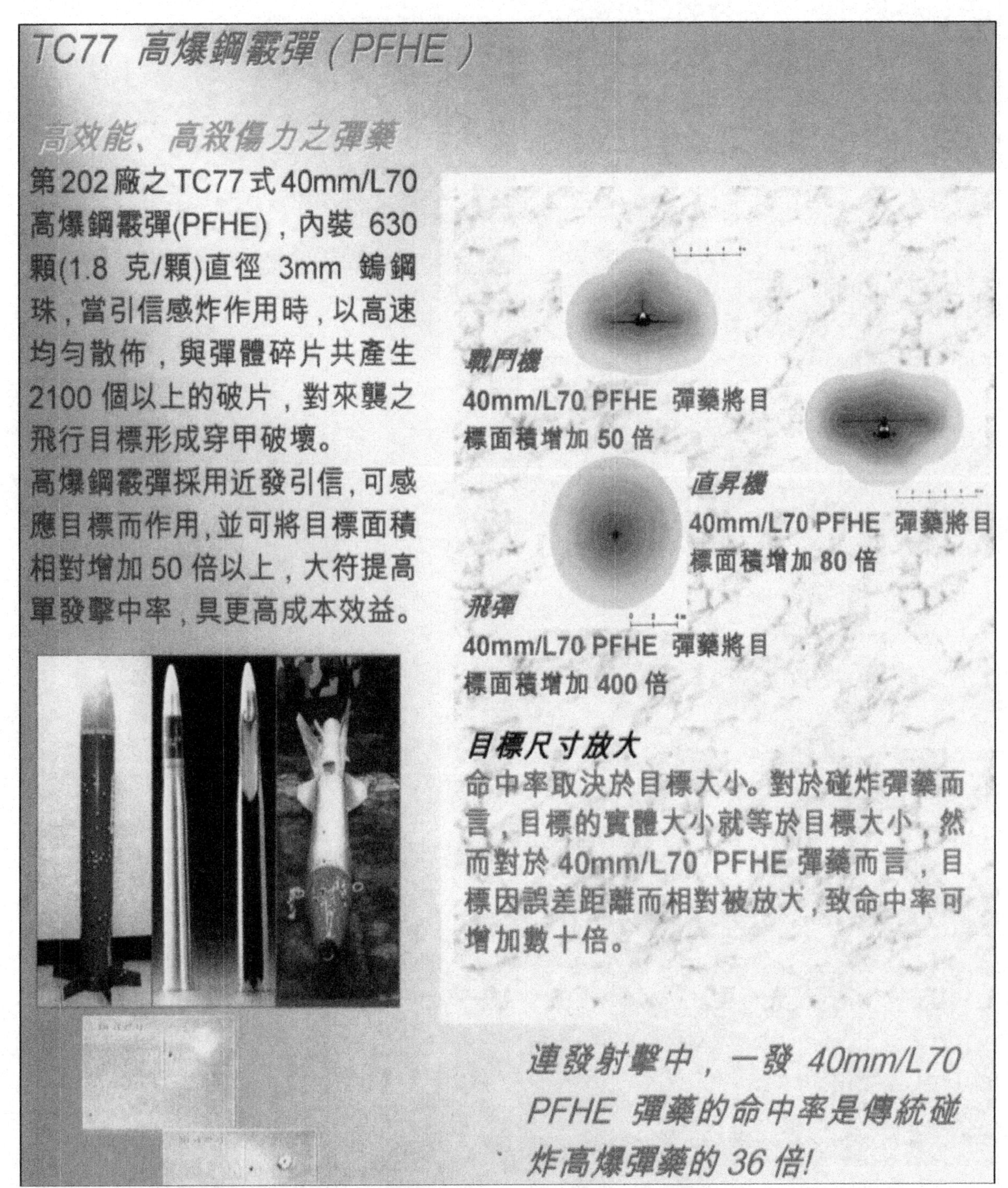

Materiel Production Center. 202nd Arsenal. TC77 Prefragmented High-Explosive (PFHE) for the 40mm/L70 T92 Air Defense Gun. Unidentified Taipei Aerospace and Defense Technology Exhibition (TADTE). 1/1 Brochure.

防空快砲是防空飛彈的最佳輔助武器
現代戰爭之空中主要威脅，來自包括戰術飛彈、戰機、攻擊直昇機、巡弋飛彈與無人搖控載具等，因其具備低空域入侵、截面積小、不易偵測與反應時間短等特性；故低空防禦之最新趨勢，為發展「彈砲合一」之低空防禦系統，而 40mm/L70 T92 防空快砲則可配合各式中、低空防空飛彈，及結合具搜索、追蹤射控雷達，成為「彈砲合一」系統。
高品質與最先進之系統
第 202 廠生產之 40mm/L70 T92 防空快砲為全天候、高效能的低空防空快砲系統，具備光電射控系統(OFCS)，可行獨立作戰，亦可搭配各類型搜索雷達及射控系統，攻擊空中、地面與海面的各種來襲目標，以滿足防空的戰術要求。

Materiel Production Center (MPC). 202nd Arsenal. Air Defense Radar for the 40mm/L70 T92 Air Defense Gun. Unidentified Taipei Aerospace and Defense Technology Exhibition (TADTE). Brochure 1/2.

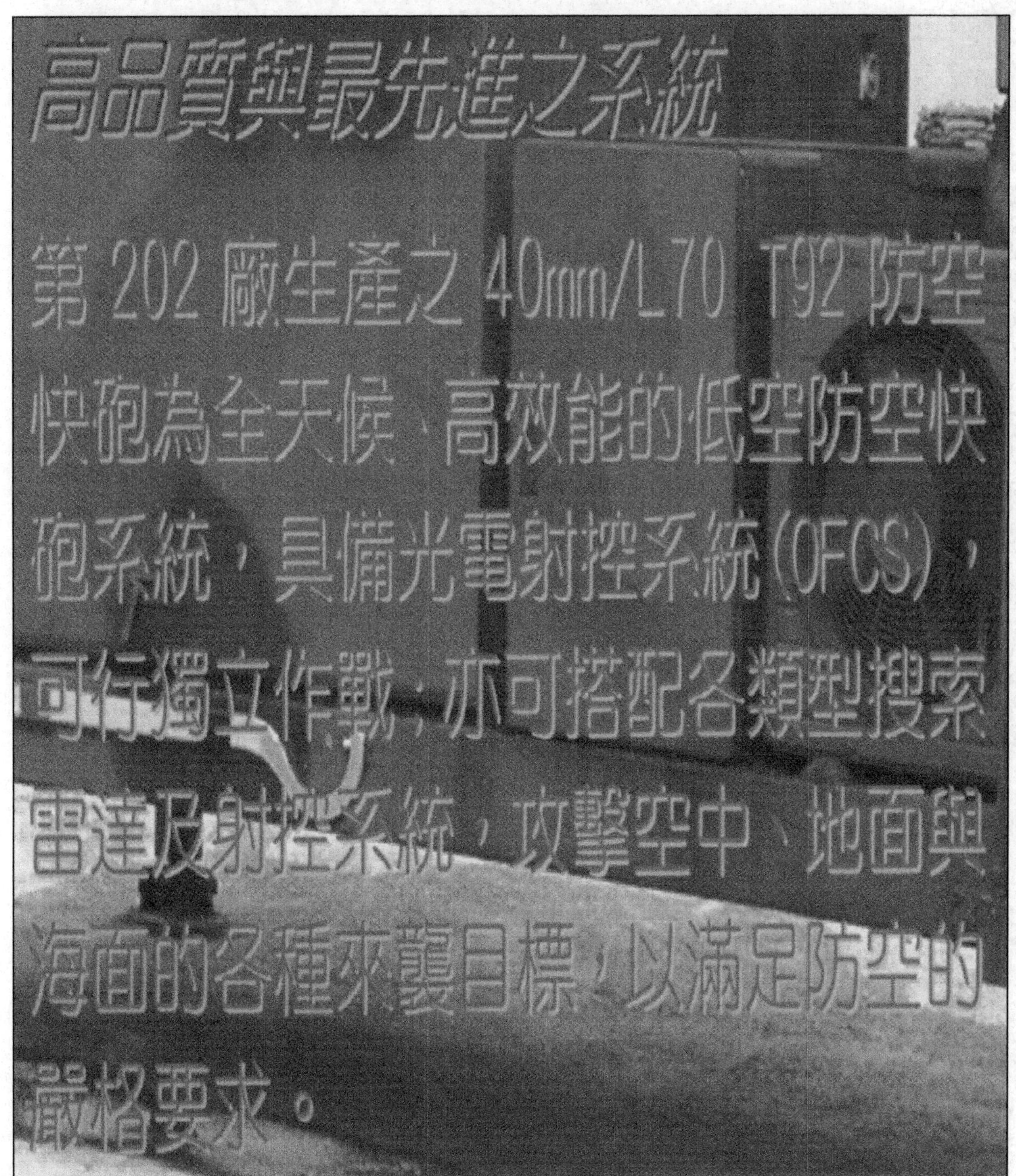

Materiel Production Center (MPC). 202nd Arsenal. Air Defense Radar for the 40mm/L70 T92 Air Defense Gun. Unidentified Taipei Aerospace and Defense Technology Exhibition (TADTE). Brochure 2/2 (CLOSE-UP).

Materiel Production Center (MPC). 205th Arsenal. 5.56/7.62mm Remote Weapons Station. 2015 Taipei Aerospace and Defense Technology Exhibition (TADTE). Brochure 1/2.

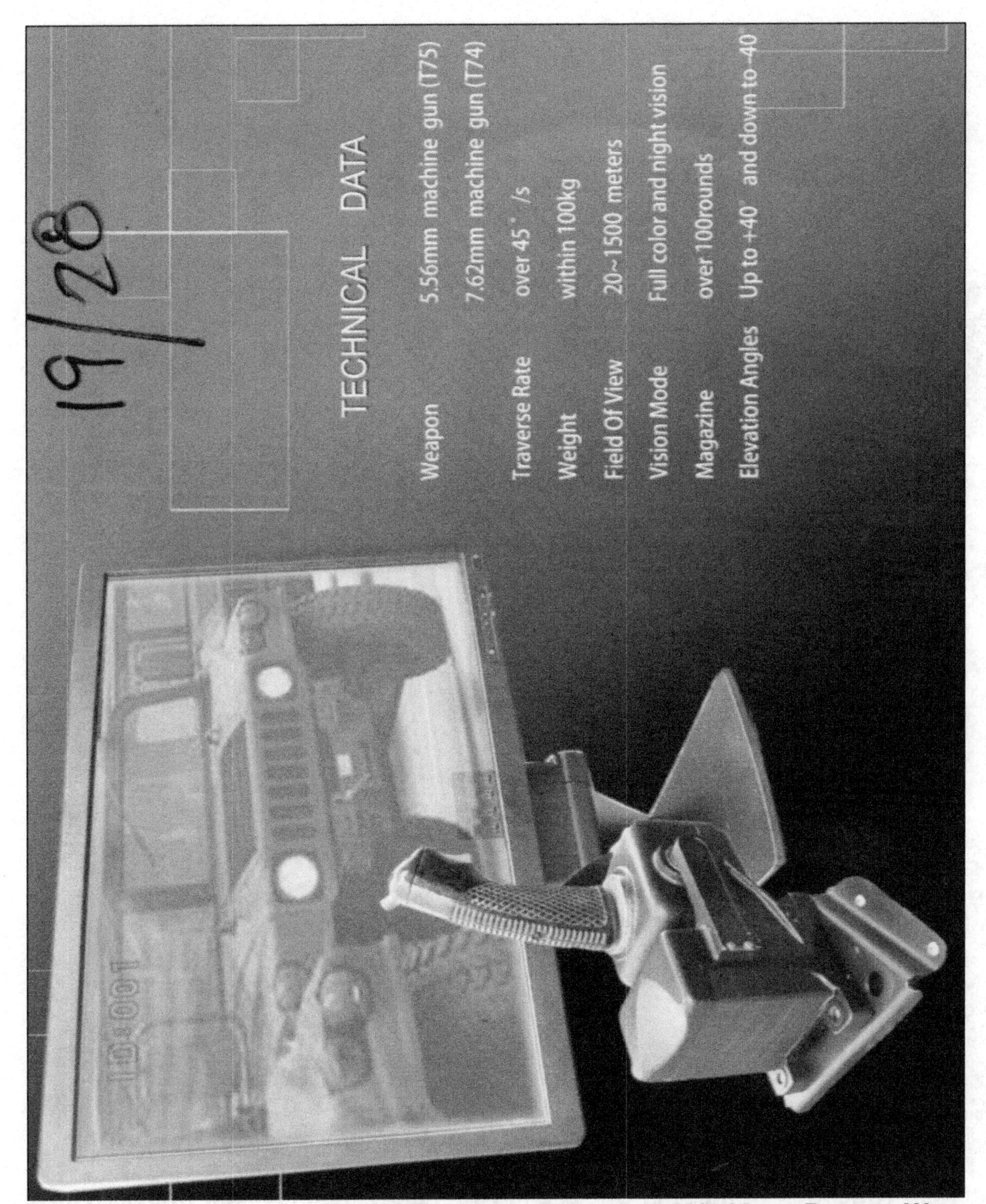

Materiel Production Center (MPC). 205th Arsenal. 5.56/7.62mm Remote Weapons Station. 2015 Taipei Aerospace and Defense Technology Exhibition (TADTE). Brochure 2/2.

Materiel Production Center (MPC). 202nd Arsenal. 7.62mm/40MM Coaxial Turret Weapon. 2019 Taipei Aerospace and Defense Technology Exhibition (TADTE). Final variant for the Clouded Leopard vehicle. Notice configuration differences. Editor Photograph.

BULLET PROOF VESTS
HELMETS
BOOTS
CAMOUFLAGE
PARACHUTES
MISCELLANEOUS

Materiel Production Center (MPC). 202nd Arsenal. Bullet Proof Vest. 2011 Taipei Aerospace and Defense Technology Exhibition (TADTE). Brochure 1/3.

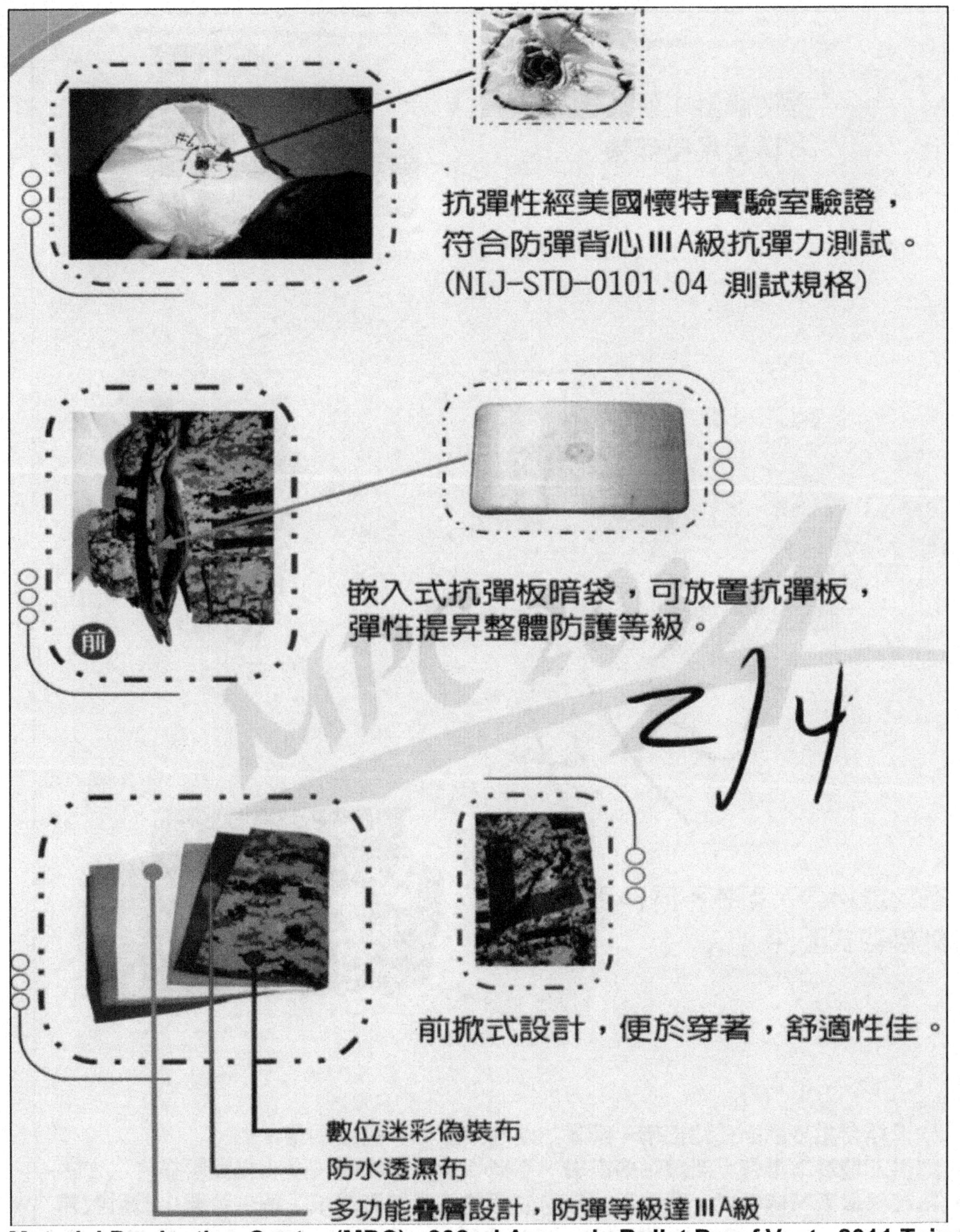

Materiel Production Center (MPC). 202nd Arsenal. Bullet Proof Vest. 2011 Taipei Aerospace and Defense Technology Exhibition (TADTE). Brochure 2/3.

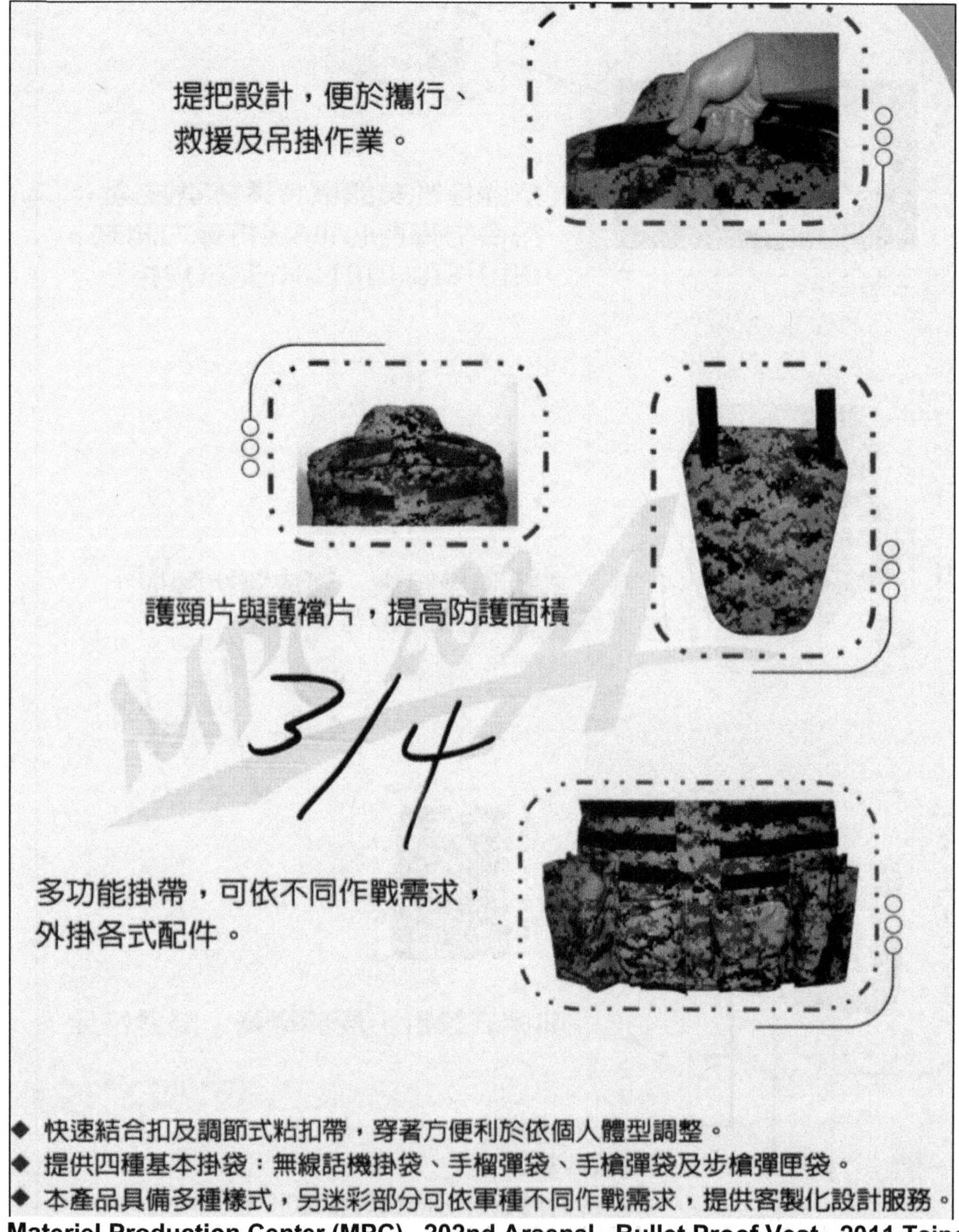

提把設計，便於攜行、
救援及吊掛作業。
護頸片與護襠片，提高防護面積
3/4
多功能掛帶，可依不同作戰需求，
外掛各式配件。
◆ 快速結合扣及調節式粘扣帶，穿著方便利於依個人體型調整。
◆ 提供四種基本掛袋：無線話機掛袋、手榴彈袋、手槍彈袋及步槍彈匣袋。
◆ 本產品具備多種樣式，另迷彩部分可依軍種不同作戰需求，提供客製化設計服務。

Materiel Production Center (MPC). 202nd Arsenal. Bullet Proof Vest. 2011 Taipei Aerospace and Defense Technology Exhibition (TADTE). Brochure 3/3.

Materiel Production Center (MPC). 202nd Arsenal. Protective Clothing. 2011 Taipei Aerospace and Defense Technology Exhibition (TADTE). Brochure 1/2.

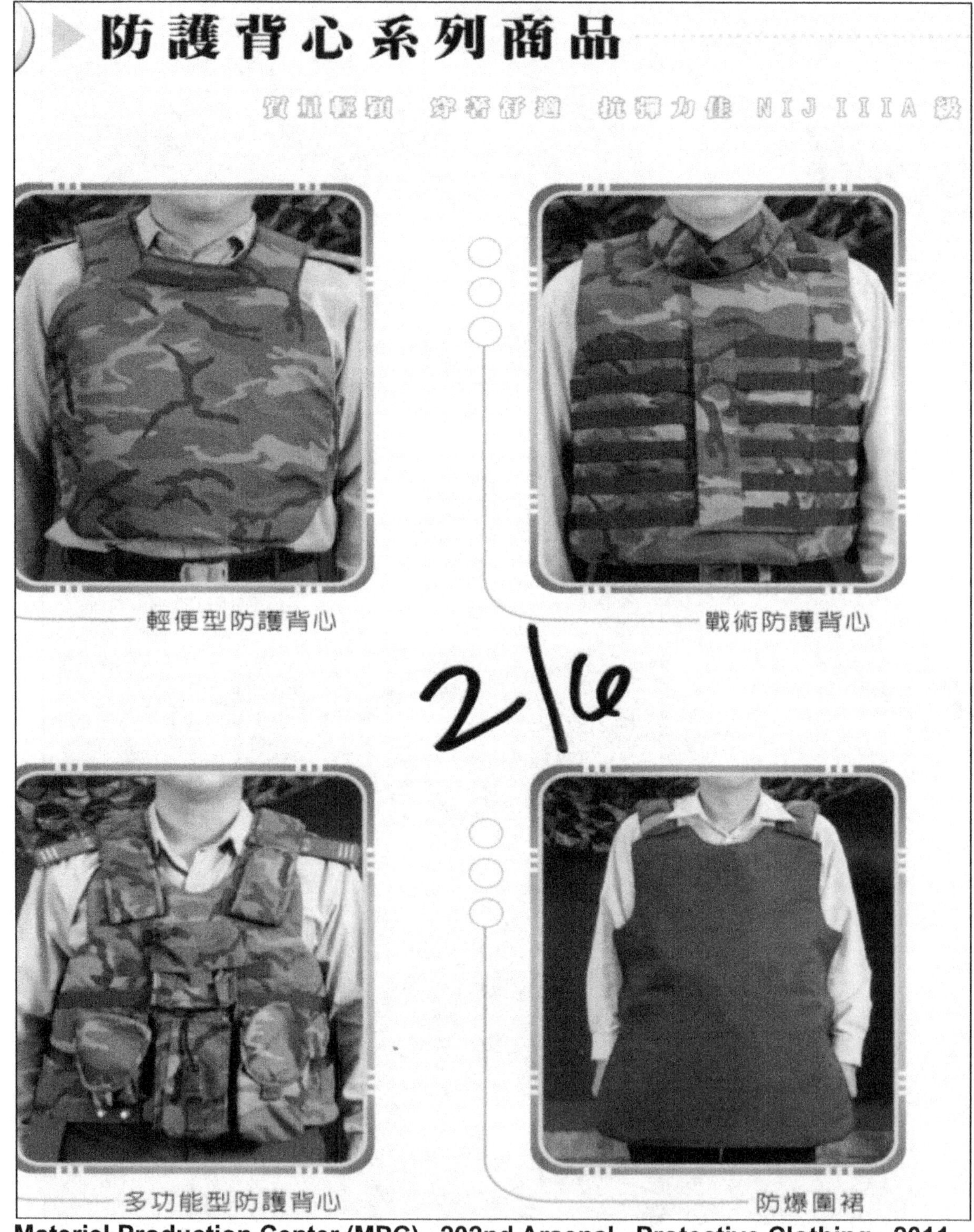

防護背心系列商品
質量輕穎 穿著舒適 抗彈力佳 NIJ IIIA 級
輕便型防護背心
戰術防護背心
多功能型防護背心
防爆圍裙

Materiel Production Center (MPC). 202nd Arsenal. Protective Clothing. 2011 Taipei Aerospace and Defense Technology Exhibition (TADTE). Brochure 2/2.

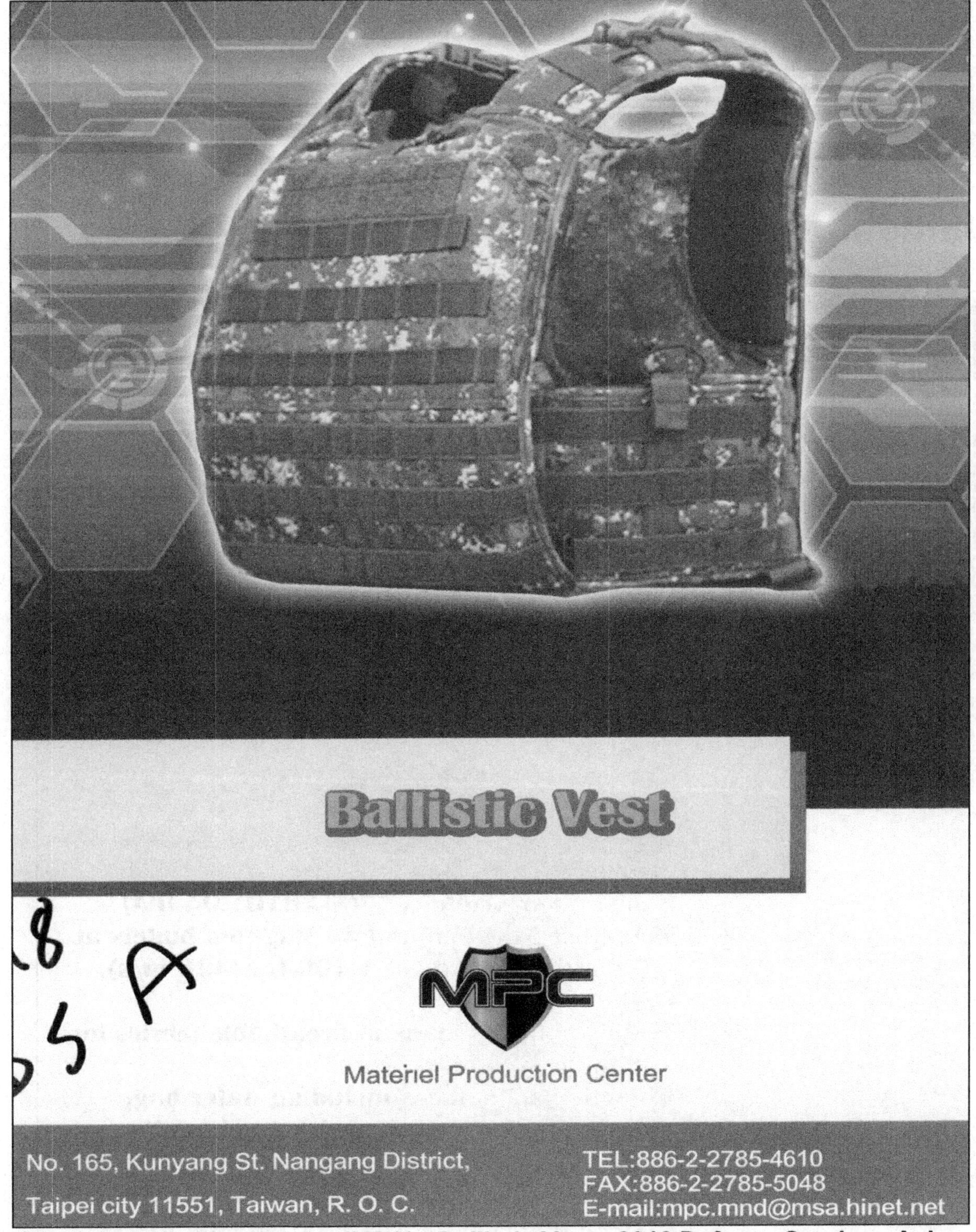

Materiel Production Center (MPC). Ballistic Vest. 2018 Defence Services Asia (DSA). Brochure 1/2.

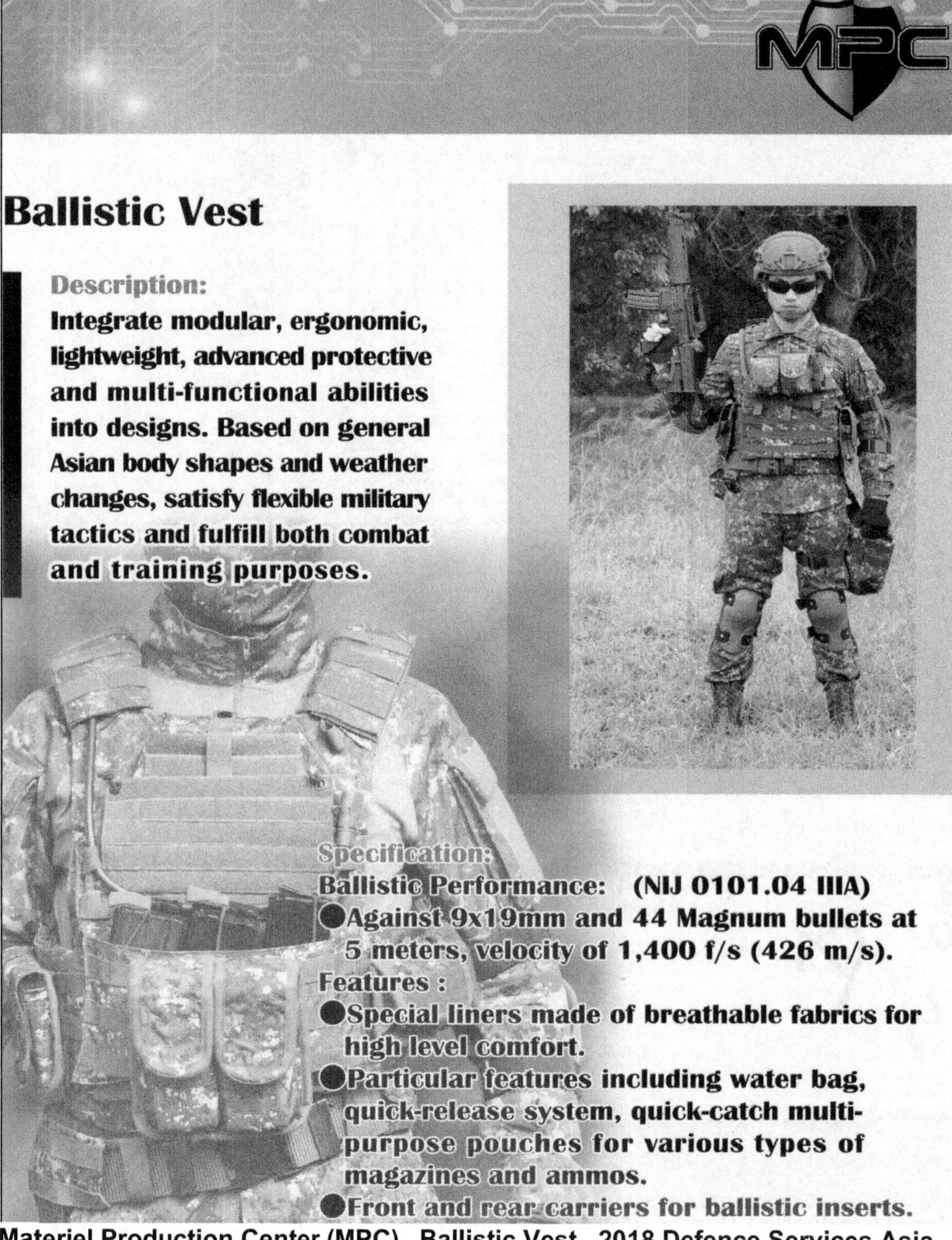

MPC
Ballistic Vest
Description:
Integrate modular, ergonomic, lightweight, advanced protective and multi-functional abilities into designs. Based on general Asian body shapes and weather changes, satisfy flexible military tactics and fulfill both combat and training purposes.
Specification:
Ballistic Performance: (NIJ 0101.04 IIIA)
●Against 9x19mm and 44 Magnum bullets at 5 meters, velocity of 1,400 f/s (426 m/s).
Features :
●Special liners made of breathable fabrics for high level comfort.
●Particular features including water bag, quick-release system, quick-catch multi-purpose pouches for various types of magazines and ammos.
●Front and rear carriers for ballistic inserts.

Materiel Production Center (MPC). Ballistic Vest. 2018 Defence Services Asia (DSA). Brochure 2/2.

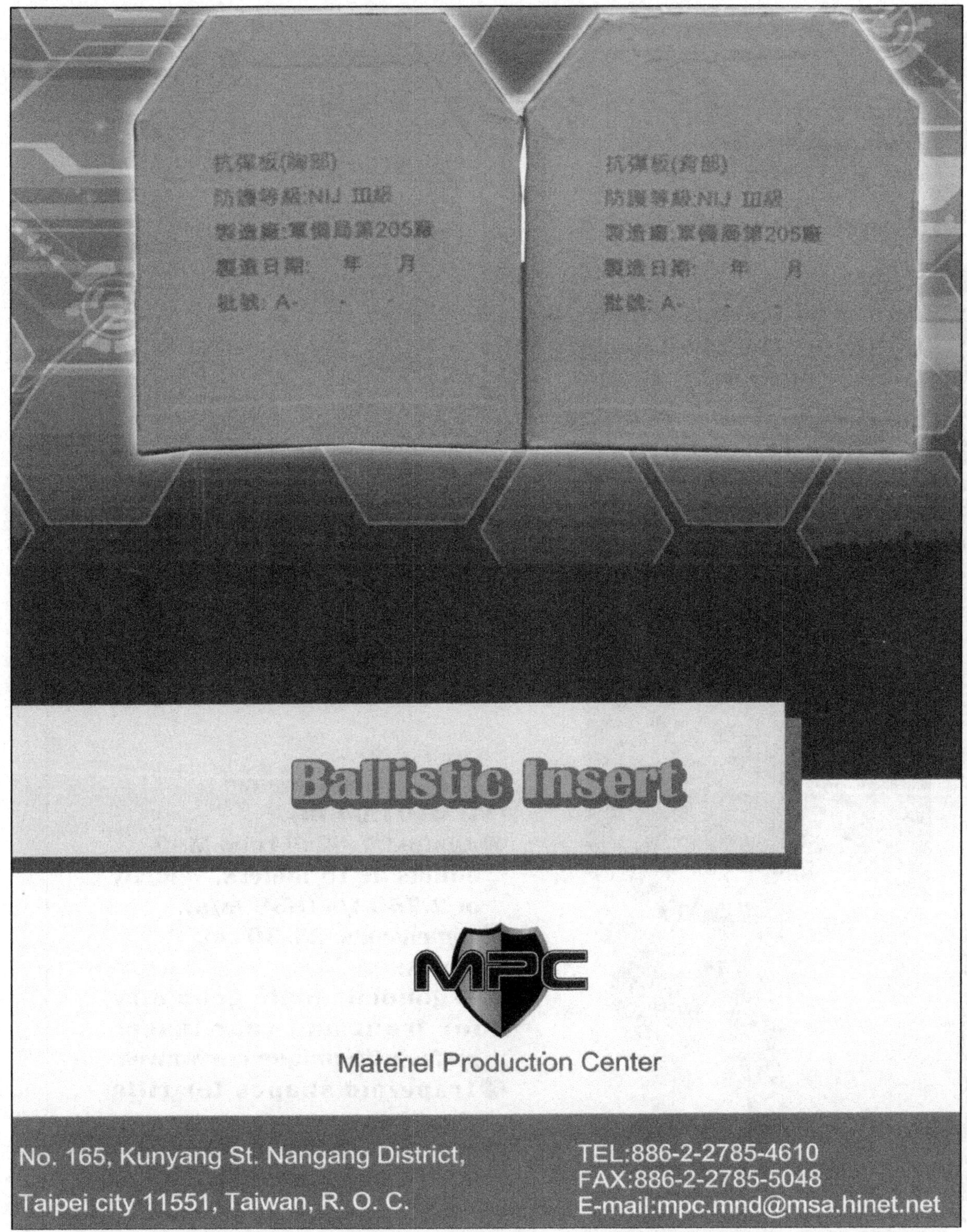

Materiel Production Center (MPC). Ballistic Insert NIJ 0101.04 III. 2018 Defence Services Asia (DSA). Brochure 1/2.

MPC

Special Forces

Ballistic Insert

Description:

The ballistic vest has achieved the level IIIA of the U.S. National Institute of Justice Standard-0101.04. In order to upgrade its ballistic protection to III, increase the ballistic inserts made of UHMWPE materials. Then it could provide protection against 7.62mm bullets.

Standard Ground Troops

Specification:

Ballistic Performance:
(NIJ 0101.04 III)

- **Against 7.62x51mm M80 bullets at 15 meters, velocity of 2,750 f/s (838 m/s).**
- **Dimensions: 25x30 cm.**

Features:

- **Ergonomic plate geometry for front and rear insert plates with unique curvatures.**
- **Trapezoid shapes for rifle shooting.**
- **Weight: 1.94kg for front and rear inserts.**

Materiel Production Center (MPC). Ballistic Insert NIJ 0101.04 III. 2018 Defence Services Asia (DSA). Brochure 2/2.

Materiel Production Center (MPC). 205th Arsenal. Modular Ballistic Vest. 2015 Taipei Aerospace and Defense Technology Exhibition (TADTE). Brochure 1/1.

Materiel Production Center (MPC). 205th Arsenal. Explosive Ordnance Disposal (EOD) EOD Ballistic Vest. 2015 Taipei Aerospace and Defense Technology Exhibition (TADTE). Brochure 1/1.

Materiel Production Center (MPC). 205th Arsenal. Ballistic Helmet. 2015 Taipei Aerospace and Defense Technology Exhibition (TADTE). Brochure 1/1.

Materiel Production Center (MPC). 202nd Arsenal. Protective Clothing. 2011 Taipei Aerospace and Defense Technology Exhibition (TADTE). Brochure 1/1.

Materiel Production Center (MPC). Modular Ballistic Helmet (MBH)/Enhanced Modular Ballistic Helmet (EMBH). 2018 Defence Services Asia (DSA). Brochure 1/2.

DSA 2018

MPC

Modular Ballistic helmet

Description:

Integrate modular, ergonomic, lightweight, advanced protective and multi-functional abilities into designs. Based on general Asian head shapes and weather changes, satisfy flexible military tactics and fulfill both combat and training purposes.

Standard Ground Troops

Special Forces

Specification:

Ballistic Performance:

- **Against 9x19mm bullets at 5 m Velocity of 1,400 f/s (426 m/s).**
- **Qualify for MIL-STD-662F V50 > 2,300ft/s.**

Features:

- **Equip Night-Vision-Telescope Frame (Aluminum) and Tactical Rail (Plastics), suitable for various gears such as a monocular night-vision-telescopes, identification lights and headlights.**

Materiel Production Center (MPC). Modular Ballistic Helmet (MBH)/Enhanced Modular Ballistic Helmet (EMBH). 2018 Defence Services Asia (DSA). Brochure 2/2.

Materiel Production Center (MPC). 205th Arsenal. CB-99 Waterproof Combat Boots. 2015 Taipei Aerospace and Defense Technology Exhibition (TADTE). Brochure 1/1.

Materiel Production Center (MPC). 205th Arsenal. CB-103 Combat Boots. 2015 Taipei Aerospace and Defense Technology Exhibition (TADTE). Brochure 1/1.

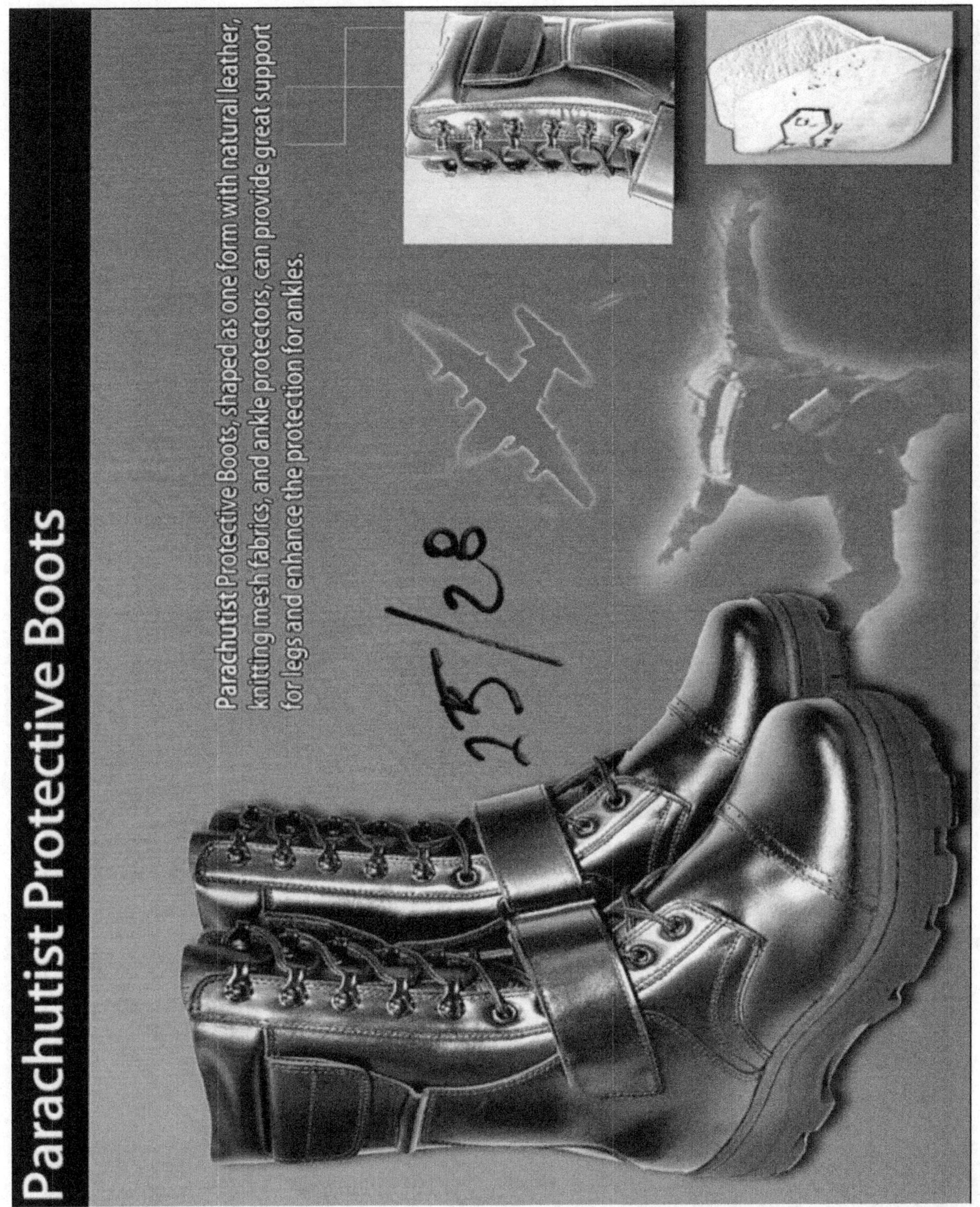

Materiel Production Center (MPC). 205th Arsenal. Parachutist Protective Boots. 2015 Taipei Aerospace and Defense Technology Exhibition (TADTE). Brochure 1/1.

Materiel Production Center (MPC). Flying Vehicle Recovering Parachute. 2018 Defence Services Asia (DSA). Brochure 1/2.

Flying Vehicle Recovering Parachute

Description:

The drone recovering parachute is designed for flying vehicle under 100kg. It provides steady drop and reduce impact when landing, minimizing the damage of the drone.

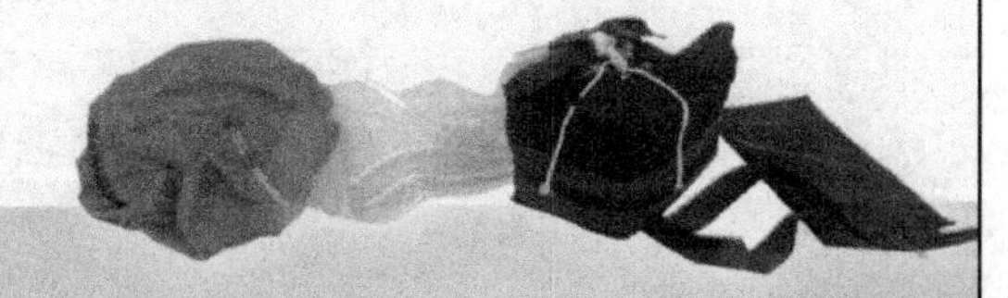

Specification:

- **Complete Assembly Weight:** 13.2lbs. (6kg)
- **Inflate Diameter:** 24Feet (7.3m)
- **Number of Suspension Lines:** 24
- **Maximum Weight Capacity:** 286lbs. (130kg)
- **Maximum Deployment Speed:** 130kts (240km/h)
- **Descent Rate:** 21ft/s (6.5m/s)

Materiel Production Center (MPC). Flying Vehicle Recovering Parachute. 2018 Defence Services Asia (DSA). Brochure 2/2.

Materiel Production Center (MPC). MC1-1B Main Parachute. 2018 Defence Services Asia (DSA). Brochure 1/2.

MC1-1B Main Parachute

Description:

Designed for precision infiltration of Airborne Forces, the MC1-1 steerable troop parachute assembly allows the jumper to maneuver the parachute towards target precisely. Mainly used in special airborne operations and training.

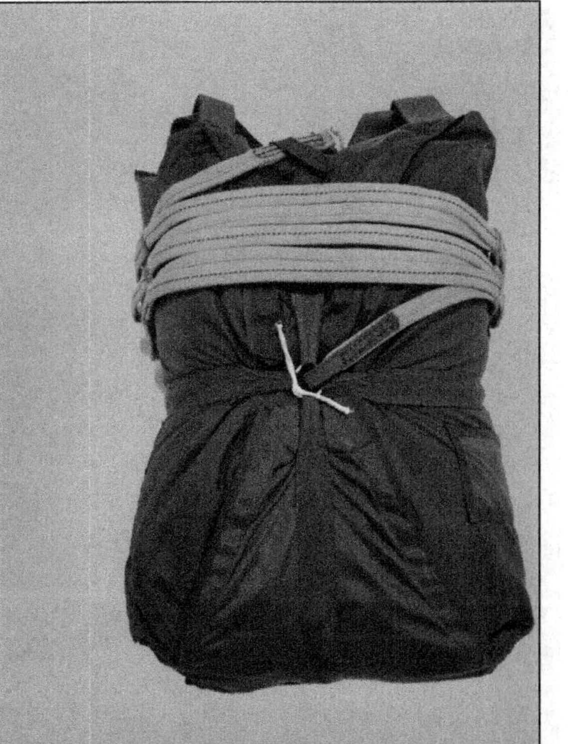

Specification:

- **Complete Assembly Weight:** 29 lbs. (13.2kg)
- **Inflate Diameter:** 35 Feet (10.67m)
- **Number of Suspension Lines:** 30
- **Maximum Weight Capacity:** 360 lbs. (163kg)
- **Descent Rate:** 12-18 ft/s (3.7-5.5m/s)

Materiel Production Center (MPC). MC1-1B Main Parachute. 2018 Defence Services Asia (DSA). Brochure 2/2.

Materiel Production Center (MPC). T-10B Main Parachute. 2018 Defence Services Asia (DSA). Brochure 1/2.

T-10B Main Parachute

Description:

The T-10B main parachute is designed for combat mass-assault airborne operations and training. It provides safety landings and has been an efficient equipment for airborne troops.

Specification:

- **Complete Assembly Weight:** 29 lbs. (13.2kg)
- **Inflate Diameter:** 35 Feet (10.7m)
- **Number of Suspension Lines:** 30
- **Maximum Weight Capacity:** 360 lbs. (163kg)
- **Descent Rate:** 12-18 ft/s (3.7-5.5m/s)

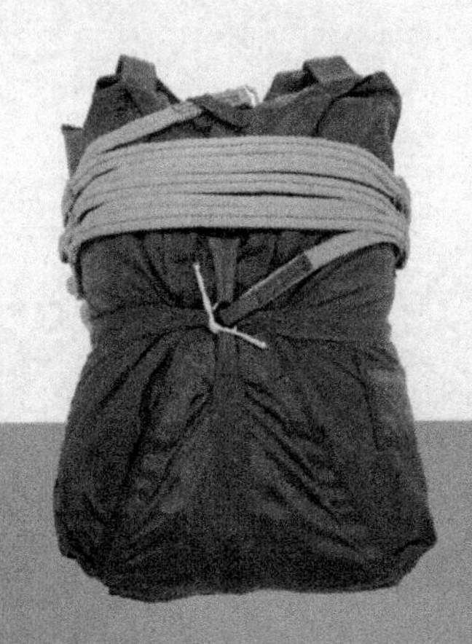

Materiel Production Center (MPC). MC1-1B Main Parachute. 2018 Defence Services Asia (DSA). Brochure 2/2.

Materiel Production Center (MPC). T-10R Reserve Parachute. 2018 Defence Services Asia (DSA). Brochure 1/2.

T-10R Reserve Parachute

Description:

T-10R Reserve Parachute is an emergency parachute designed to be used in the event of a malfunction of the primary back type parachute.

Specification:

- **Complete Assembly Weight:** 13 lbs (6kg)
- **Inflate Diameter:** 24 Feet (7.3m)
- **Number of Suspension Lines:** 24
- **Maximum Weight Capacity:** 300 lbs (136kg)
- **Descent Rate:** 12-18 ft/s (3.7-5.5m/s)

Materiel Production Center (MPC). T-10R Reserve Parachute. 2018 Defence Services Asia (DSA). Brochure 2/2.

Army Aviation and Special Forces Command (陸軍航空特戰指揮部). Skydiving Demonstration Team (中華民國陸軍神龍小組). The members are from the parachute training instructor group of the Army Airborne Training Center (大武營「陸軍空降訓練中心」). 1/3.

Army Aviation and Special Forces Command (陸軍航空特戰指揮部). **Skydiving Demonstration Team** (中華民國陸軍神龍小組). **The members are from the parachute training instructor group of the Army Airborne Training Center** (大武營「陸軍空降訓練中心」). **2/3.**

Cold War Era patch. Skydiving Demonstration Team (中華民國陸軍神龍小組). 3/3.

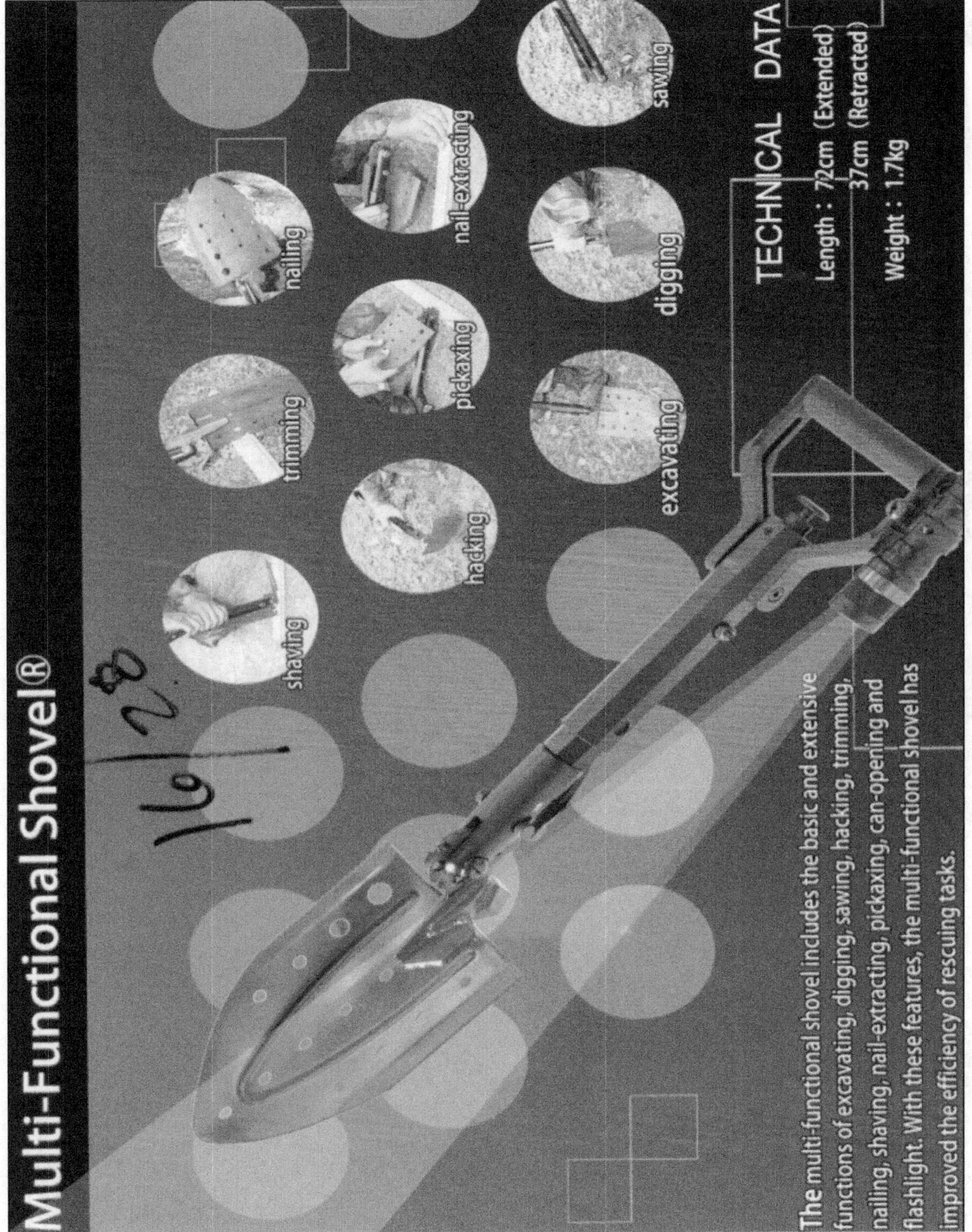

Materiel Production Center (MPC). 205th Arsenal. Multi-Functional Shovel. 2015 Taipei Aerospace and Defense Technology Exhibition (TADTE). Brochure 1/1.

Materiel Production Center (MPC). 202nd Arsenal. Camouflage. 2011 Taipei Aerospace and Defense Technology Exhibition (TADTE). Brochure 1/2.

防護用品相關資訊

防護背心

質量輕，採人因工程設計，製供國軍最佳適體性，穿著舒適。

抗彈性能：達美國司法委員會抗彈標準NIJ IIIA級，可抵擋9mm全金屬包覆圓頭子彈(FMJRN)，彈頭重量8.0克，彈速427m/s(1400ft/s)以下之射擊及抵擋44麥格農手槍包覆軟質彈頭(JSP)，彈頭重量15.6克，彈速427m/s(1400ft/s)以下之射擊。

多功能防護背心

適用於不同演訓及作戰需求，外掛各式配件。

輕便型防護背心

適用於一般衛哨勤務。

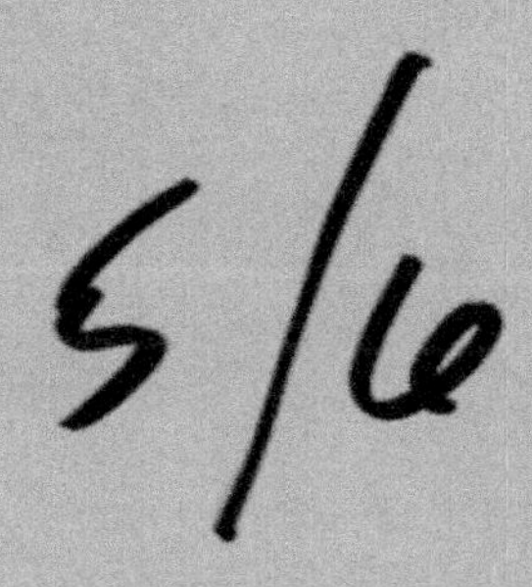

戰術防護背心

並依不同演訓及作戰需求，外掛各式配件。

防爆圍裙

重量：L37kg XL40kg 依MIL-STD-662E法測試，V50値測試大於每秒1650呎以上。

高階防護頭盔

1.符合國際防護材料性能精進趨勢與滿足未來單兵作戰要求，
提昇國軍戰力與士官兵於戰場之生存安全。

2.抗彈性能：
依MIL-STD-662F測試法及MIL-H-44099A測試標準，
V50値達2200呎/秒(含)以上；另依NIJ-STD-0106.01頭盔抗彈測試法達NIJ-STD-0108.01
防護材料抗彈標準IIIA 級 9mm FMJ 彈及 44Magnum(麥格農)Lead SWC彈之測試標準。

3.「重量輕」、「抗彈性高」、「防護面積大」，「符合頭型尺寸」。

偽裝網

使敵人不易辨別或誤判經偽裝後之工事裝備，達有效欺敵之目的。

功能特性：1.組合搭配方便，人員操作簡易，可視偽裝標的物大小調整。
2.接近背景之可見光及近紅外線偽裝，有效降低敵人辨識，達欺敵功能。
3.以近紅外線觀測鏡實測，防近紅外線功效良好。

Materiel Production Center (MPC). 202nd Arsenal. Camouflage. 2011 Taipei Aerospace and Defense Technology Exhibition (TADTE). Brochure 2/2.

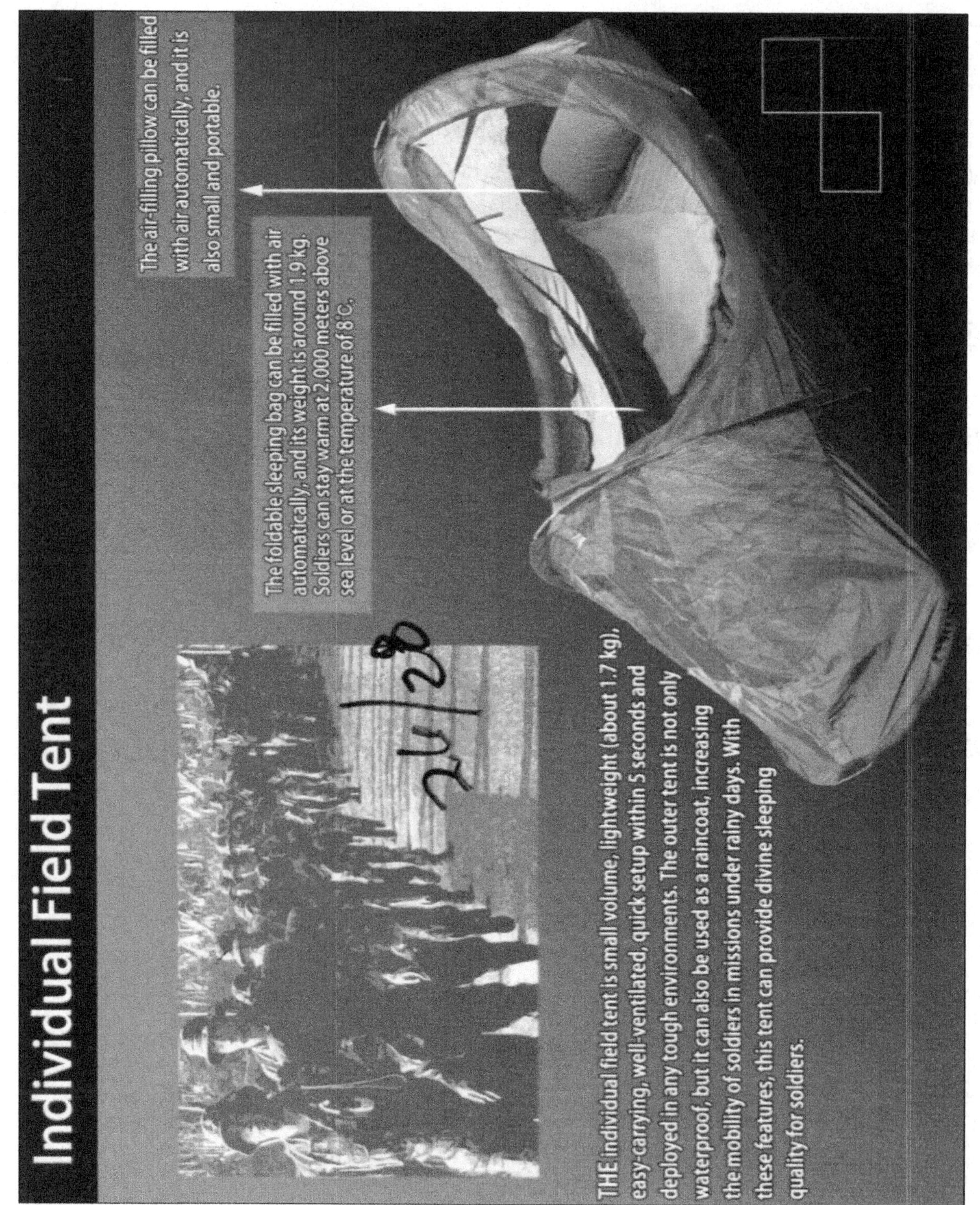

Materiel Production Center (MPC). 205th Arsenal. Individual Field Tent. 2015 Taipei Aerospace and Defense Technology Exhibition (TADTE). Brochure 1/1.

Materiel Production Center (MPC). 204th Arsenal. T4-98 Field Movable Shower. 2011 Taipei Aerospace and Defense Technology Exhibition (TADTE). Brochure 1/2.

Materiel Production Center (MPC). 204th Arsenal. T4-98 Field Movable Shower. 2011 Taipei Aerospace and Defense Technology Exhibition (TADTE). Brochure 2/2.

煙光求救信號筒

◎煙光求救信號筒以單手擊發後，手持發煙或發光之信號筒，適用於飛行員或登山者遇難時，對搜救人員發出求救信號，白天發橙煙，夜間發紅光。

◎本產品已獲得中華民國專利新型第185180號。

Materiel Production Center (MPC). 204th Arsenal. Hand Signal Flare. 2011 Taipei Aerospace and Defense Technology Exhibition (TADTE). Brochure 1/2.

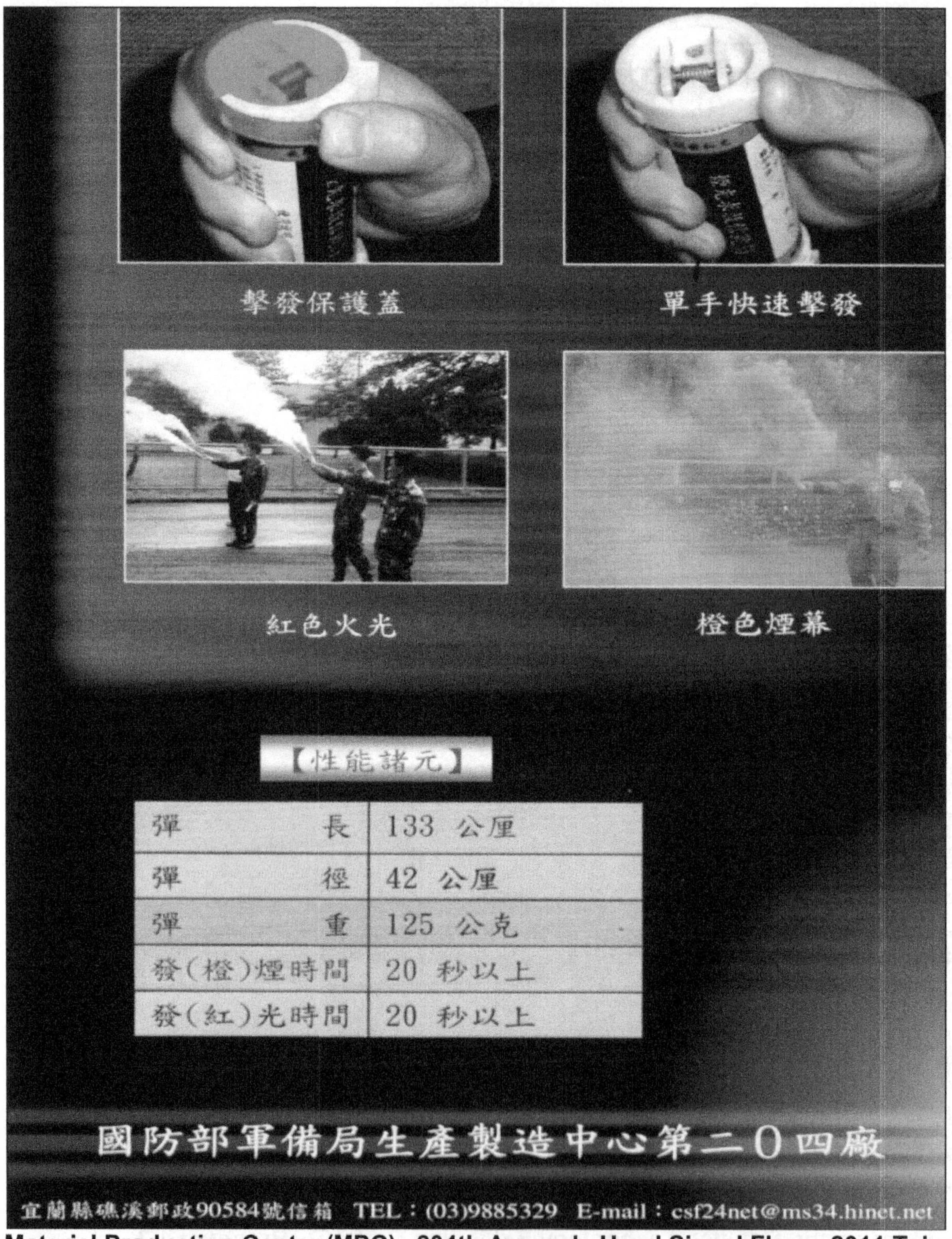

彈長	133 公厘
彈徑	42 公厘
彈重	125 公克
發(橙)煙時間	20 秒以上
發(紅)光時間	20 秒以上

Materiel Production Center (MPC). 204th Arsenal. Hand Signal Flare. 2011 Taipei Aerospace and Defense Technology Exhibition (TADTE). Brochure 2/2.

TC91個人用信號包

◎TC91個人用信號包，包含一拋棄式發射器及五枚信號彈，信號彈可依需求選擇紅色、黃色、綠色或白色。

◎體積小，重量輕，使用於單兵作戰中小區域信號傳遞；並可提供飛行員或海上、登山活動人員，遭遇急難時，發出求救信號。

◎發射器具有可快速拆裝之彈藥結合座，可在短時間內以單手完成裝彈與擊發動作，增加被搜救人員發現機會。

Materiel Production Center (MPC). 204th Arsenal. TC91 Hand Signal Flare. 2011 Taipei Aerospace and Defense Technology Exhibition (TADTE). Brochure 1/2.

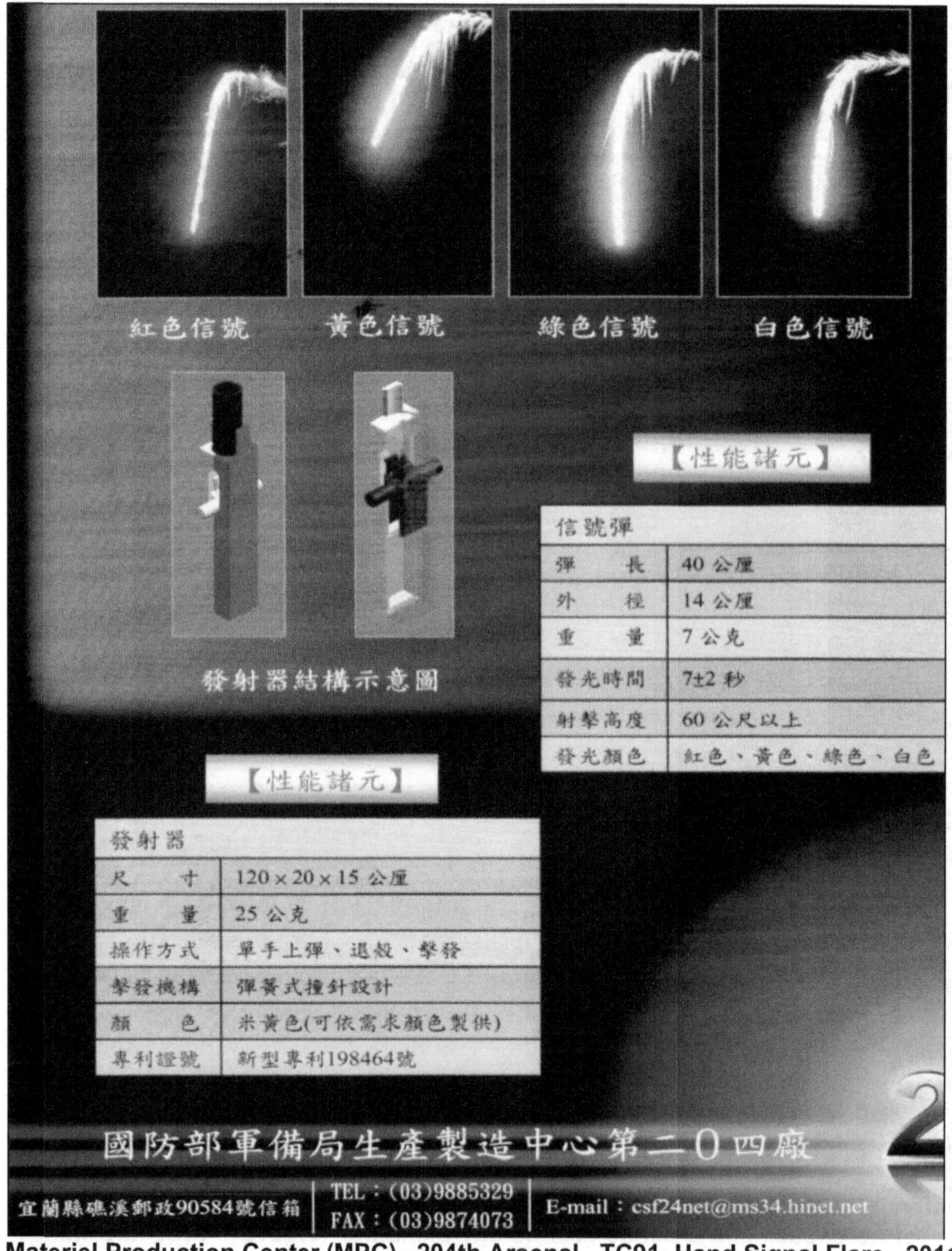

信號彈	
彈　長	40 公厘
外　徑	14 公厘
重　量	7 公克
發光時間	7±2 秒
射擊高度	60 公尺以上
發光顏色	紅色、黃色、綠色、白色

發射器	
尺　寸	120×20×15 公厘
重　量	25 公克
操作方式	單手上彈、退殼、擊發
擊發機構	彈簧式撞針設計
顏　色	米黃色(可依需求顏色製供)
專利證號	新型專利198464號

Materiel Production Center (MPC). 204th Arsenal. TC91 Hand Signal Flare. 2011 Taipei Aerospace and Defense Technology Exhibition (TADTE). Brochure 2/2.

Materiel Production Center (MPC). 204th Arsenal. TC96 Smoke Canister. 2011 Taipei Aerospace and Defense Technology Exhibition (TADTE). Brochure 1/2.

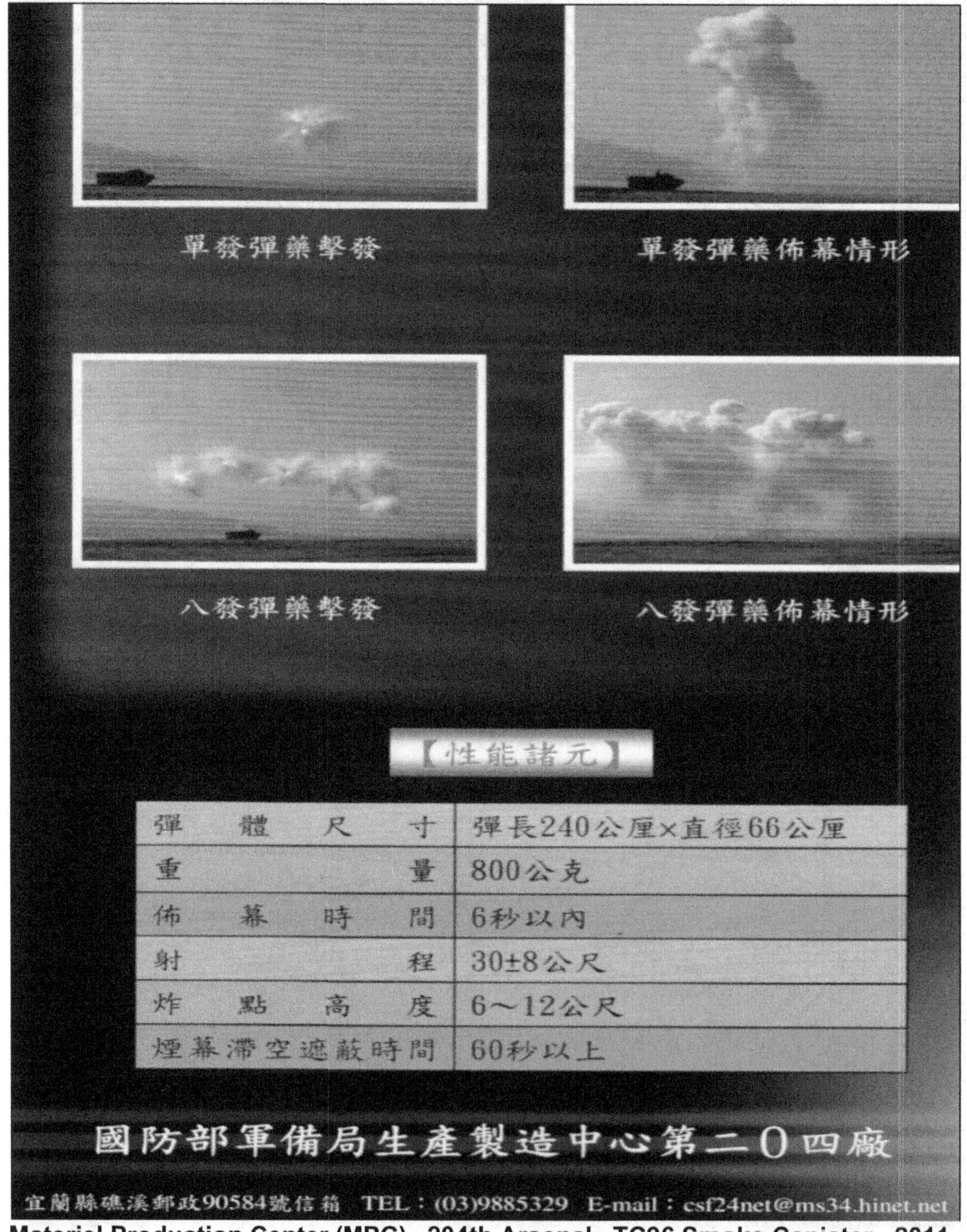

Materiel Production Center (MPC). 204th Arsenal. TC96 Smoke Canister. 2011 Taipei Aerospace and Defense Technology Exhibition (TADTE). Brochure 2/2.

防一氧化碳緊急照明逃生面罩箱

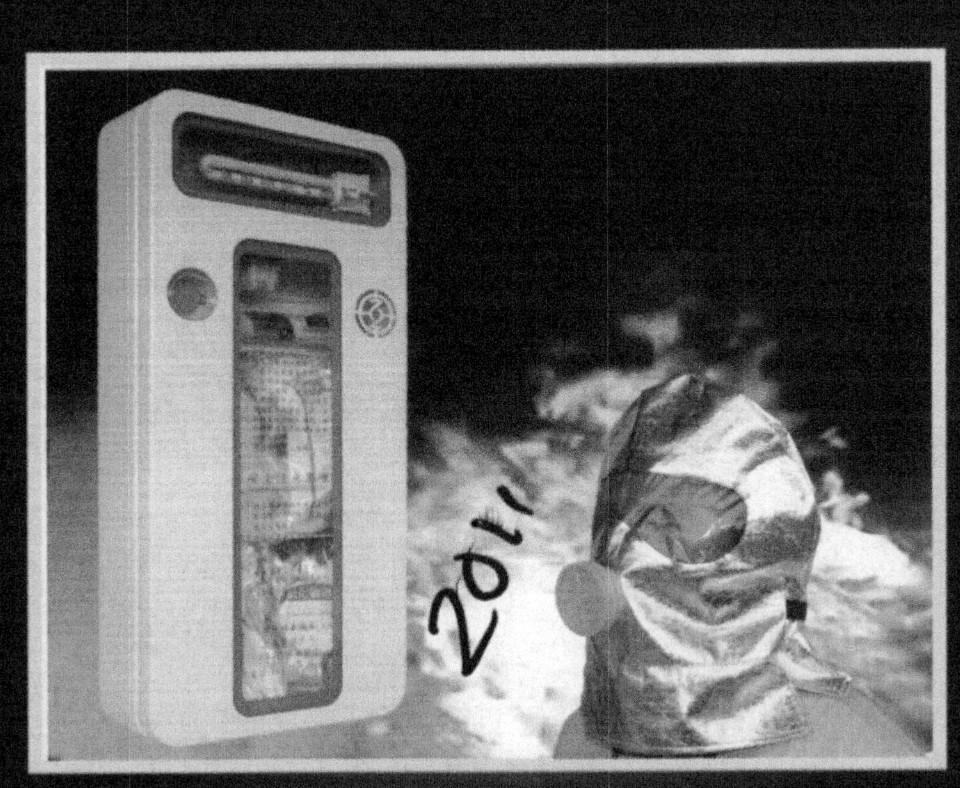

◎防一氧化碳逃生面罩可供人員於火災現場第一時間逃生時保護頭部(包括呼吸器官、眼睛、臉及頸部)，避免遭受火場高熱、濃煙及一氧化碳的傷害，為火場逃生時最佳的個人防護裝備。

◎防一氧化碳濾罐採用高效能濾材，可濾除火場濃煙中有害粒子，並以高科技蜂巢結構裝填奈米級觸媒，可有效轉換一氧化碳，保障人員生命安全。

◎一氧化碳防護效能符合CNS11981 Z2064及歐盟EN403規範。

◎面罩箱具備自動照明功能，於火災現場停電時，面罩箱可自動照明1小時以上，箱內配置防一氧化碳逃生面罩4只、求生口哨及手電筒，提供火場人員緊急逃生使用，增加火場救災搜救功能。

◎本產品已獲得中華民國專利新型第198275號及第M321787號。

Materiel Production Center (MPC). 204th Arsenal. Fire Protection Equipment. 2011 Taipei Aerospace and Defense Technology Exhibition (TADTE). Brochure 1/2.

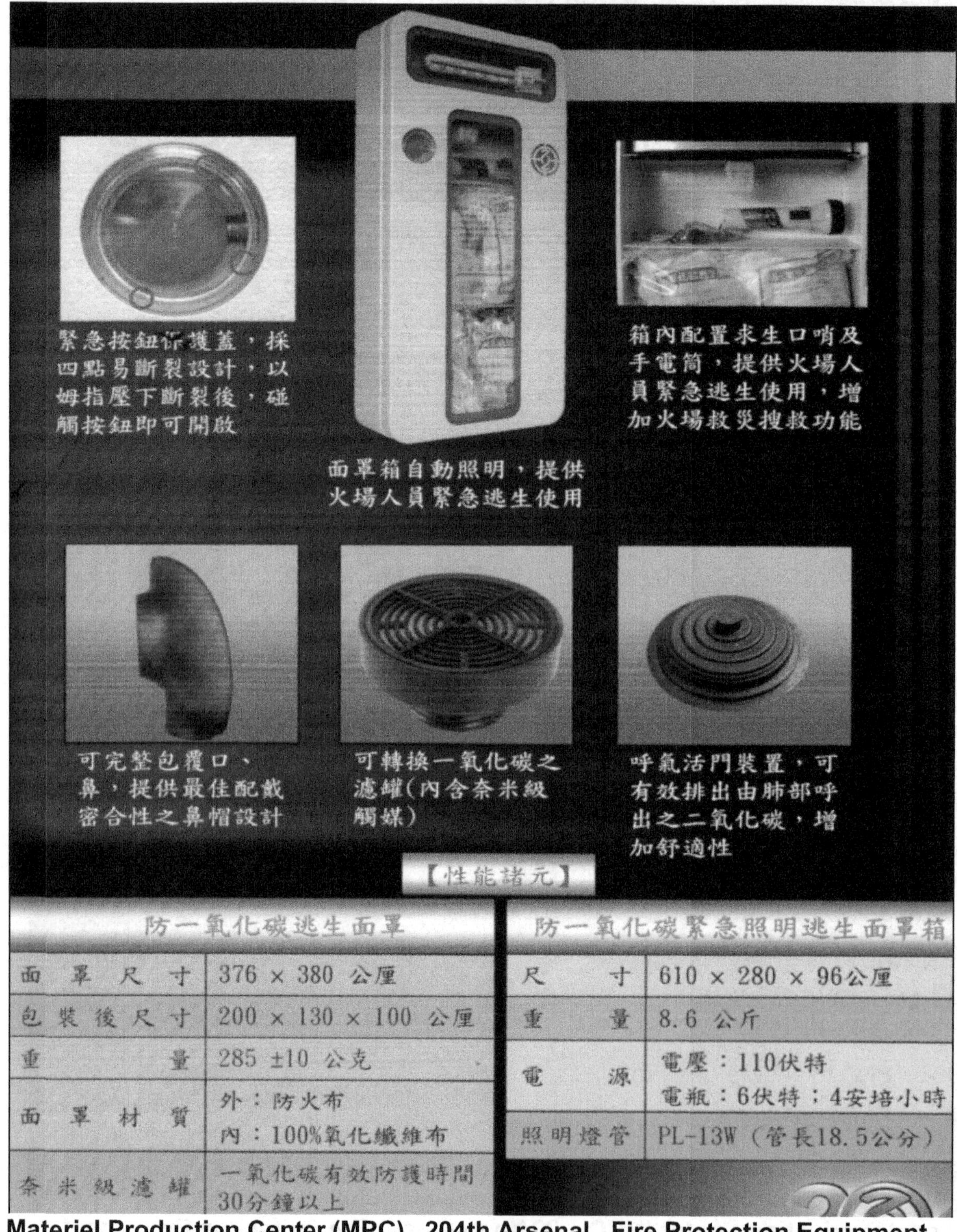

防一氧化碳逃生面罩	
面罩尺寸	376 × 380 公厘
包裝後尺寸	200 × 130 × 100 公厘
重量	285 ±10 公克
面罩材質	外：防火布 內：100%氧化纖維布
奈米級濾罐	一氧化碳有效防護時間30分鐘以上

防一氧化碳緊急照明逃生面罩箱	
尺寸	610 × 280 × 96公厘
重量	8.6 公斤
電源	電壓：110伏特 電瓶：6伏特；4安培小時
照明燈管	PL-13W（管長18.5公分）

Materiel Production Center (MPC). 204th Arsenal. Fire Protection Equipment. 2011 Taipei Aerospace and Defense Technology Exhibition (TADTE). Brochure 2/2.

防一氧化碳緊急逃生面罩

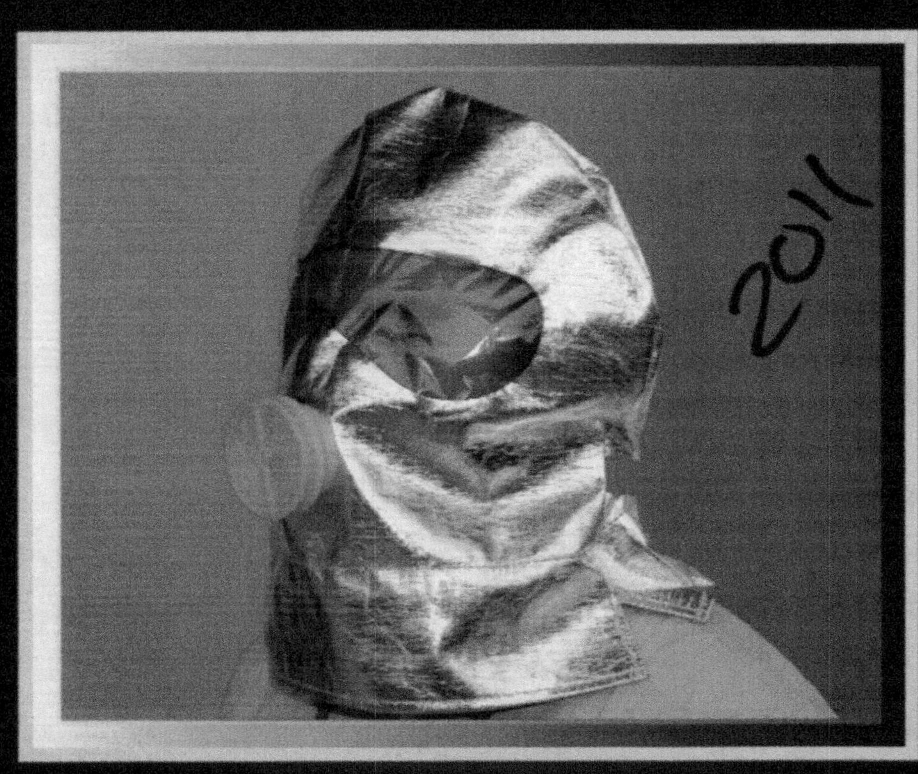

◎本產品可供人員於火災現場第一時間逃生時保護頭部(包括呼吸器官、眼睛、臉及頸部)，避免遭受火場高熱、濃煙及一氧化碳的傷害，為火場逃生時最佳的個人防護裝備。

◎濾毒罐採用高效能濾材，可濾除火場濃煙中有害粒子，並以高科技蜂巢結構裝填奈米級觸媒，可有效轉換一氧化碳，保障人員生命安全。

◎本產品結合工研院奈米合成材料技術(美國專利US 6,911,413 B2)，一氧化碳防護效能符合CNS 11981 Z2064及歐盟EN403規範。

◎面罩內部鼻帽設計可完整包覆口、鼻，提供最佳之配戴密合性及舒適性，同時具有呼氣活門裝置，可優化呼吸空氣品質，視窗採用耐高溫材質及寬闊視野設計，有利於逃離火場。

Materiel Production Center (MPC). 204th Arsenal. Fire Protection Equipment. 2011 Taipei Aerospace and Defense Technology Exhibition (TADTE). Brochure 1/2.

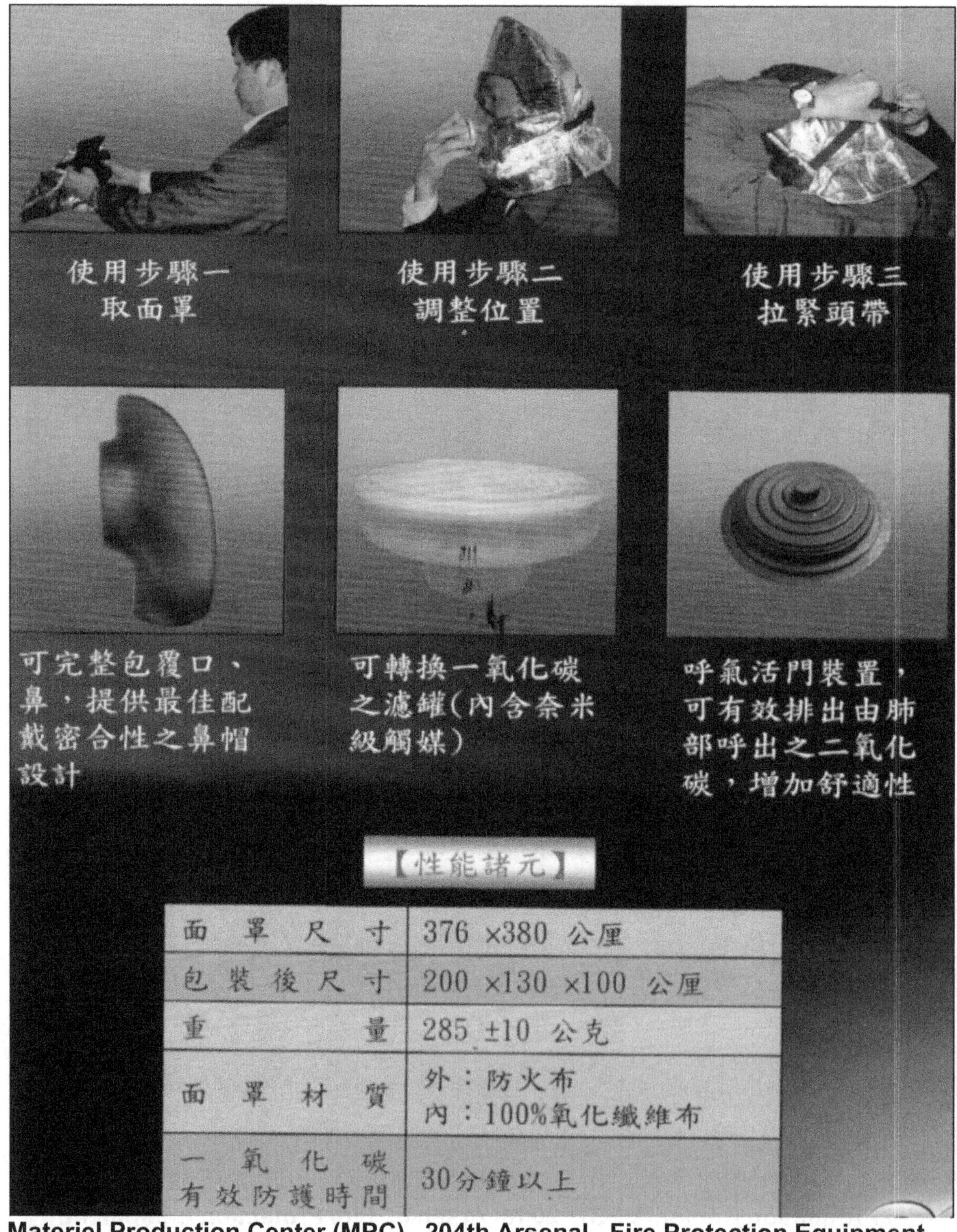

使用步驟一
取面罩

使用步驟二
調整位置

使用步驟三
拉緊頭帶

可完整包覆口、鼻，提供最佳配戴密合性之鼻帽設計

可轉換一氧化碳之濾罐(內含奈米級觸媒)

呼氣活門裝置，可有效排出由肺部呼出之二氧化碳，增加舒適性

【性能諸元】

面罩尺寸	376 ×380 公厘
包裝後尺寸	200 ×130 ×100 公厘
重量	285 ±10 公克
面罩材質	外：防火布 內：100%氧化纖維布
一氧化碳有效防護時間	30分鐘以上

Materiel Production Center (MPC). 204th Arsenal. Fire Protection Equipment. 2011 Taipei Aerospace and Defense Technology Exhibition (TADTE). Brochure 2/2.

97式濾毒罐、62式濾毒罐

◎可搭配國軍T3-75防護面具、72式戰車人員防護面具、T3-92 防護面具及T3-94 防護面具使用，可有效過濾神經性、糜爛性、血液性、窒息性及催淚性等化學戰劑。

◎濾毒罐螺紋採用 NATO 40MM規範，可共用於各國軍用及民用相同螺紋之防護面具。

Materiel Production Center (MPC). 204th Arsenal. Gas Mask Filter. 2011 Taipei Aerospace and Defense Technology Exhibition (TADTE). Brochure 1/2.

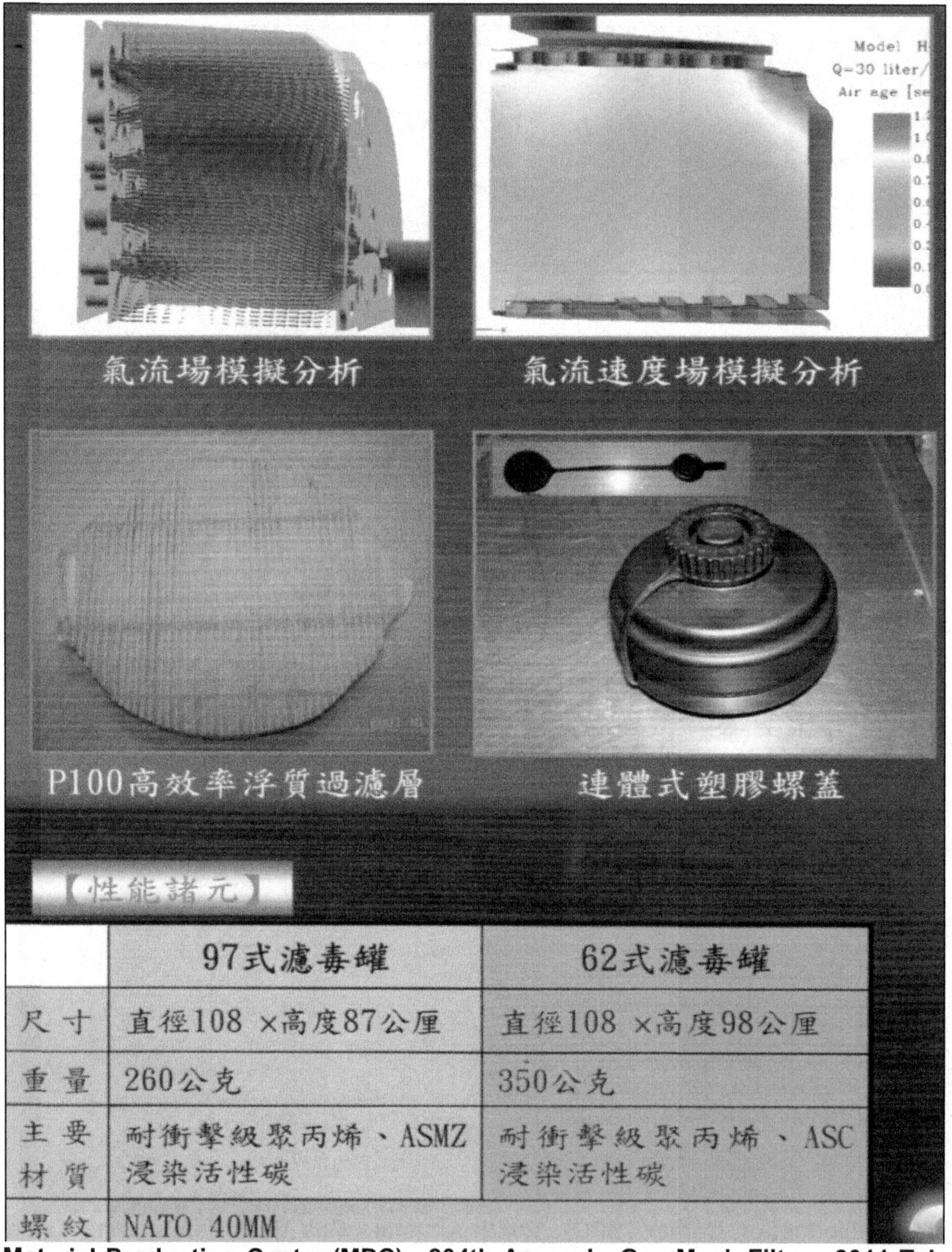

	97式濾毒罐	62式濾毒罐
尺寸	直徑108 ×高度87公厘	直徑108 ×高度98公厘
重量	260公克	350公克
主要材質	耐衝擊級聚丙烯、ASMZ浸染活性碳	耐衝擊級聚丙烯、ASC浸染活性碳
螺紋	NATO 40MM	

Materiel Production Center (MPC). 204th Arsenal. Gas Mask Filter. 2011 Taipei Aerospace and Defense Technology Exhibition (TADTE). Brochure 2/2.

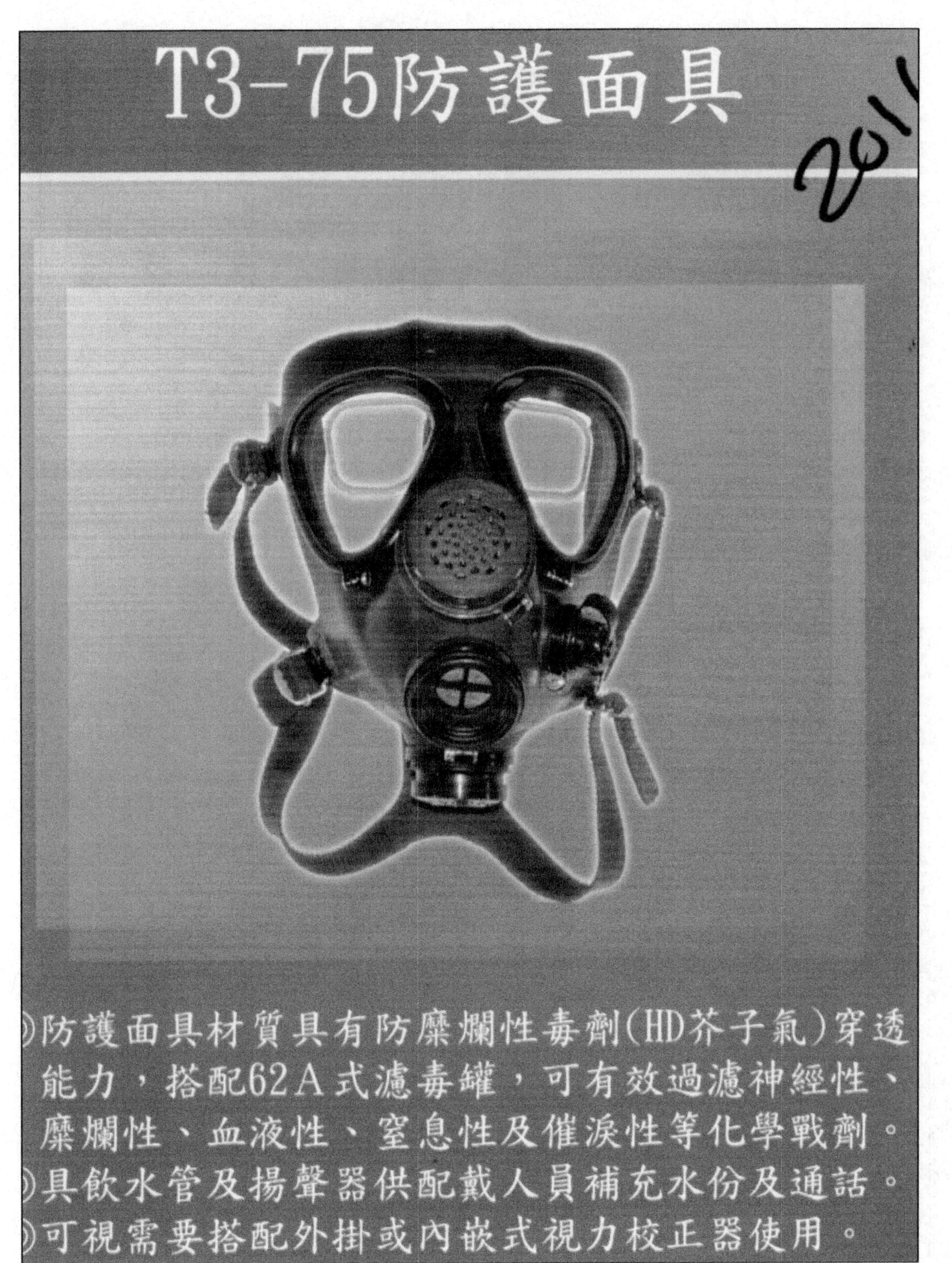

Materiel Production Center (MPC). 204th Arsenal. T3-75 Gas Mask. 2011 Taipei Aerospace and Defense Technology Exhibition (TADTE). Brochure 1/2.

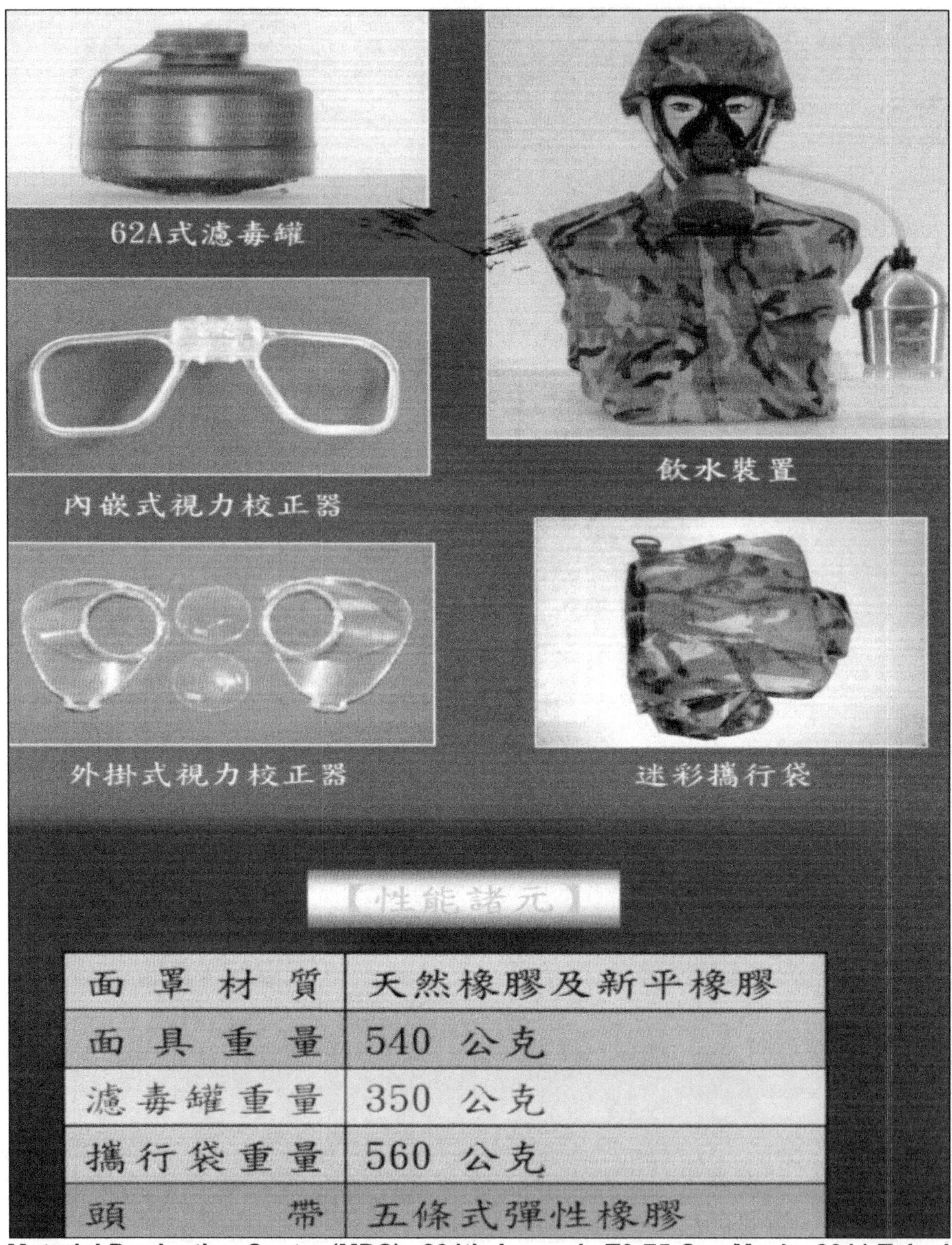

Materiel Production Center (MPC). 204th Arsenal. T3-75 Gas Mask. 2011 Taipei Aerospace and Defense Technology Exhibition (TADTE). Brochure 2/2.

T3-92防護面具

- 面罩材質具有優異之防糜爛性毒劑(ＨＤ芥子氣)穿透能力，搭配62Ａ式濾毒罐，可有效過濾神經性、糜爛性、血液性、窒息性及催淚性等化學戰劑，儲存壽限長達15年。
- 面具依據臉型統計數據設計構型，提供最佳之配戴密合性及舒適性。同時具有飲水、揚聲、麥克風插座及視力校正等功能。
- 旋轉式濾毒罐座可依需求調整至左、中、右三種位置，已獲得中華民國新型專利第224418號及美國專利US 7,130,439 B2號。

Materiel Production Center (MPC). 204th Arsenal. T3-92 Gas Mask. 2011 Taipei Aerospace and Defense Technology Exhibition (TADTE). Brochure 1/2.

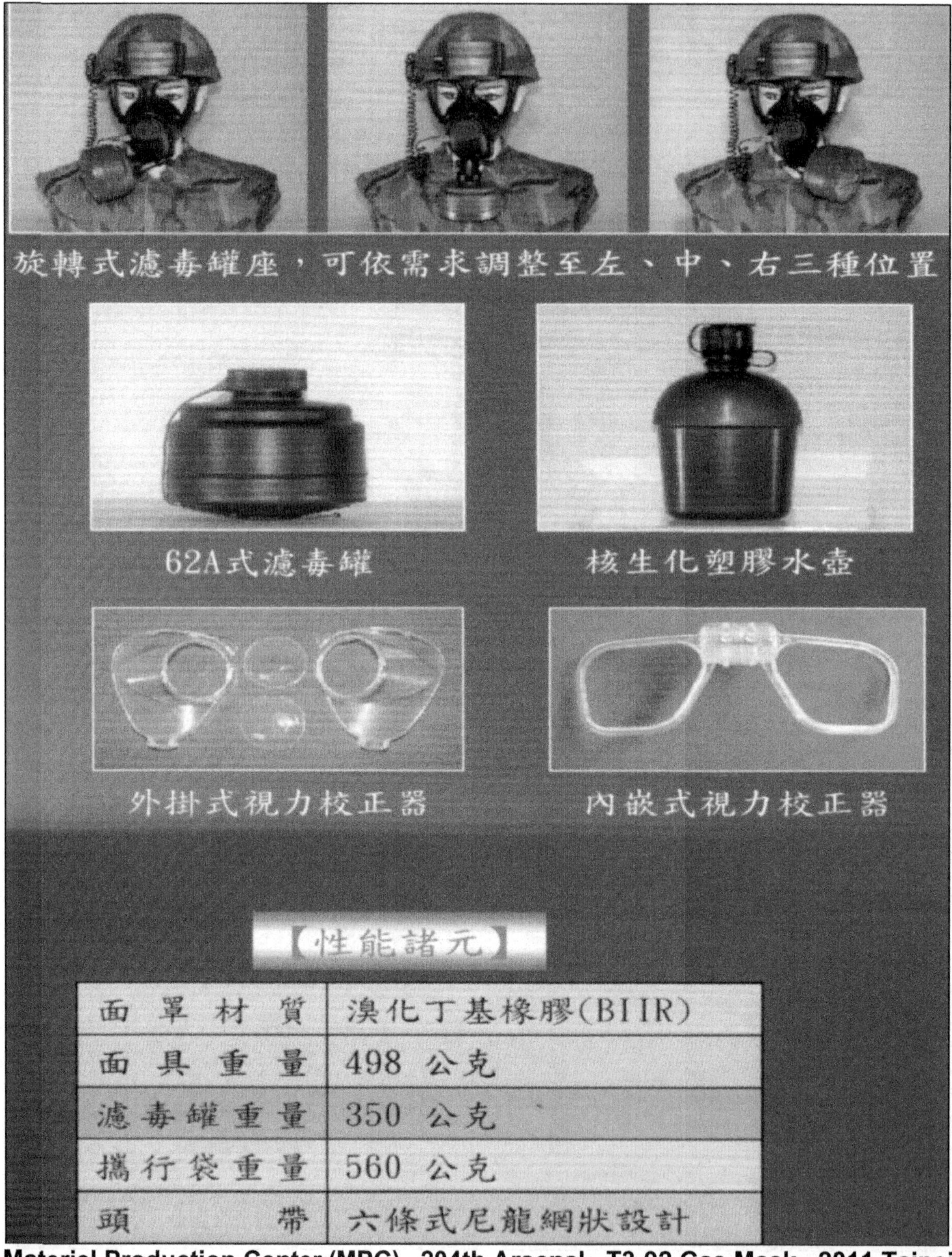

面罩材質	溴化丁基橡膠(BIIR)
面具重量	498 公克
濾毒罐重量	350 公克
攜行袋重量	560 公克
頭帶	六條式尼龍網狀設計

Materiel Production Center (MPC). 204th Arsenal. T3-92 Gas Mask. 2011 Taipei Aerospace and Defense Technology Exhibition (TADTE). Brochure 2/2.

T3-94防護面具

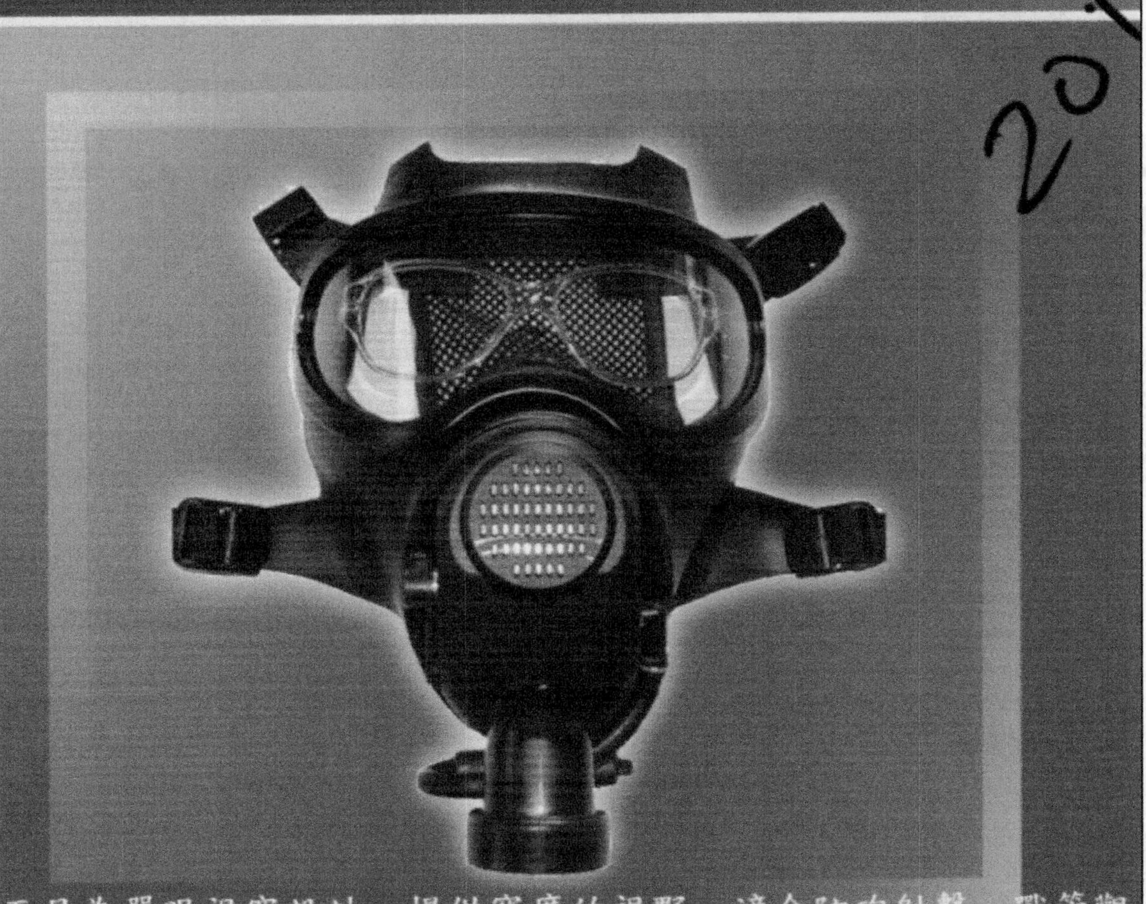

◎面具為單眼視窗設計，提供寬廣的視野，適合防砲射擊、戰管觀測、艦上勤務與空軍地勤等任務之執行。

◎面罩材質具有優異之防糜爛性毒劑(HD芥子氣) 穿透能力，搭配62A式濾毒罐，可有效過濾神經性、糜爛性、血液性、窒息性及催淚性等化學戰劑，儲存壽限長達15年。

◎面具依據臉型統計數據設計構型，提供最佳之配戴密合性及舒適性。同時具有飲水、揚聲、麥克風插座及視力校正等功能。

◎旋轉式濾毒罐座可依需求調整至左、中、右三種位置，已獲得中華民國新型專利第224418號及美國專利US 7,130,439 B2號。

Materiel Production Center (MPC). 204th Arsenal. T3-94 Gas Mask. 2011 Taipei Aerospace and Defense Technology Exhibition (TADTE). Brochure 1/2.

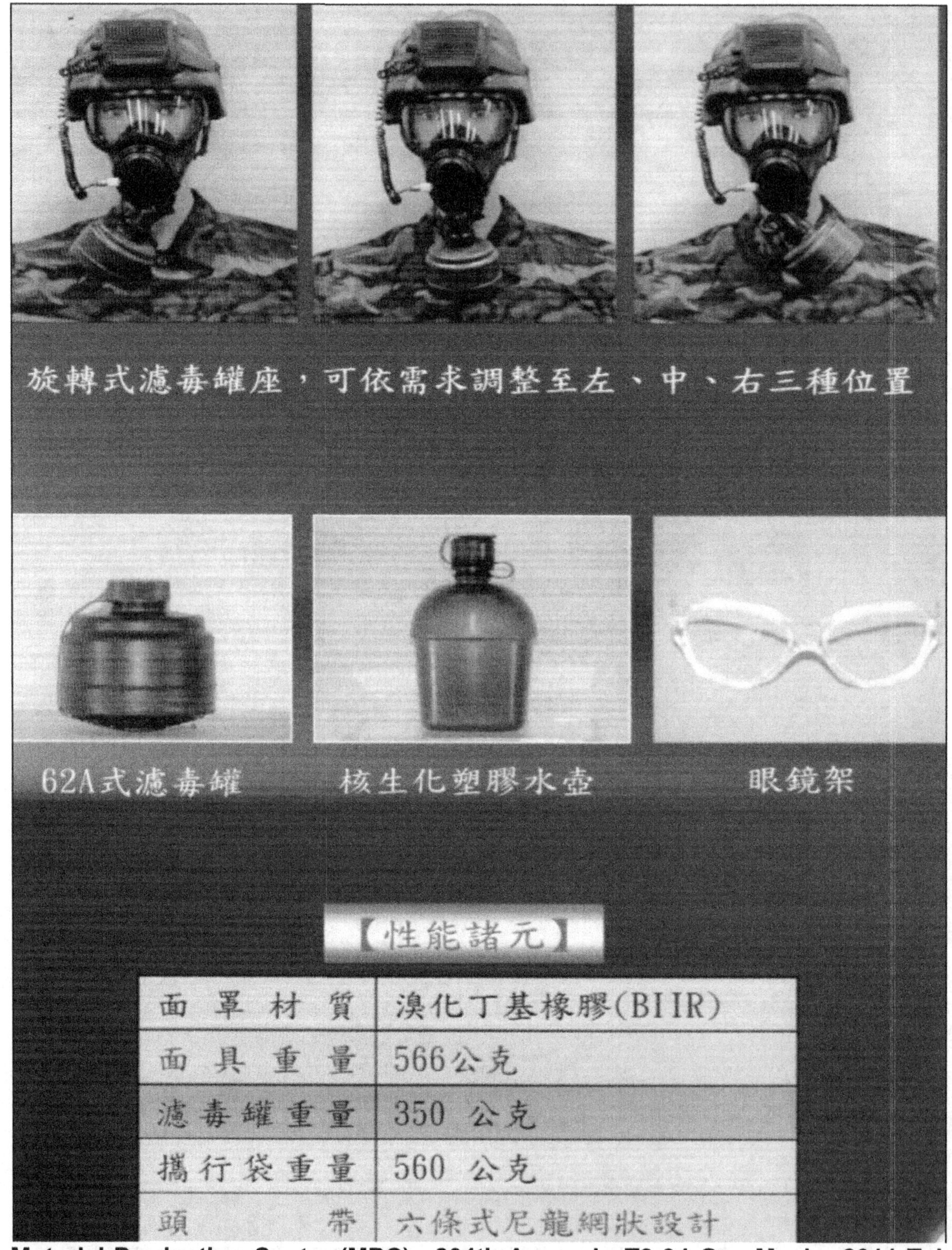

旋轉式濾毒罐座，可依需求調整至左、中、右三種位置

62A式濾毒罐

核生化塑膠水壺

眼鏡架

【性能諸元】

面罩材質	溴化丁基橡膠(BIIR)
面具重量	566公克
濾毒罐重量	350 公克
攜行袋重量	560 公克
頭帶	六條式尼龍網狀設計

Materiel Production Center (MPC). 204th Arsenal. T3-94 Gas Mask. 2011 Taipei Aerospace and Defense Technology Exhibition (TADTE). Brochure 2/2.

T4-86輕型消毒器

◎自備電力、動力，附提把供四人搬運，適合野外操作，可產生高溫霧狀或高壓及霧狀水柱，供執行消除、清除與消毒作業。

◎軍事用途：適用於房舍、車輛、裝備或人員受核生化戰劑污染時消除與清除，於野外搭配淋浴架，可提供熱水供大量人員淋浴使用。

◎一般用途：可清洗房舍、車輛及飛機，亦可配合各式藥劑實施環境消毒。

Materiel Production Center (MPC). 204th Arsenal. T4-86 NBC Shower Equipment. 2011 Taipei Aerospace and Defense Technology Exhibition (TADTE). Brochure 1/2.

運用於戰機之消除清洗作業　運用於車輛之消除清洗作業

淋浴架設計，提供人員清洗消除　勤務箱及各式另配件

【性能諸元】

尺　　寸	95 × 75 × 75 公分
重　　量	170 公斤
電　　源	自備電力系統(12伏特電瓶)
操作溫度(可調式)	100℃(熱水消除)、150℃(蒸氣消除)
操作壓力(可調式)	20 kg/cm^2 (熱水消除、淋浴使用) 9～10 kg/cm^2 (蒸氣消除)

Materiel Production Center (MPC). 204th Arsenal. T4-86 NBC Shower Equipment. 2011 Taipei Aerospace and Defense Technology Exhibition (TADTE). Brochure 2/2.

Materiel Production Center (MPC). T4-103 Light Duty Decontaminating System. 2018 Defence Services Asia (DSA). Brochure 1/2.

T4-103 Light Duty
Decontaminating System
Description:
The decontaminating system is equipped with digital control panel, engine, and battery. It could generate hot steam or high pressure water which is suitable to be utilized for building, vehicle, equipment, ground, and personnel sterilization or decontamination. It can also provide hot water for personnel bath.
Specification:
Weight: 260kg
Maximum outflow: 42L/min
Maximum outflow(hot water): 35L/min
Maximum water pressure: 100kg/cm2

Materiel Production Center (MPC). T4-103 Light Duty Decontaminating System. 2018 Defence Services Asia (DSA). Brochure 2/2.

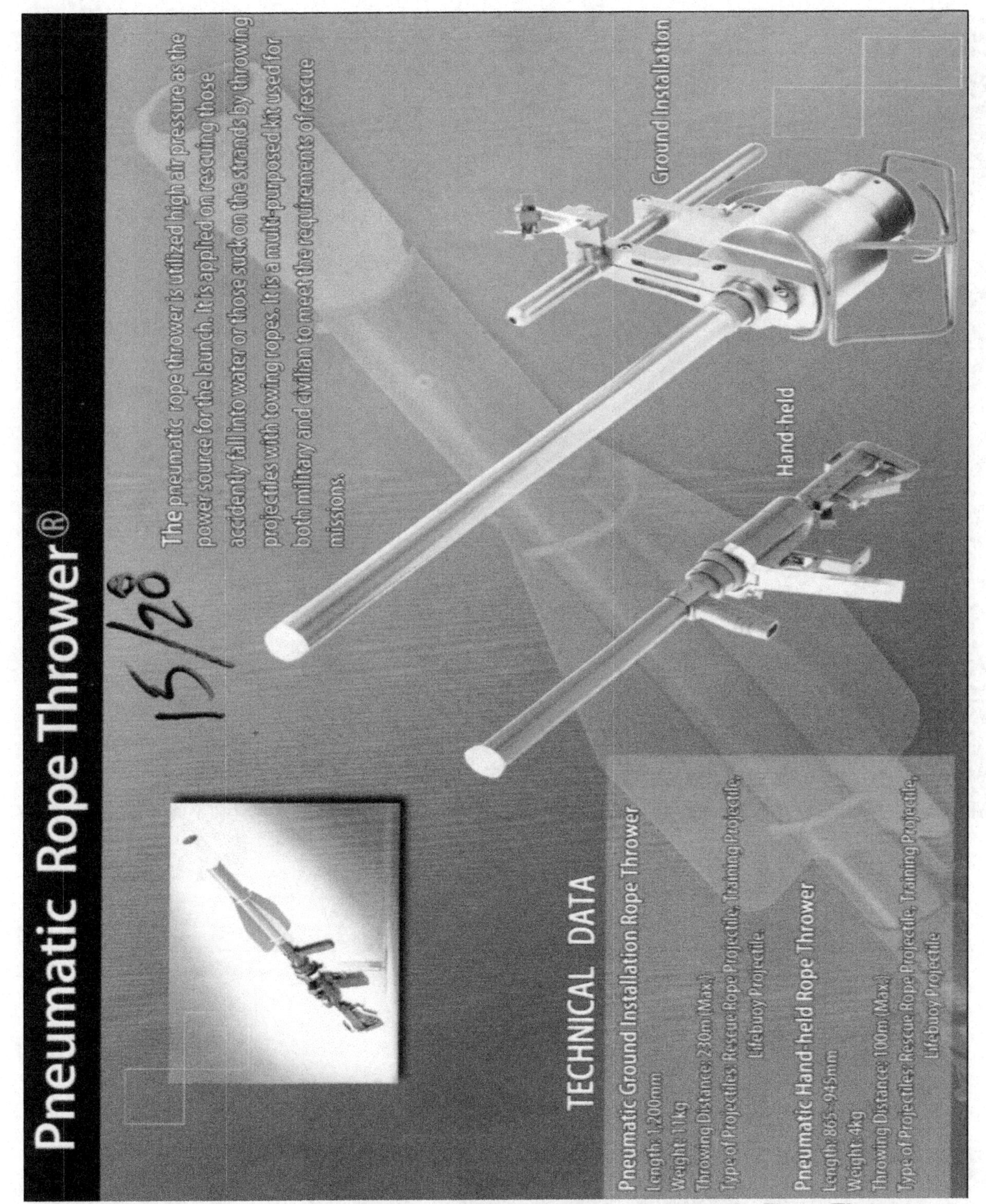

Materiel Production Center (MPC). 205th Arsenal. Pneumatic Rope Thrower. 2015 Taipei Aerospace and Defense Technology Exhibition (TADTE). Brochure 1/1.

Materiel Production Center (MPC). Pneumatic Rope Thrower. 2018 Defence Services Asia (DSA). Brochure 1/2.

Pneumatic Rope Thrower

Description:

The pneumatic rope thrower is utilized high pressure air as the power source for launch. It is applied on water rescue and stranded person rescue by throwing projectiles with towing ropes. It is a multi-purposed kit used for both military and civilian to meet the requirements for rescue missions.

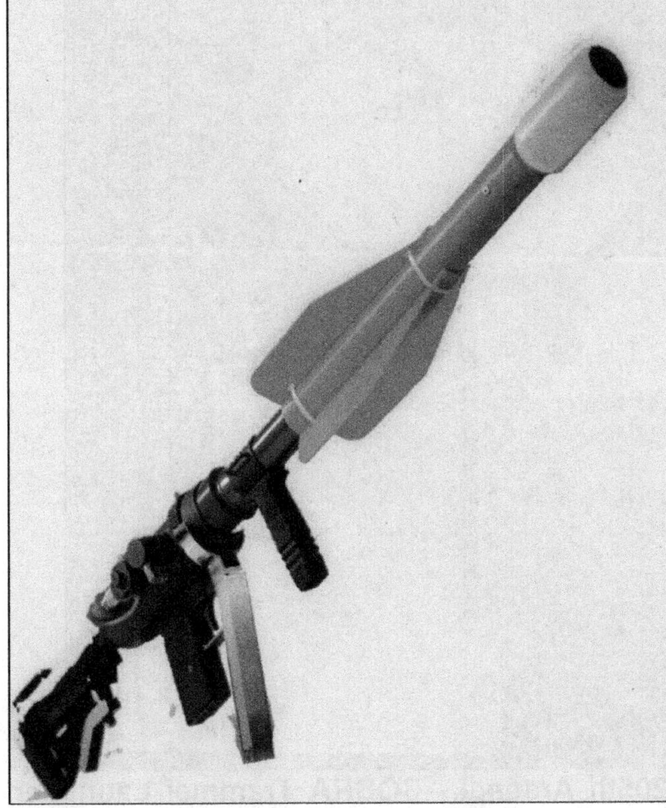

Specification:

- **Pneumatic system to keep the operate safety.**
- **Easy for Maintenance.**
- **Light-weight and Long-distance.**

Hand-held:

- **Length:** 865-945mm
- **Weight:** 4kg (without projectiles)
- **Throwing Distance:** 110m (Max.)

Ground Installation:

- **Length:** 1,200mm
- **Weight:** 11kg (without projectiles)
- **Throwing Distance:** 230m (Max.)

Materiel Production Center (MPC). Pneumatic Rope Thrower. 2018 Defence Services Asia (DSA). Brochure 2/2.

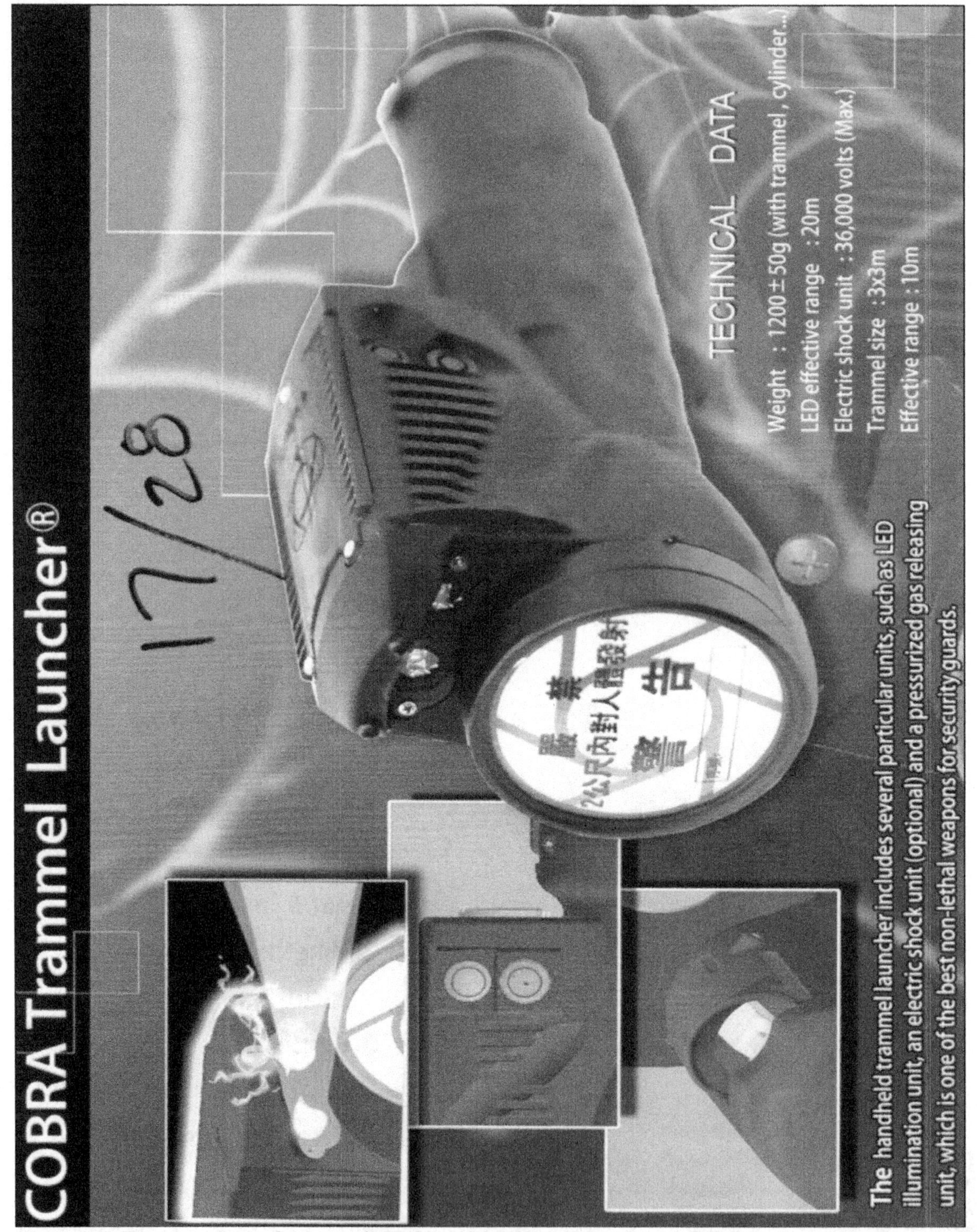

Materiel Production Center (MPC). 205th Arsenal. COBRA Trammel Launcher. 2015 Taipei Aerospace and Defense Technology Exhibition (TADTE). Brochure 1/1.

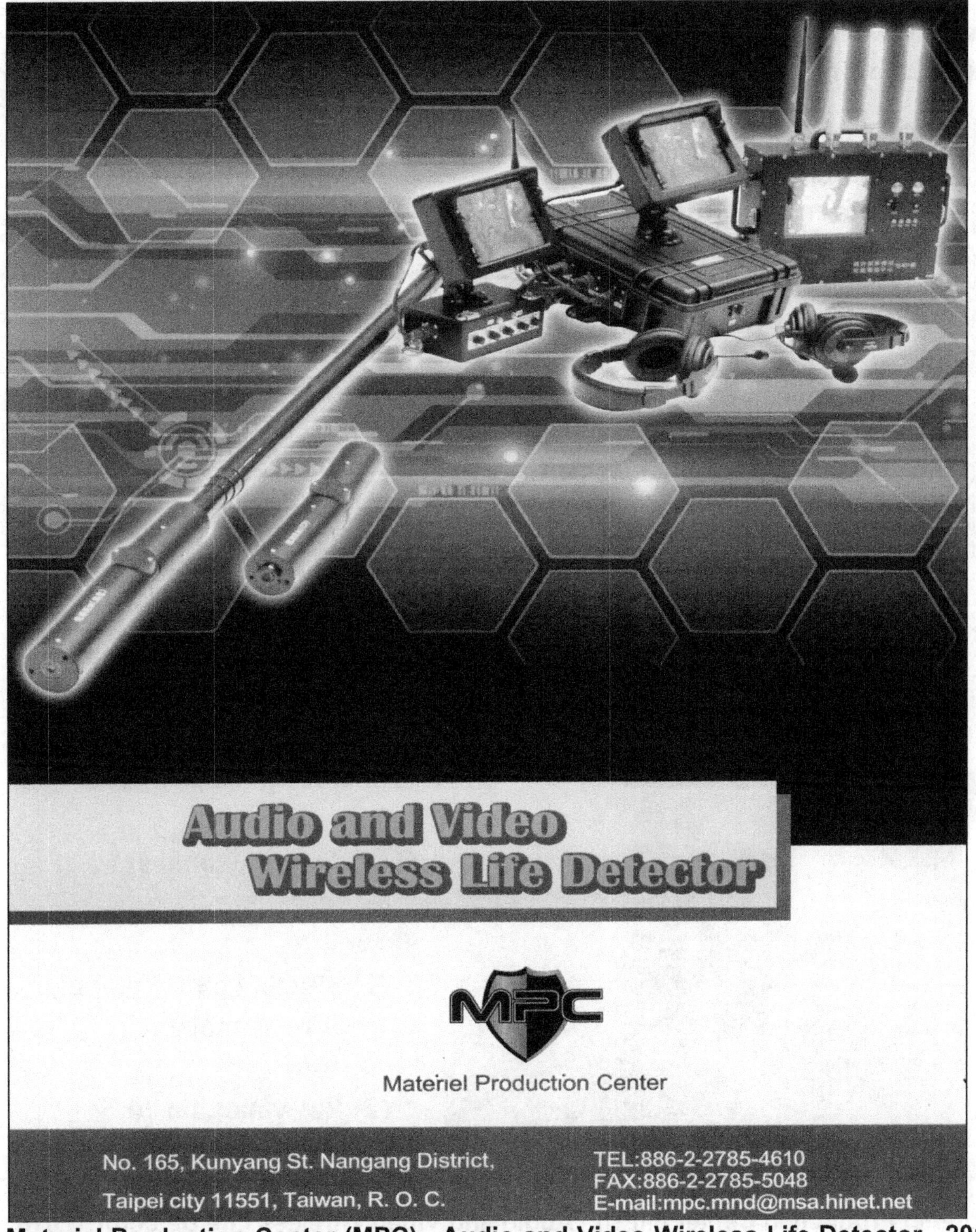

Materiel Production Center (MPC). Audio and Video Wireless Life Detector. 2018 Defence Services Asia (DSA). Brochure 1/2.

Audio and Video Wireless Life Detector

Description:

The modular design allows the equipment to subdivide into camera module, telescoping pole, system control module and monitor relay module. The camera module not only produces thermal or visible images but also provides audio fuction with a built-in microphone and speaker. Both system control module and monitor relay module transmit imaging signals wirelessly, which enables real-time images be displayed on a commander center's screen.

Specification:

- **Camera module：**
  - **(1)CCD camera : 640 x 480 pixels**
  - **(2)Thermal camera : 320 x 240 pixels**
- **Carbon fiber telescoping pole (Maximum length up to 3m)**
- **Operating time：2 hours**
- **6.5 inch diagonal TFT LCD monitors.**
- **Wireless：**
  - **(1) System control module to relay monitor : 40m**
  - **(2) Relay monitor to command center : 200m**

Materiel Production Center (MPC). Audio and Video Wireless Life Detector. 2018 Defence Services Asia (DSA). Brochure 2/2.

MUNITIONS

Materiel Production Center (MPC)

Materiel Production Center (MPC). 205th Arsenal. 2015 Taipei Aerospace and Defense Technology Exhibition (TADTE). Brochure 1/1.

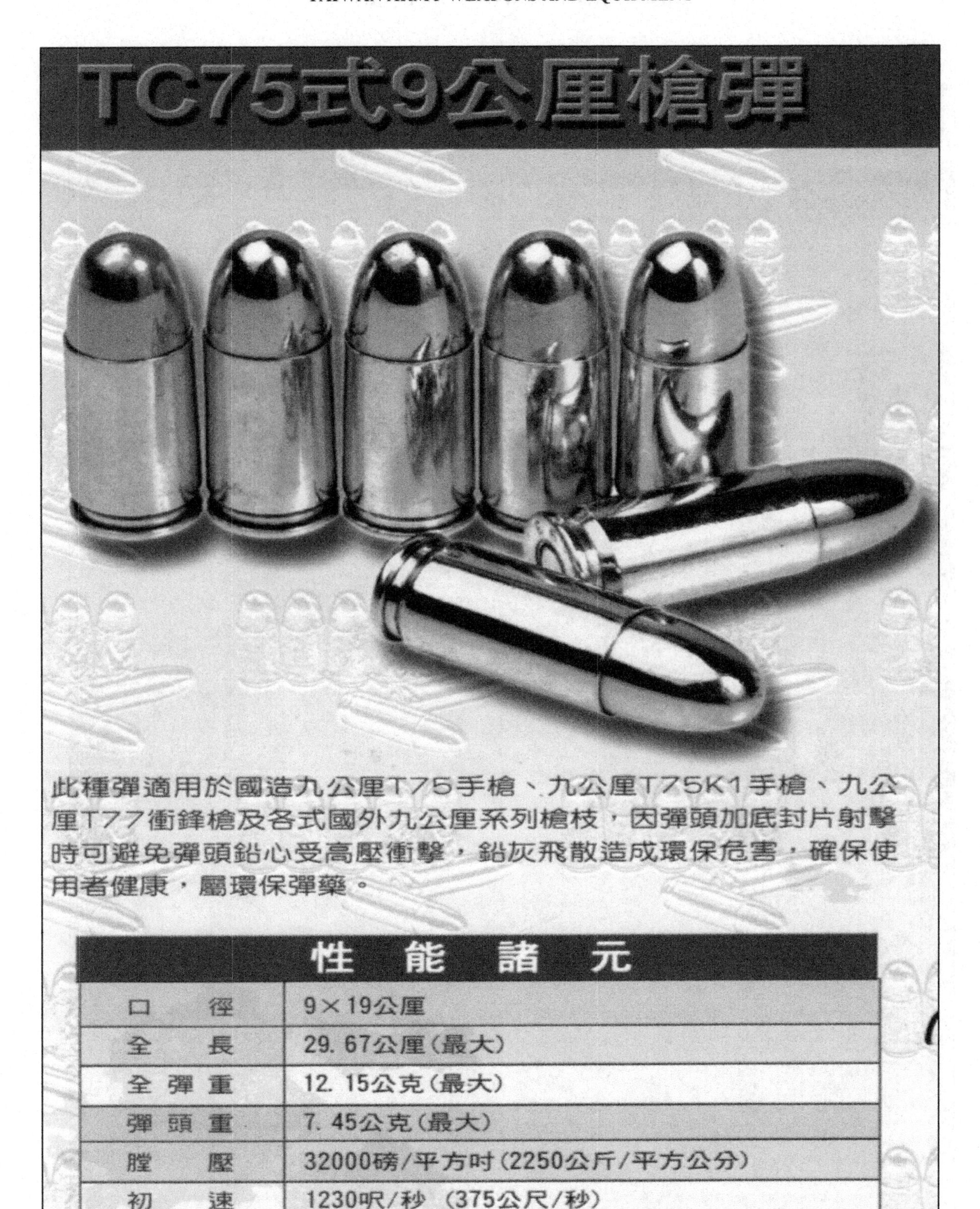

TC75式9公厘槍彈

此種彈適用於國造九公厘T75手槍、九公厘T75K1手槍、九公厘T77衝鋒槍及各式國外九公厘系列槍枝，因彈頭加底封片射擊時可避免彈頭鉛心受高壓衝擊，鉛灰飛散造成環保危害，確保使用者健康，屬環保彈藥。

性能諸元	
口徑	9×19公厘
全長	29.67公厘(最大)
全彈重	12.15公克(最大)
彈頭重	7.45公克(最大)
膛壓	32000磅/平方吋(2250公斤/平方公分)
初速	1230呎/秒 (375公尺/秒)

Materiel Production Center (MPC). 205th Arsenal. TC75 9mm bullets. 2011 Taipei Aerospace and Defense Technology Exhibition (TADTE). Brochure 1/1.

Materiel Production Center (MPC). 205th Arsenal. 9mm TC89 Marking Cartridge. 2011 Taipei Aerospace and Defense Technology Exhibition (TADTE). Brochure 1/3.

性能諸元 (TECHNICAL DATA)	
口徑 (Caliber)	9×19公厘 (9×19 mm)
初速 (Muzzle Velocity)	150±30公尺/秒 (150 ± 30 m/s)
彈頭材質 (Projectile Material)	塑膠 (Plastic)
色膏 (Marking Compound)	無毒、可用水洗及非刺激性之化合物 (Non-toxic, washable, inert)
終端效應 (Terminal Effect)	假如皮膚無適當防護，會造成非致命性的瘀青或瘀痛 (Non-lethal bruises or welts if skin is unprotected)

Materiel Production Center (MPC). 205th Arsenal. 9mm TC89 Marking Cartridge. 2011 Taipei Aerospace and Defense Technology Exhibition (TADTE). Brochure 2/3.

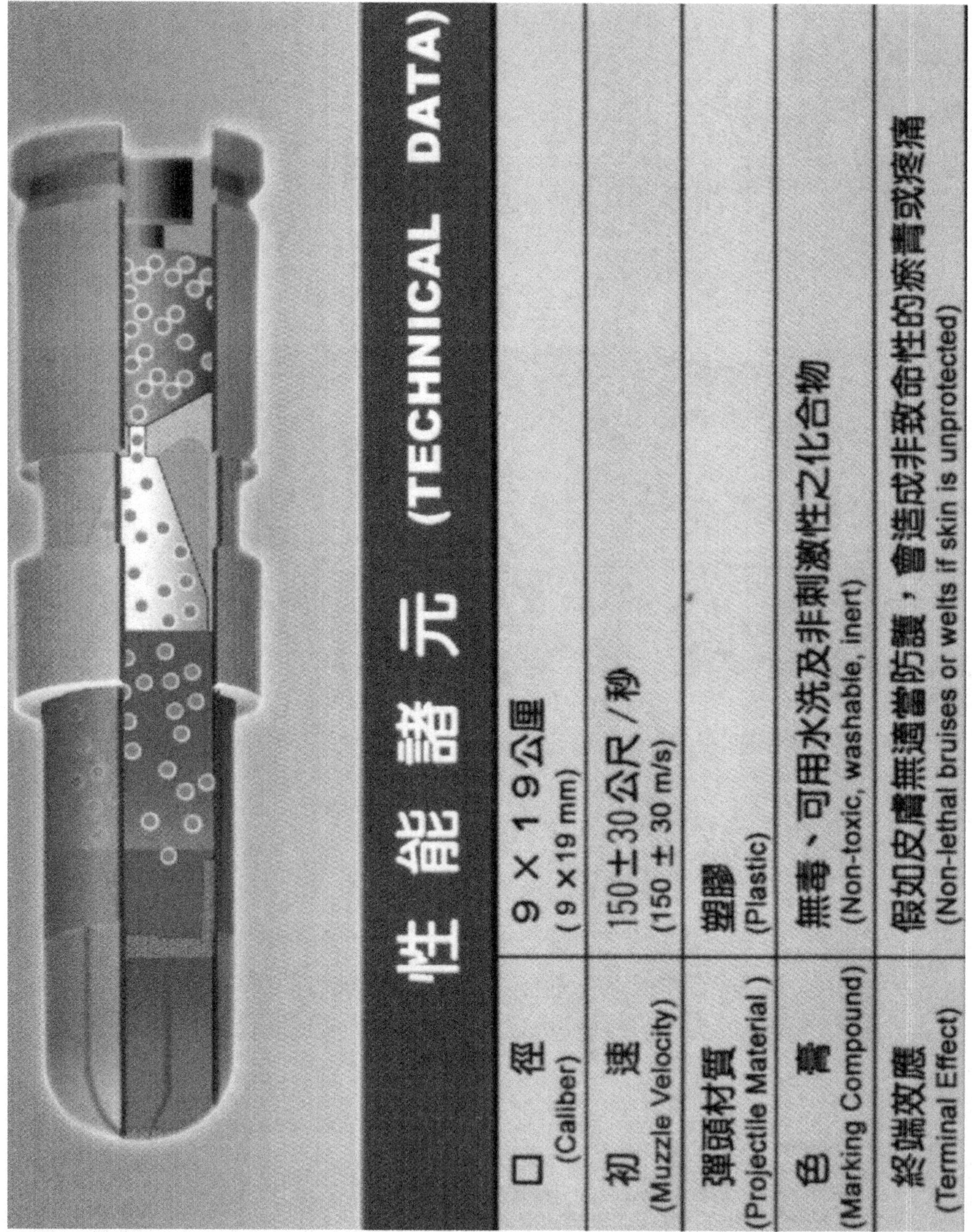

性能諸元 (TECHNICAL DATA)

口徑 (Caliber)	9×19公厘 (9×19 mm)
初速 (Muzzle Velocity)	150±30公尺/秒 (150 ± 30 m/s)
彈頭材質 (Projectile Material)	塑膠 (Plastic)
色膏 (Marking Compound)	無毒、可用水洗及非刺激性之化合物 (Non-toxic, washable, inert)
終端效應 (Terminal Effect)	假如皮膚無適當防護，會造成非致命性的瘀青或痠痛 (Non-lethal bruises or welts if skin is unprotected)

Materiel Production Center (MPC). 205th Arsenal. 9mm TC89 Marking Cartridge. 2011 Taipei Aerospace and Defense Technology Exhibition (TADTE). Brochure 3/3 (CLOSE-UP).

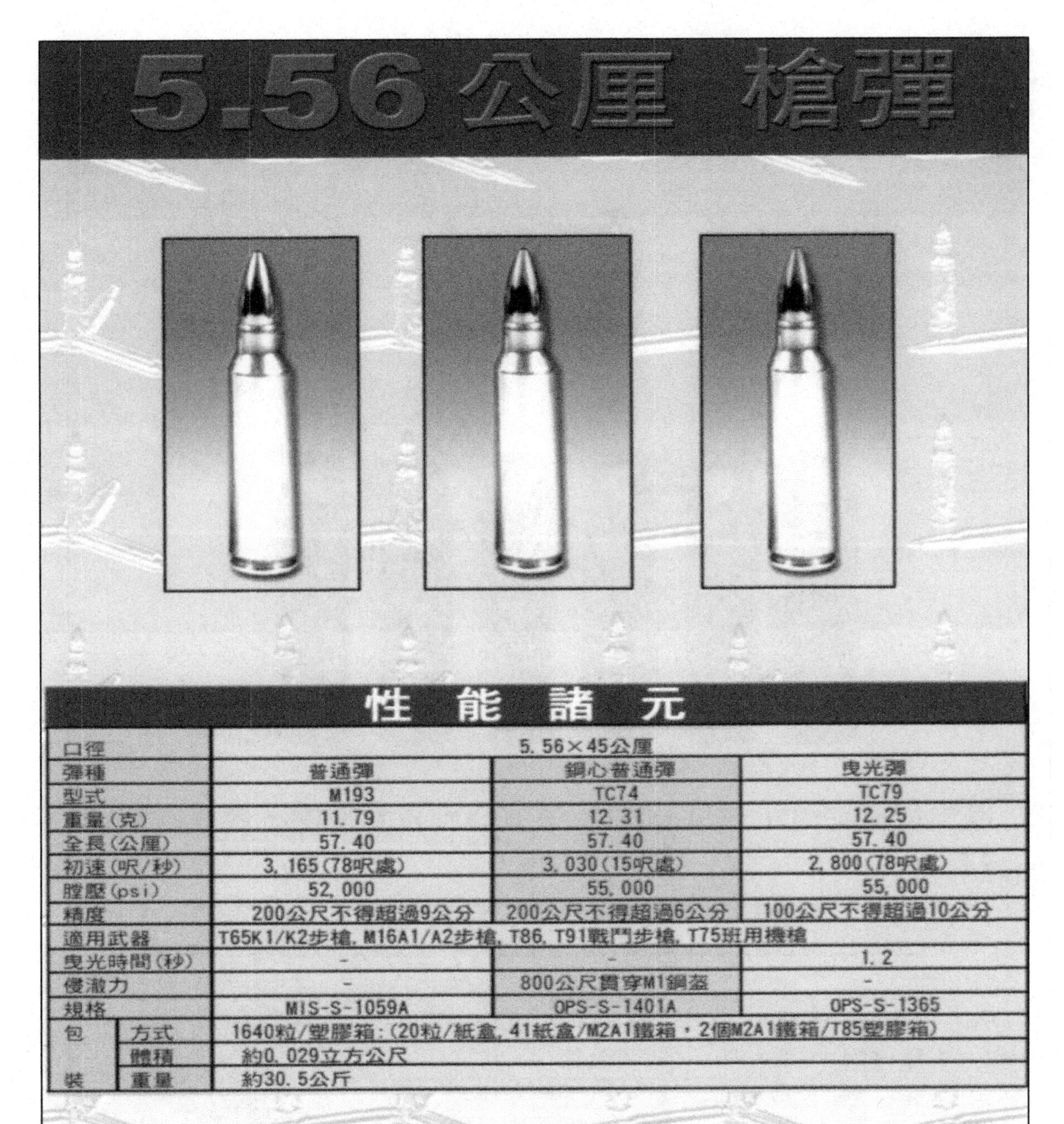

5.56公厘 槍彈

性能諸元

口徑		5. 56×45公厘		
彈種		普通彈	鋼心普通彈	曳光彈
型式		M193	TC74	TC79
重量(克)		11. 79	12. 31	12. 25
全長(公厘)		57. 40	57. 40	57. 40
初速(呎/秒)		3, 165(78呎處)	3, 030(15呎處)	2, 800(78呎處)
膛壓(psi)		52, 000	55, 000	55, 000
精度		200公尺不得超過9公分	200公尺不得超過6公分	100公尺不得超過10公分
適用武器		T65K1/K2步槍, M16A1/A2步槍, T86, T91戰鬥步槍, T75班用機槍		
曳光時間(秒)		-	-	1. 2
侵澈力		-	800公尺貫穿M1鋼盔	-
規格		MIS-S-1059A	OPS-S-1401A	OPS-S-1365
包裝	方式	1640粒/塑膠箱:(20粒/紙盒, 41紙盒/M2A1鐵箱，2個M2A1鐵箱/T85塑膠箱)		
	體積	約0. 029立方公尺		
	重量	約30. 5公斤		

5.56×45公厘槍彈是一種精確且具高效益的彈藥，適用於T65K1/K2步槍、美製M16A1/A2步槍、T86、T91戰鬥步槍、T75班用機槍。5.56公厘彈種計有M193式普通彈、TC74式鋼心普通彈及TC79 式曳光彈，可作不同的排列組裝應用。TC74 式鋼心普通彈 (綠色彈尖)， 可於800公尺距離處貫穿美製M1 式防護鋼盔；TC79 式曳光彈 (紅色彈尖)，可清楚地標示彈頭射擊方向，提昇射擊準確度。

Materiel Production Center (MPC). 205th Arsenal. 5.56mm. 2011 Taipei Aerospace and Defense Technology Exhibition (TADTE). Brochure 1/1.

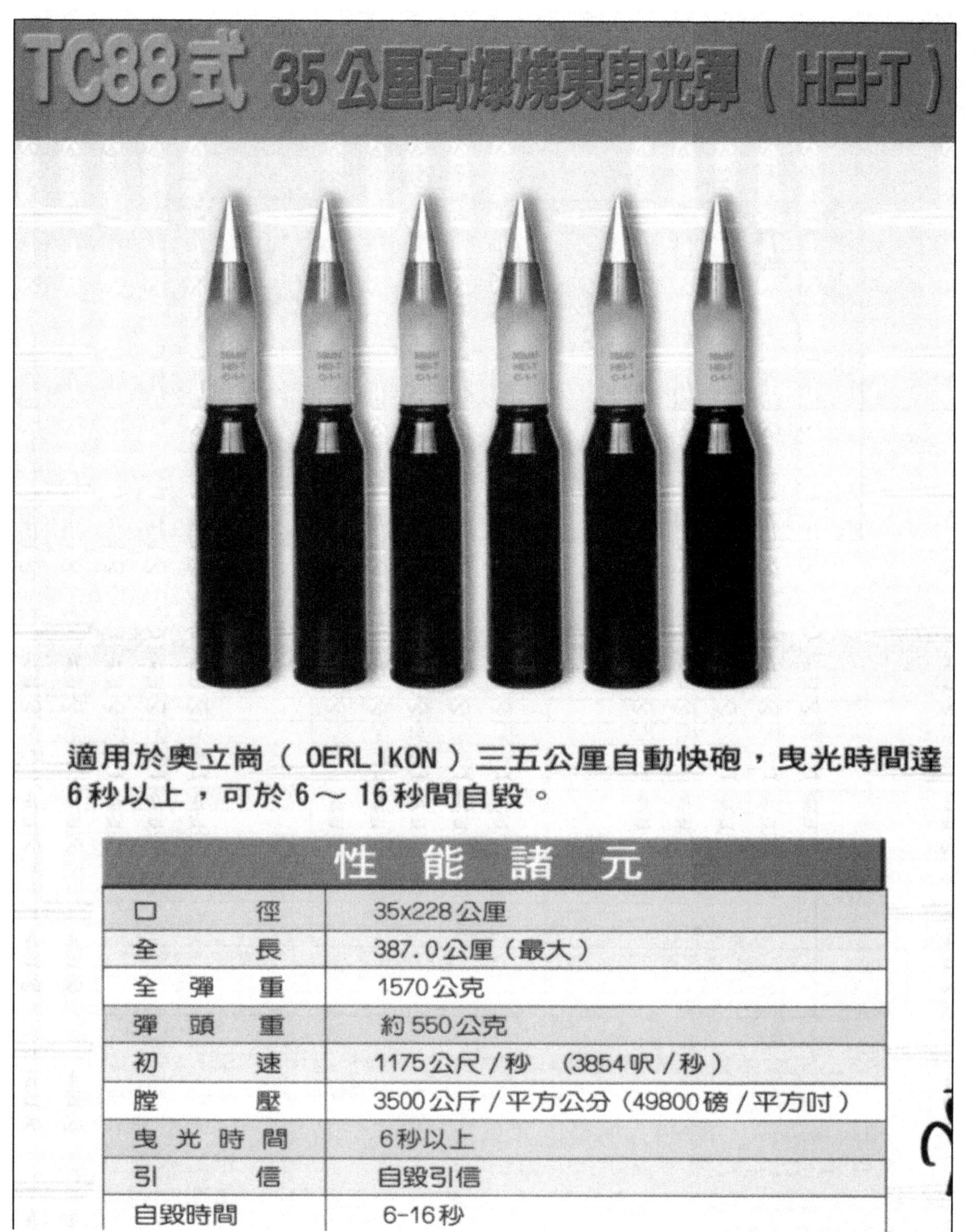

TC88式 35公厘高爆燒夷曳光彈（HEI-T）

適用於奧立崗（OERLIKON）三五公厘自動快砲，曳光時間達6秒以上，可於6～16秒間自毀。

性能諸元	
口徑	35x228公厘
全長	387.0公厘（最大）
全彈重	1570公克
彈頭重	約550公克
初速	1175公尺／秒　(3854呎／秒)
膛壓	3500公斤／平方公分（49800磅／平方吋）
曳光時間	6秒以上
引信	自毀引信
自毀時間	6-16秒

Materiel Production Center (MPC). 205th Arsenal. TC88 High Explosive Incendiary Tracer (HEI-T) 35mm - Oerlikon. 2011 Taipei Aerospace and Defense Technology Exhibition (TADTE). Brochure 1/1.

OPTICS
and
NIGHT VISION

Materiel Production Center (MPC). 401st Factory. Optoelectronics Business – Red Dot Sight. 2018 Kaohsiung International Maritime and Defense Exhibition. Brochure 1/2.

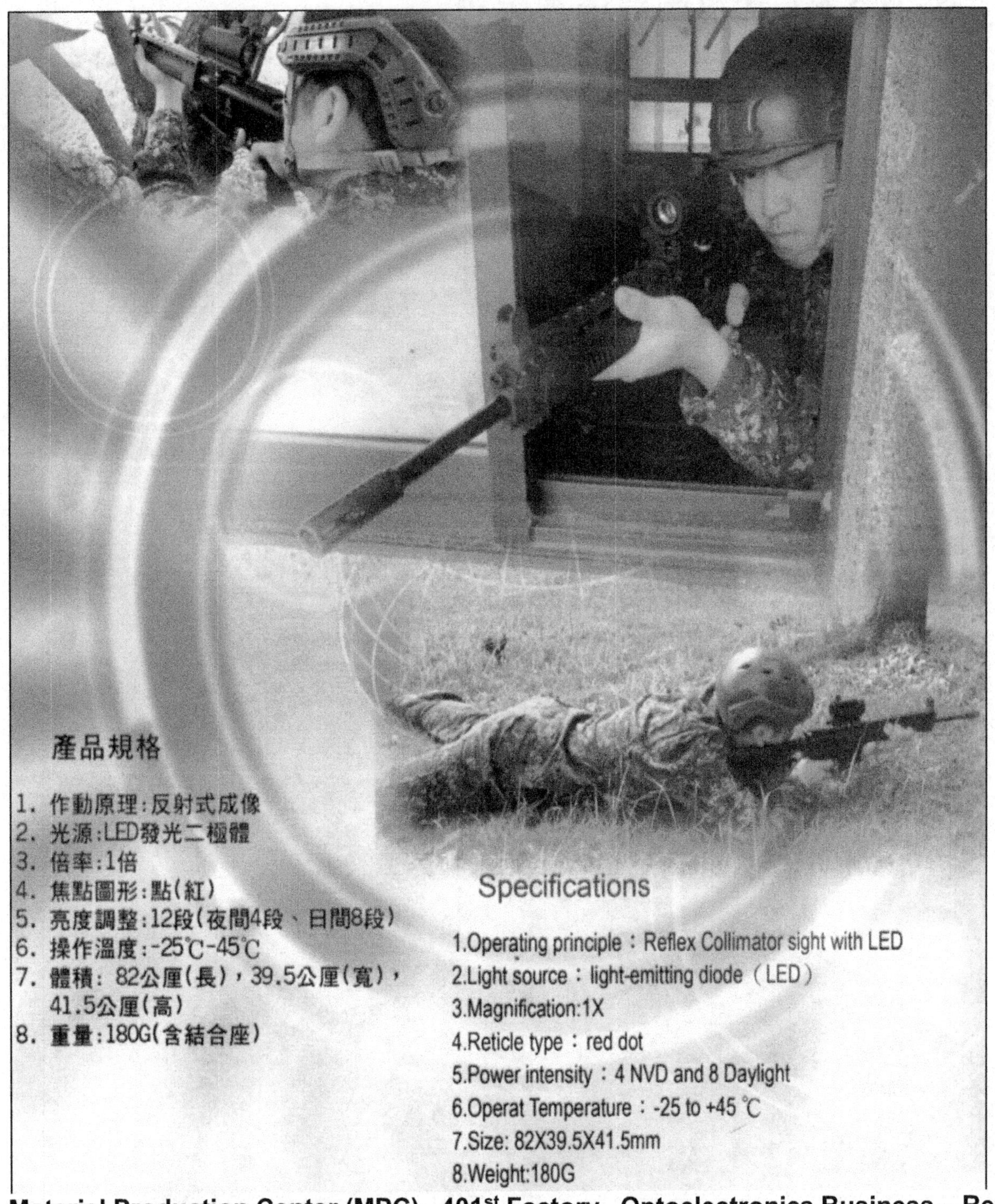

Materiel Production Center (MPC). 401st Factory. Optoelectronics Business – Red Dot Sight. 2018 Kaohsiung International Maritime and Defense Exhibition. Brochure 2/2.

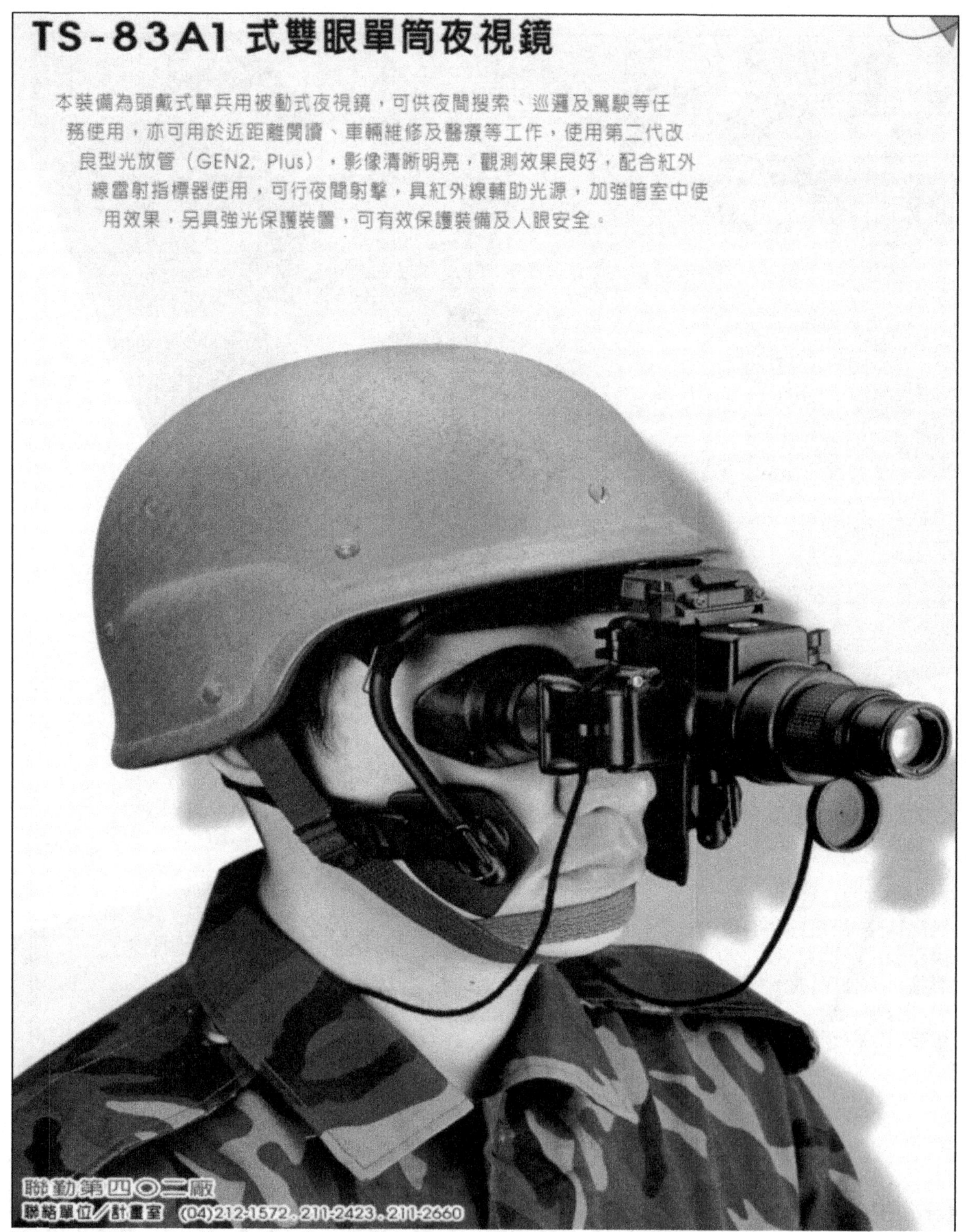

Materiel Production Center (MPC). 402nd Factory. TS-83A1 Night Vision Goggle. Unidentified Taipei Aerospace and Defense Technology Exhibition (TADTE). Brochure 1/3.

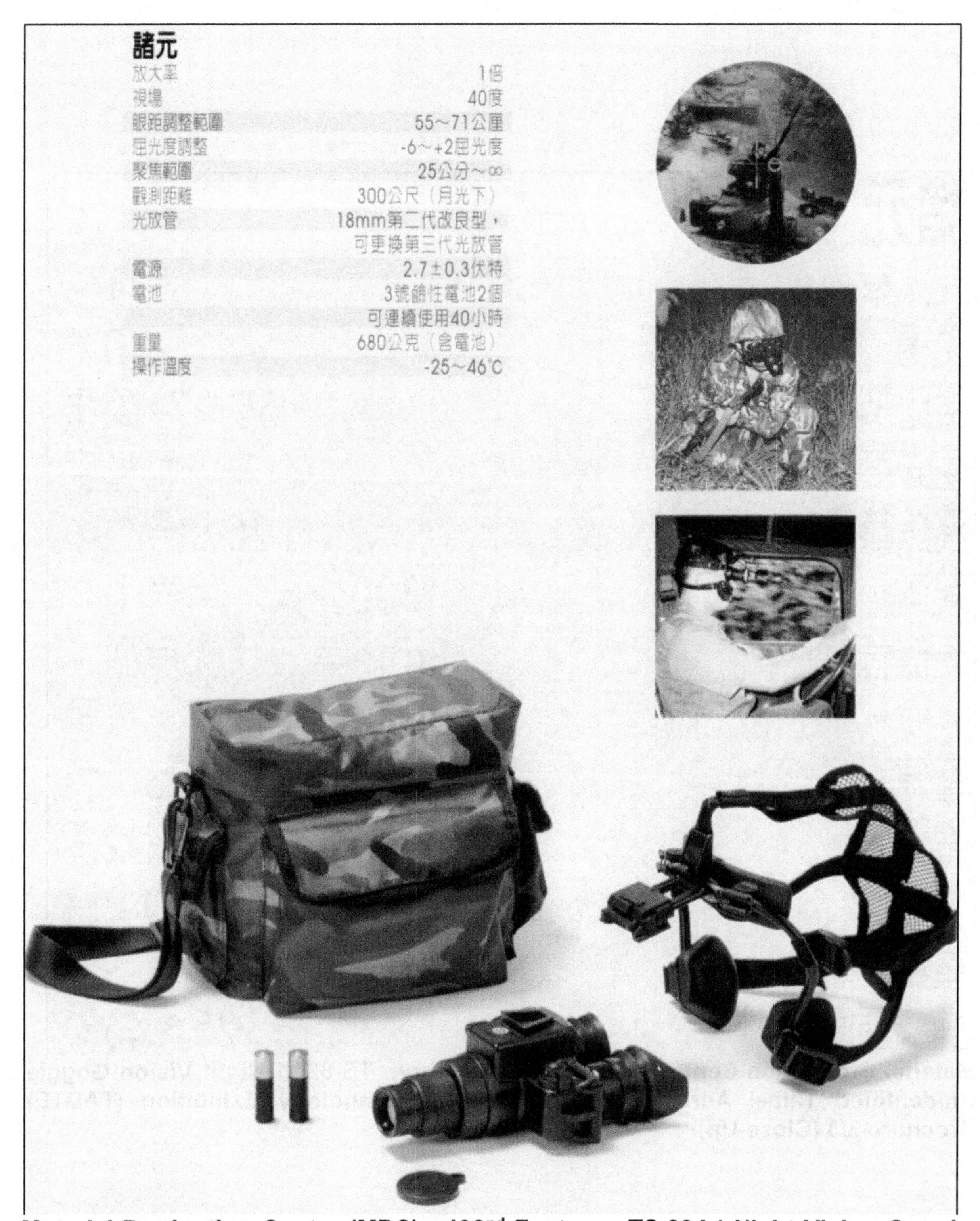

Materiel Production Center (MPC). 402nd Factory. TS-83A1 Night Vision Goggle. Unidentified Taipei Aerospace and Defense Technology Exhibition (TADTE). Brochure 2/3.

諸元	
放大率	1倍
視場	40度
眼距調整範圍	55～71公厘
屈光度調整	-6～+2屈光度
聚焦範圍	25公分～∞
觀測距離	300公尺（月光下）
光放管	18mm第二代改良型，可更換第三代光放管
電源	2.7±0.3伏特
電池	3號鹼性電池2個 可連續使用40小時
重量	680公克（含電池）
操作溫度	-25～46℃

Materiel Production Center (MPC). 402nd Factory. TS-83A1 Night Vision Goggle. Unidentified Taipei Aerospace and Defense Technology Exhibition (TADTE). Brochure 3/3 (Close-Up).

TS-84式步機槍夜視鏡

本裝備係專門針對國造65K2步槍及T74排用機槍而設計之夜視瞄準鏡，經特殊設計，方便貼腮瞄準射擊與拆卸更換武器。分劃板為十字刻劃（K2與T74共用），亮度可調，射擊瞄準方便且準確度高。因放大倍率高（四倍），使用第二代改良型光放管或第三代光放管，影像清晰明亮，易於搜尋目標。

Materiel Production Center (MPC). 402nd Factory. TS-84 Night Vision Scope. Unidentified Taipei Aerospace and Defense Technology Exhibition (TADTE). Brochure 1/3.

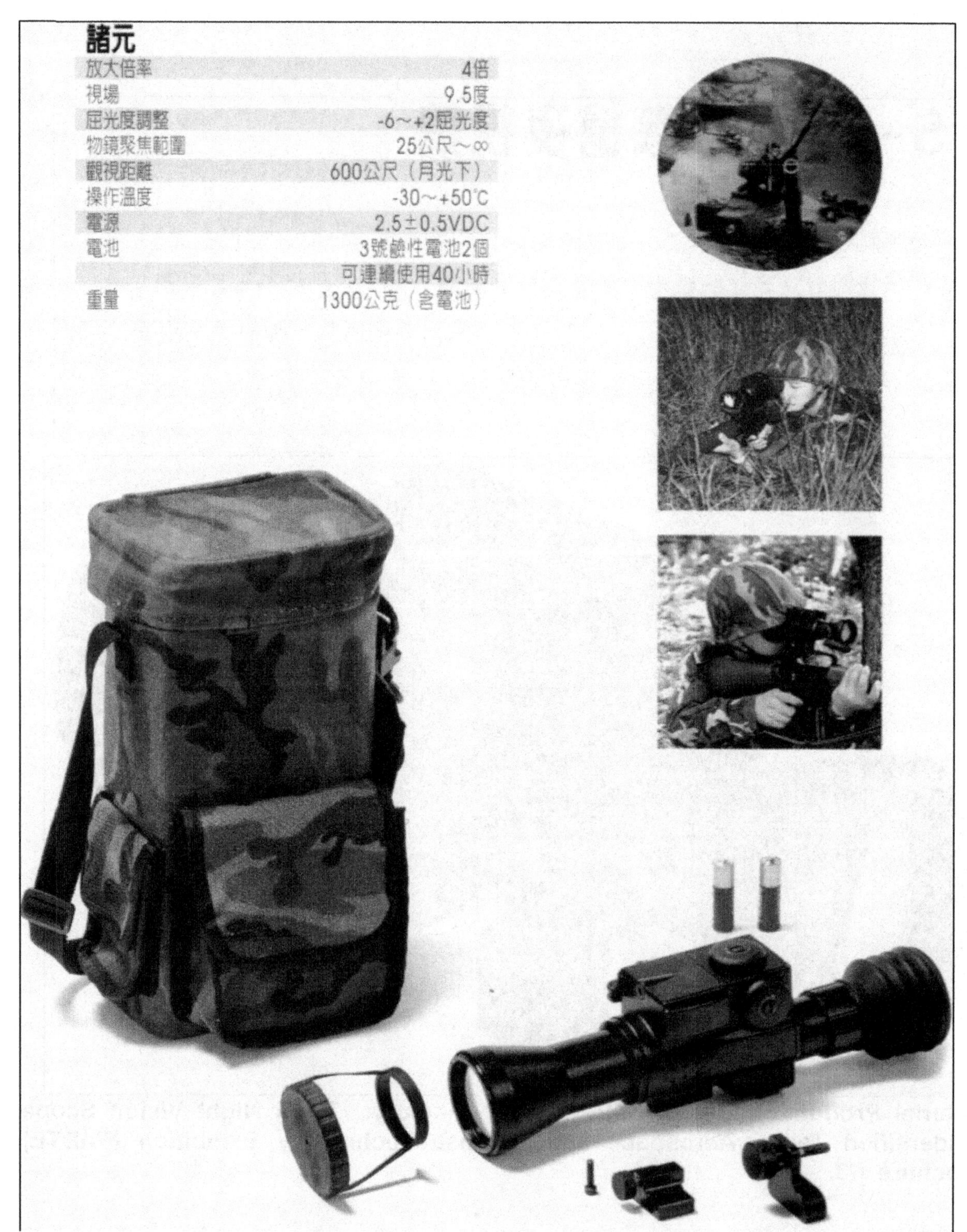

諸元

放大倍率	4倍
視場	9.5度
屈光度調整	-6～+2屈光度
物鏡聚焦範圍	25公尺～∞
觀視距離	600公尺（月光下）
操作溫度	-30～+50℃
電源	2.5±0.5VDC
電池	3號鹼性電池2個
	可連續使用40小時
重量	1300公克（含電池）

Materiel Production Center (MPC). 402nd Factory. TS-84 Night Vision Scope. Unidentified Taipei Aerospace and Defense Technology Exhibition (TADTE). Brochure 2/3.

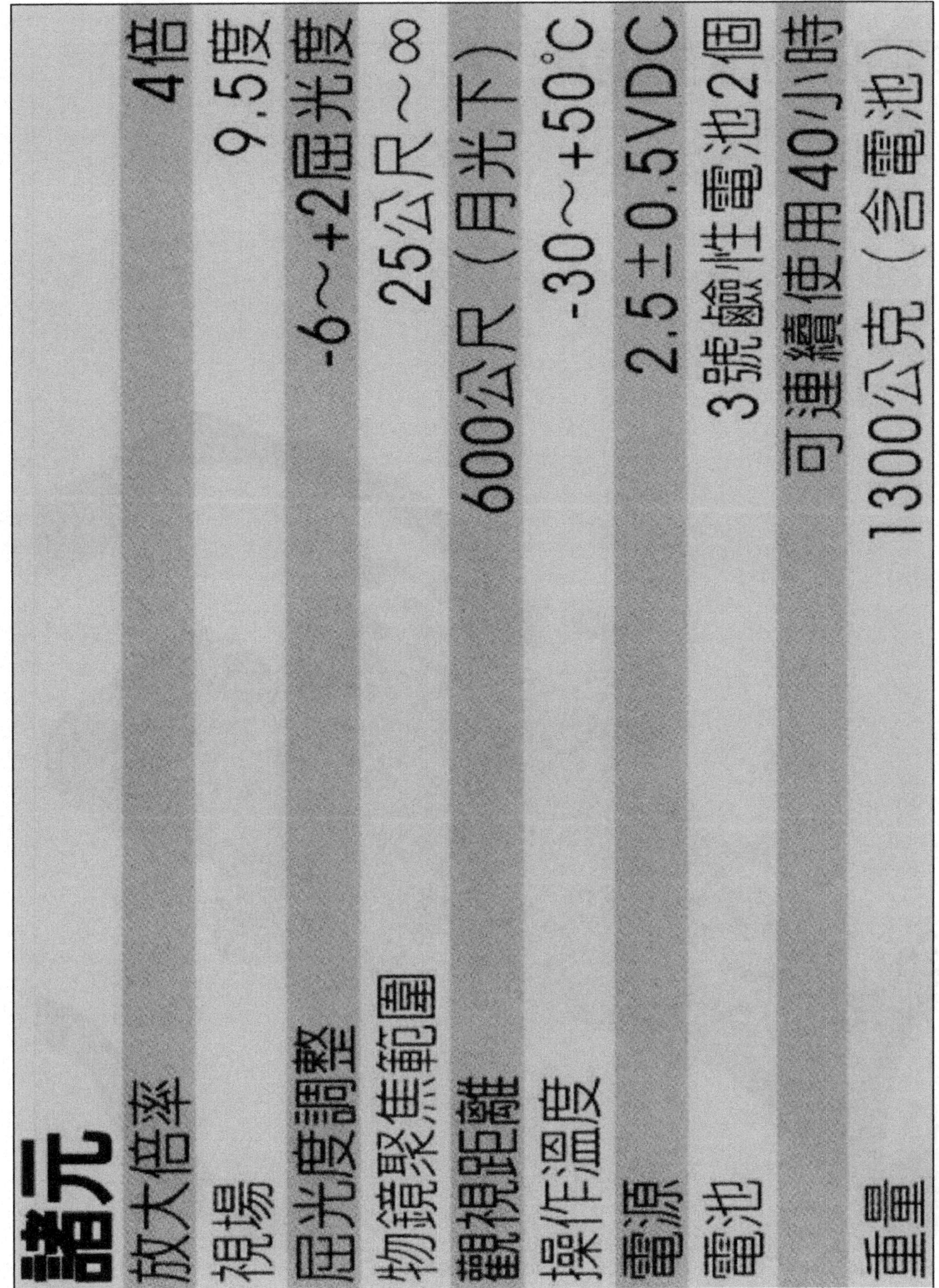

諸元	
放大倍率	4倍
視場	9.5度
屈光度調整	-6～+2屈光度
物鏡聚焦範圍	25公尺～∞
觀視距離	600公尺（月光下）
操作溫度	-30～+50°C
電源	2.5±0.5VDC
電池	3號鹼性電池2個
	可連續使用40小時
重量	1300公克（含電池）

Materiel Production Center (MPC). 402nd Factory. TS-84 Night Vision Scope. Unidentified Taipei Aerospace and Defense Technology Exhibition (TADTE). Brochure 3/3 (Close-Up).

TS-84A式夜視望遠鏡

本裝備為被動式手持夜視望遠鏡，於暗夜月光下行中距離觀測，觀測距離可達一公里，內附十字刻線鏡，可簡易估測目標物距離，適合部隊指揮官於夜間陣地指揮與戰況掌握，亦可作中距離觀測與監視搜索用。使用第二代改良型或第三代光放管，影像清晰明亮，觀測效果佳，另具強光保護裝置，可有效保護裝備及人眼之安全。

Materiel Production Center (MPC). 402nd Factory. TS-84A Night Vision Goggle. Unidentified Taipei Aerospace and Defense Technology Exhibition (TADTE). Brochure 1/3.

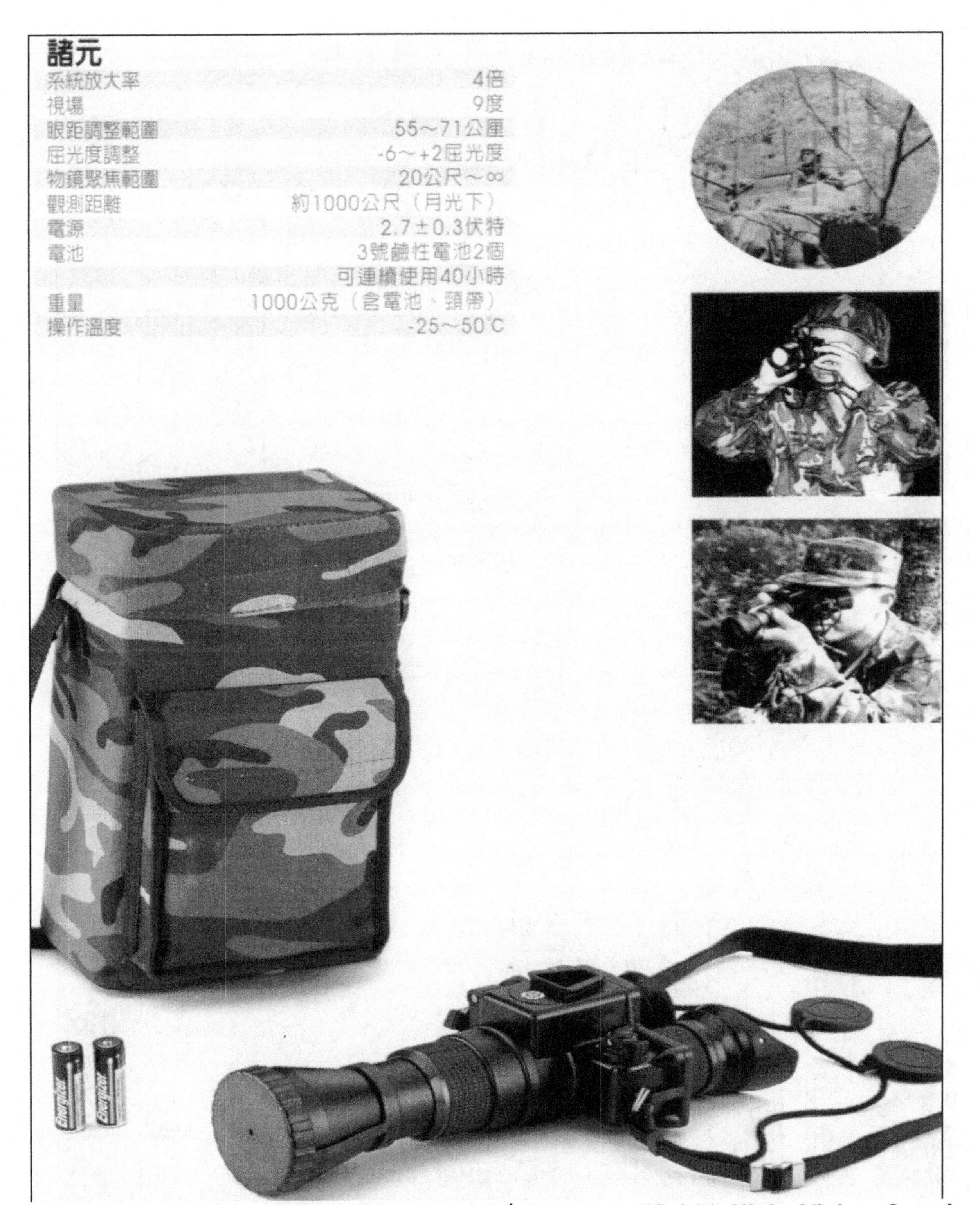

Materiel Production Center (MPC). 402nd Factory. TS-84A Night Vision Goggle. Unidentified Taipei Aerospace and Defense Technology Exhibition (TADTE). Brochure 2/3.

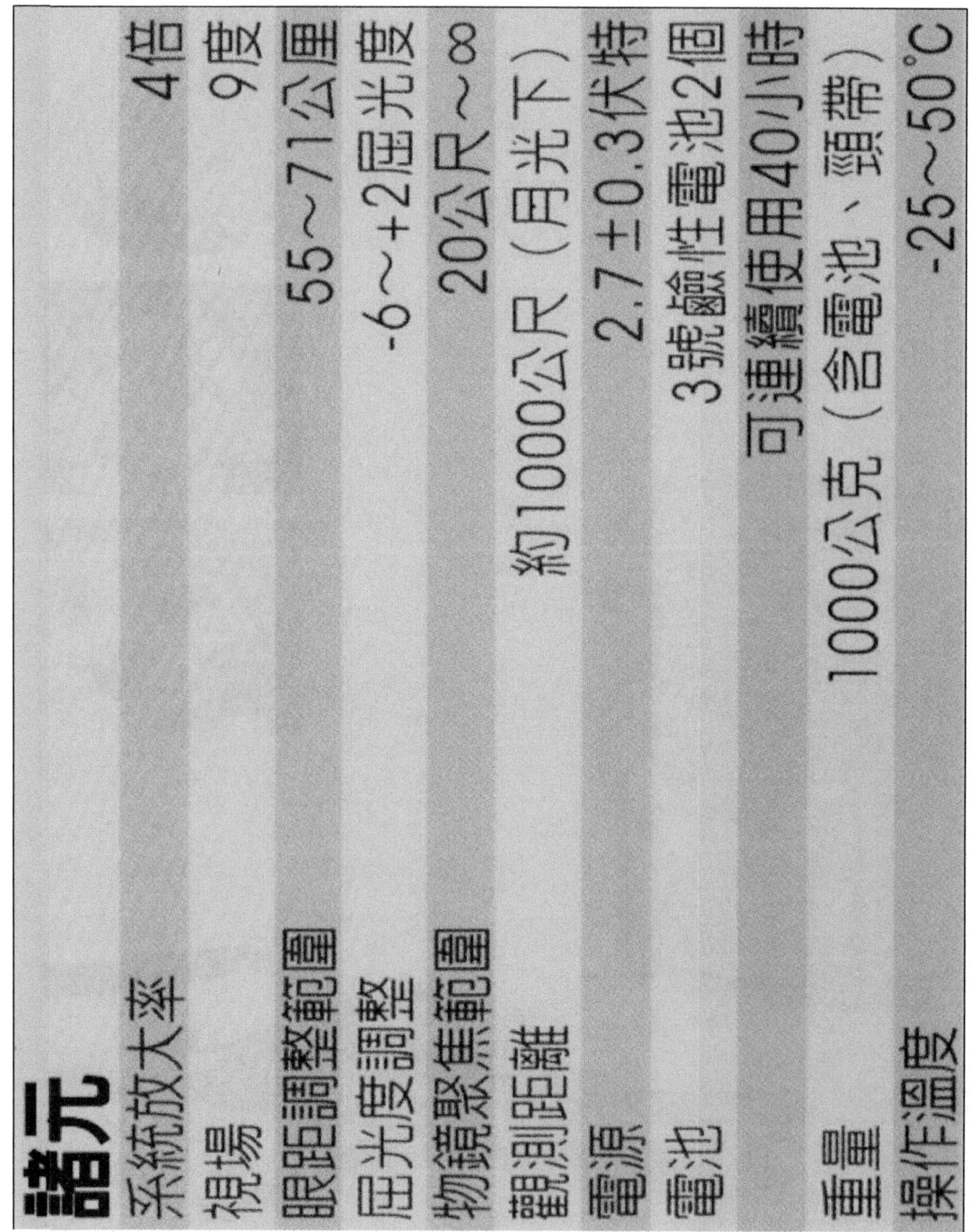

諸元	
系統放大率	4倍
視場	9度
眼距調整範圍	55~71公厘
屈光度調整	-6~+2屈光度
物鏡聚焦範圍	20公尺~∞
觀測距離	約1000公尺（月光下）
電源	2.7±0.3伏特
電池	3號鹼性電池2個
	可連續使用40小時
重量	1000公克（含電池、頭帶）
操作溫度	-25~50°C

Materiel Production Center (MPC). 402nd Factory. TS-84A Night Vision Goggle. Unidentified Taipei Aerospace and Defense Technology Exhibition (TADTE). Brochure 3/3 (Close-Up).

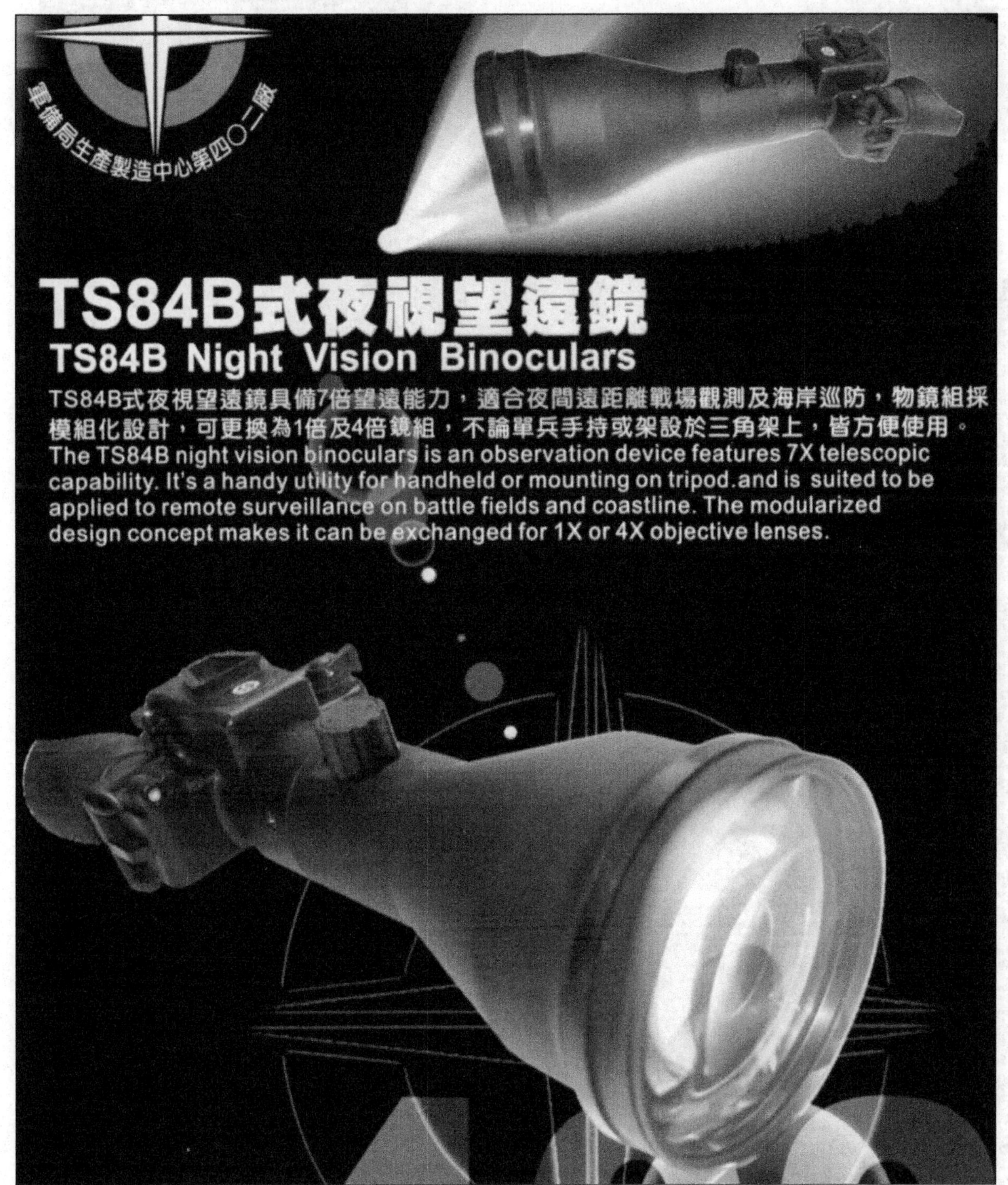

Materiel Production Center (MPC). 402nd Factory. TS-84B Night Vision Binoculars. Unidentified Taipei Aerospace and Defense Technology Exhibition (TADTE). Brochure 1/3.

Materiel Production Center (MPC). 402nd Factory. TS-84B Night Vision Binoculars. Unidentified Taipei Aerospace and Defense Technology Exhibition (TADTE). Brochure 2/3.

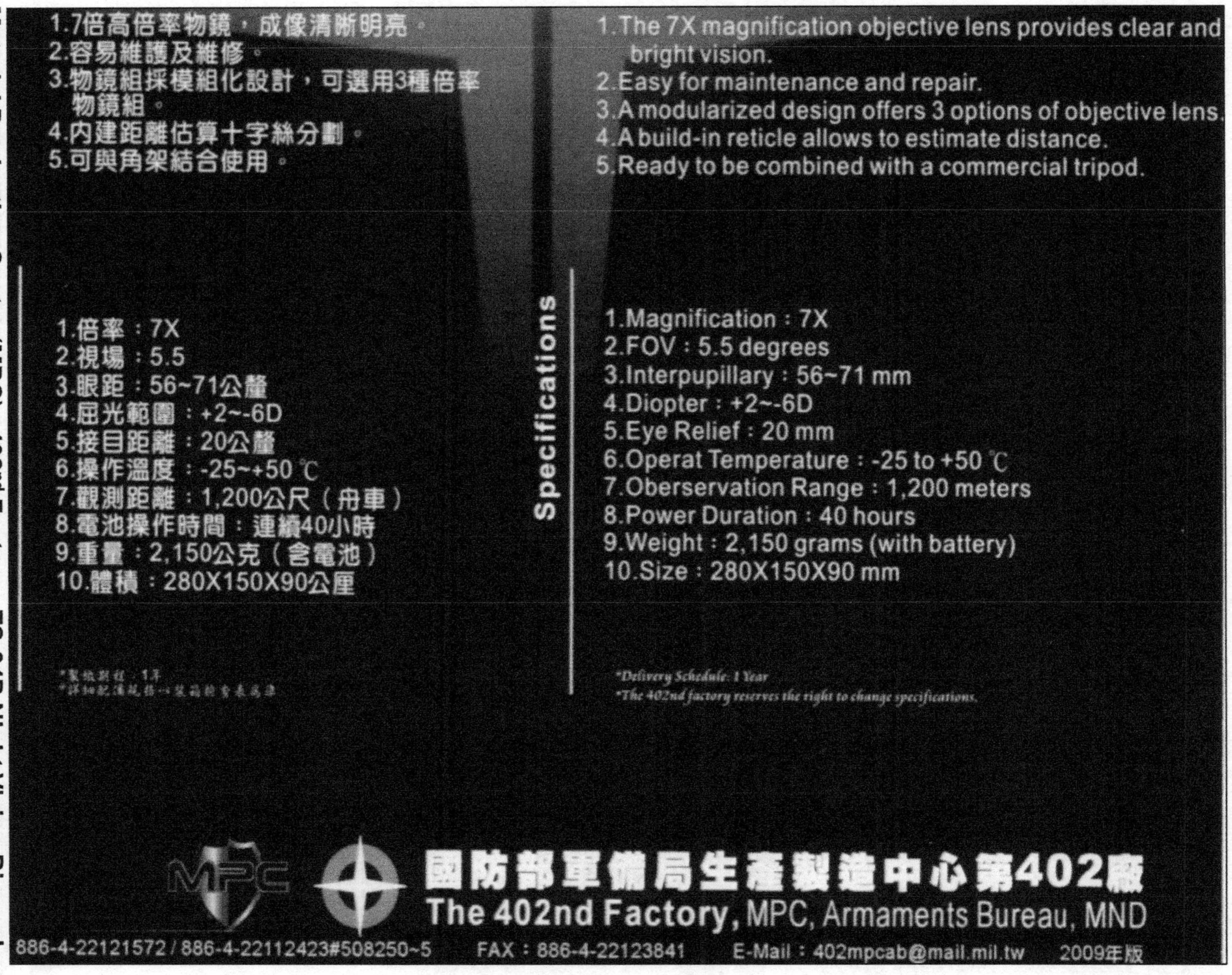

Materiel Production Center (MPC). 402nd Factory. TS-84B Night Vision Binoculars. Unidentified Taipei Aerospace and Defense Technology Exhibition (TADTE). Brochure 3/3 (CLOSE-UP).

Materiel Production Center (MPC). 401st Factory. Optoelectronics Business – TS-91B Rifle Sight. 2018 Kaohsiung International Maritime and Defense Exhibition. Brochure 1/2.

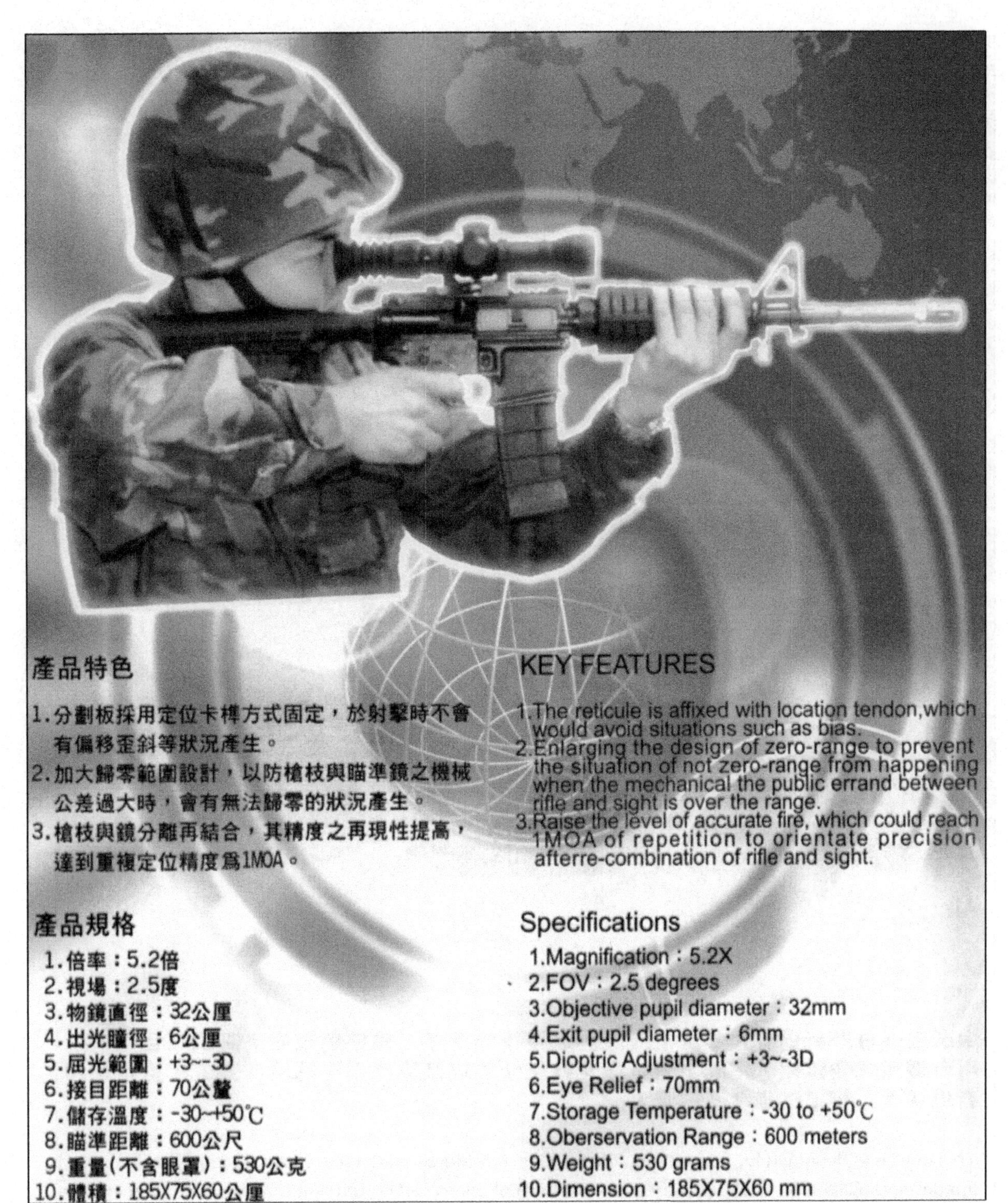

產品特色
1.分劃板採用定位卡榫方式固定，於射擊時不會有偏移歪斜等狀況產生。
2.加大歸零範圍設計，以防槍枝與瞄準鏡之機械公差過大時，會有無法歸零的狀況產生。
3.槍枝與鏡分離再結合，其精度之再現性提高，達到重複定位精度爲1MOA。
KEY FEATURES
1.The reticule is affixed with location tendon,which would avoid situations such as bias.
2.Enlarging the design of zero-range to prevent the situation of not zero-range from happening when the mechanical the public errand between rifle and sight is over the range.
3.Raise the level of accurate fire, which could reach 1MOA of repetition to orientate precision afterre-combination of rifle and sight.
產品規格
1.倍率：5.2倍
2.視場：2.5度
3.物鏡直徑：32公厘
4.出光瞳徑：6公厘
5.屈光範圍：+3~-3D
6.接目距離：70公釐
7.儲存溫度：-30~+50℃
8.瞄準距離：600公尺
9.重量(不含眼罩)：530公克
10.體積：185X75X60公厘
Specifications
1.Magnification：5.2X
2.FOV：2.5 degrees
3.Objective pupil diameter：32mm
4.Exit pupil diameter：6mm
5.Dioptric Adjustment：+3~-3D
6.Eye Relief：70mm
7.Storage Temperature：-30 to +50℃
8.Oberservation Range：600 meters
9.Weight：530 grams
10.Dimension：185X75X60 mm

Materiel Production Center (MPC). 401st Factory. Optoelectronics Business – TS-91B Rifle Sight. 2018 Kaohsiung International Maritime and Defense Exhibition. Brochure 2/2.

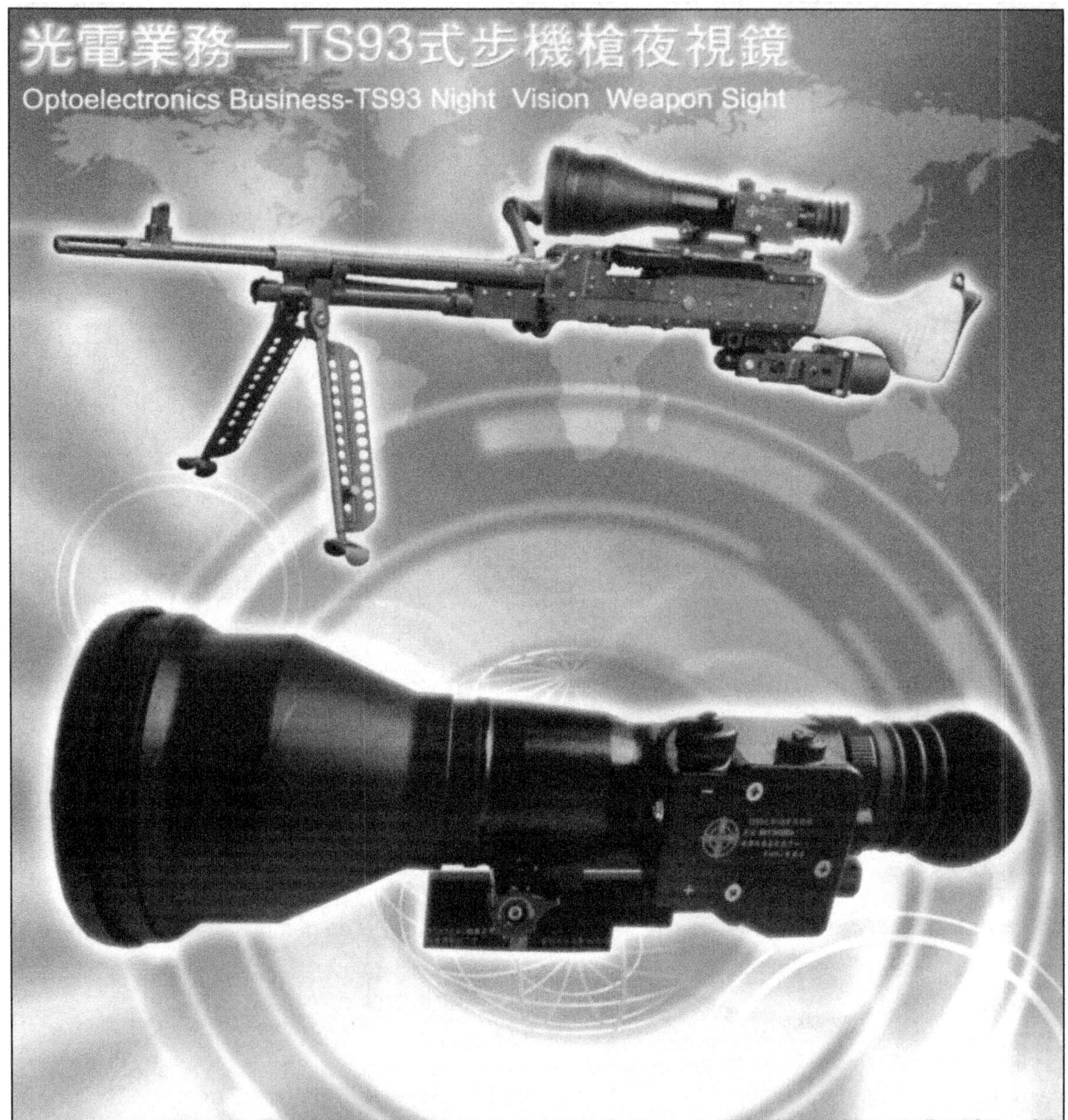

本裝備係專門針對各式步機槍所設計之夜視瞄準鏡，爲使射擊者清晰瞄準目標，採用可調焦式物鏡組，搭配本廠戰術滑軌結合座以及更改十字絲調整方式，提供使用者更快速、精準的瞄準及射擊。

This device is designed for use with multiform rifles. It features adjustable focus node to acquire sharp image, and utilizes our multi-country patented tactical rail mounting bracket that provides users with a rapid and accurate shooting experience.

Materiel Production Center (MPC). 401st Factory. Optoelectronics Business – TS-93 Night Vision Weapon Sight. 2018 Kaohsiung International Maritime and Defense Exhibition. Brochure 1/2.

產品特色
1.4倍高倍率物鏡，成像清晰明亮。
2.容易維護及維修。
3.物鏡組採模組化設計，可選用3種倍率物鏡組。
4.內建距離估算十字絲分劃。
5.可與各式步槍結合使用。

KEY FEATURES
1.The 4X power objective lens provides clear and bright vision.
2.Easy for maintenance and repair.
3.A modularized design offers 3 options of objective power.
4.A build-in reticle allows to distance estimation.
5.It could be mounted on multiform rifles.

產品規格
1.倍率：6X
2.視場：5.6度
3.調焦範圍：25公尺~∞
4.屈光範圍：+2~-5D
5.接目距離：45公釐
6.操作溫度： -30~+50℃
7.觀測距離：1200(600)公尺（人型）
8.解析度：3.56 lp/mm
9.電池操作時間：連續20小時
10.重量：1800/1350公克（含電池）
11.體積：300X90X70公厘

Specifications
1.Magnification：6X
2.FOV：5.6 degrees
3.Focus Range：25m ~ ∞
4.Dioptric Adjustment：+2~-5D
5.Eye Relief：45mm
6.Operat Temperature：-30 to +50 ℃
7.Oberservation Range：1200(600) meters
8.Resolution：3.56 lp/mm
9.Battery operation Duration：20 hours
10.Weight：1,350 grams (with battery)
11.Dimension：300X90X70 mm

Materiel Production Center (MPC). 401st Factory. Optoelectronics Business – TS-93 Night Vision Weapon Sight. 2018 Kaohsiung International Maritime and Defense Exhibition. Brochure 2/2.

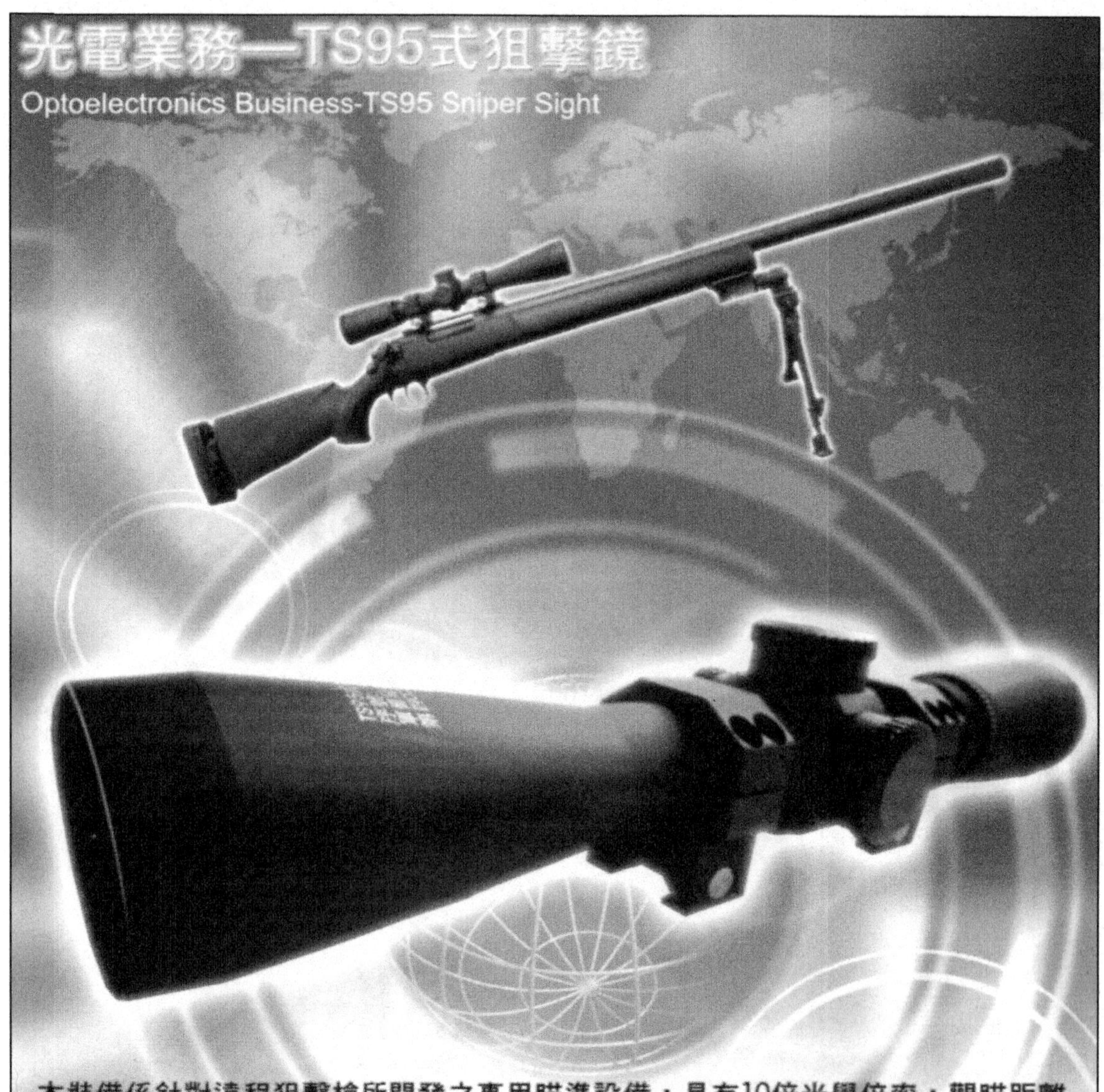

本裝備係針對遠程狙擊槍所開發之專用瞄準設備，具有10倍光學倍率，觀瞄距離可達1000公尺；分劃板採用精密光學技術製作，可完全匹配狙擊槍彈道，並具有密位（MIL）標記，可針對目標實施距離量測評估；同時配賦戰術滑軌結合座，可應用於各種不同型式之狙擊槍。

This device is designed for long-distance rifle. Its magnification is 10. The capability of aiming target is within 1000 meters. The reticle is manufactured from advanced optic technique. It conforms completely to rifle's trajectory and also using MIL as a unit. It can simply measure the target distance. With tactical rail mounting bracket, it is capable of being collocate with any types of rifles.

Materiel Production Center (MPC). 401st Factory. Optoelectronics Business – TS-95 Sniper Sight. 2018 Kaohsiung International Maritime and Defense Exhibition. Brochure 1/2.

Materiel Production Center (MPC). 401st Factory. Optoelectronics Business – TS-95 Sniper Sight. 2018 Kaohsiung International Maritime and Defense Exhibition. Brochure 2/2.

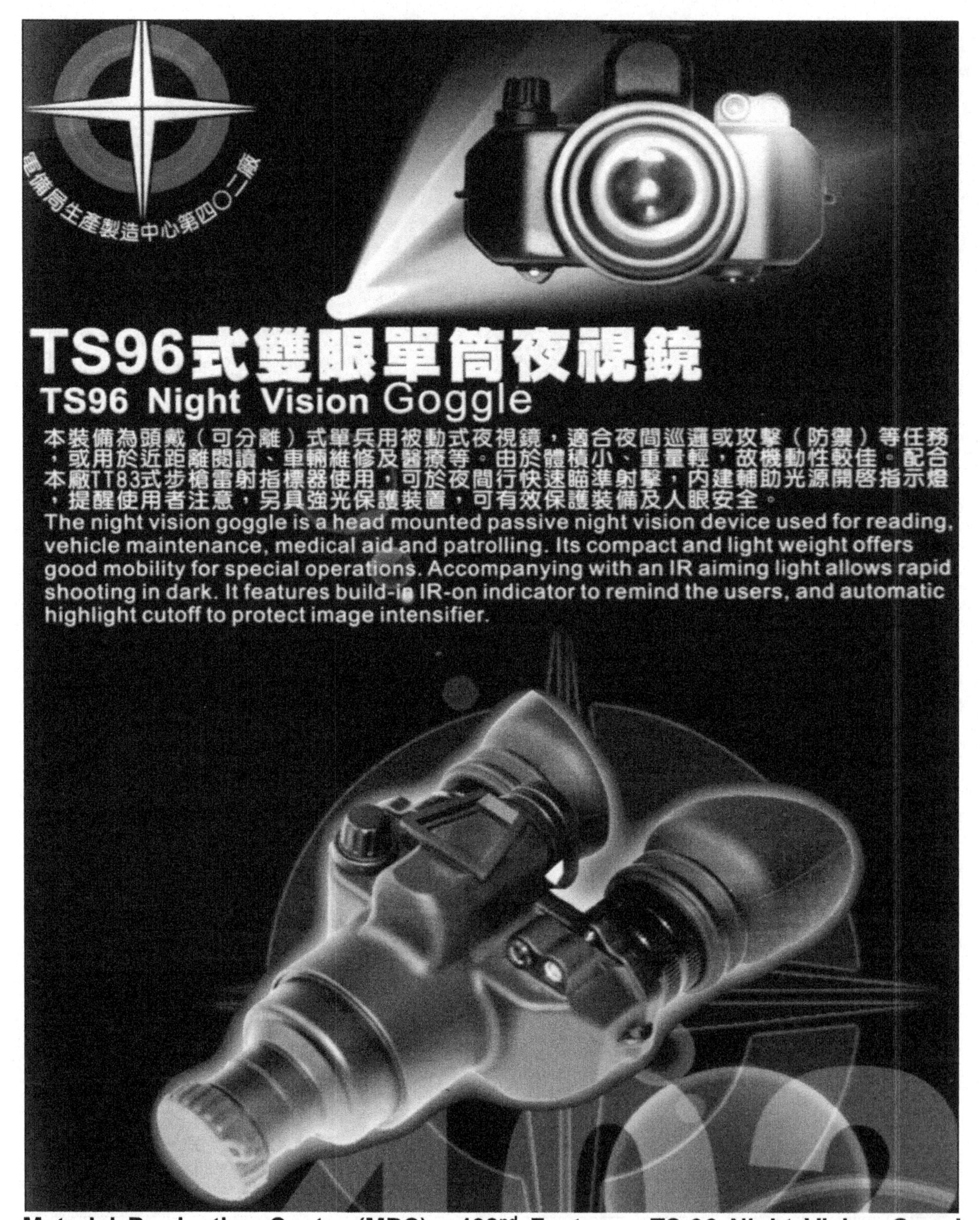

Materiel Production Center (MPC). 402nd Factory. TS-96 Night Vision Goggle. Unidentified Taipei Aerospace and Defense Technology Exhibition (TADTE). Brochure 1/3.

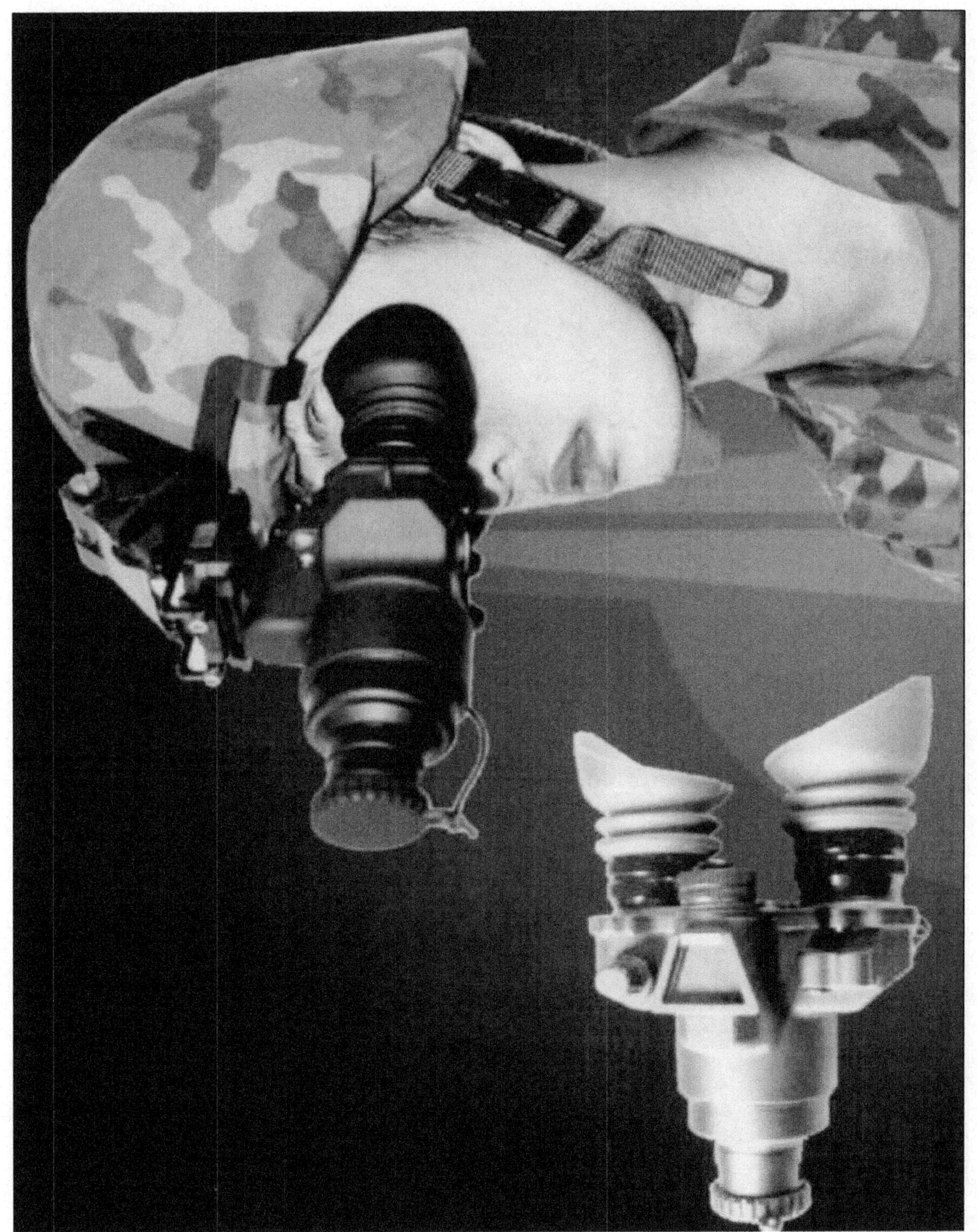

Materiel Production Center (MPC). 402nd Factory. TS-96 Night Vision Goggle. Unidentified Taipei Aerospace and Defense Technology Exhibition (TADTE). Brochure 2/3.

KEY FEATURES

1. A helmet mounted or head strap for hand-free operation.
2. The quick-release lever to allow one-handed attach or detach.
3. A modulized design to offer 2 kinds of objective power.
4. Automatic high light cutoff to protect user's eyes and the intensified tube.
5. The build-in indicators to warn user attention.

Specifications

1. Magnification : 1X
2. FOV : 40 degrees
3. Interpupillary Adjustment : 55~71mm
4. Focus Adjustment Range : 25cm ~ ∞
5. Diopter : +2~-5D
6. Eye Relief : 20mm
7. Operat Temperature : -25 to +45 ℃
8. Oberservation Range : 300 meters
9. Power Duration : 20 hours
10. Weight : less than 540 grams (with battery)
11. Size : less than 155X110X80 mm

產品特色

1.可選用頭戴或盔帶2種免持操作方式。
2.可單手快速拆裝結合。
3.物鏡組採模組化設計，可選用2種倍率物鏡組。
4.強光自動斷電裝置保護人眼與裝備。
5.內建光源開啟及弱電警示燈。

1.倍率：1X
2.視場：40度
3.眼距：55~71公釐
4.調焦範圍：25公分~∞
5.屈光範圍：+2~-5D
6.接目距離：20公釐
7.操作溫度：-25~+45℃
8.觀測距離：300公尺
9.電池操作時間：連續20小時
10.重量：小於540公克（含電池）
11.體積：155X110X80公厘

Materiel Production Center (MPC). 402nd Factory. TS-96 Night Vision Goggle. Unidentified Taipei Aerospace and Defense Technology Exhibition (TADTE). Brochure 3/3 (Close-Up).

本裝備爲頭戴（可分離）式單兵用被動式夜視鏡，適合夜間巡邏或攻擊等任務，或用於近距離閱讀、車輛維修及醫療等。本裝備光學系統是針對國人眼距55至71mm而設計，可讓單兵在未經調整眼距的情況下，直接配戴使用，改善眼距調整不當所造成暈眩情形，提高使用者效率。

The night vision monocular is a head mounted passive night vision device applied for reading, vehicle maintenance, medical aid and patrolling. This off-axis optical system is designed for Asian soldiers with eyes interval between 55~71 mm to put on the NVG directly without adjusting. Therefore, the soldiers fighting efficiency and battlefield survivability can be highly improved.

Materiel Production Center (MPC). 401st Factory. Optoelectronics Business. TS-96 Night Vision Goggle. 2018 Kaohsiung International Maritime and Defense Exhibition. Brochure 1/2.

產品特色

1. 可選用頭戴或盔帶2種免持操作方式。
2. 可單手快速拆裝結合。
3. 物鏡組採模組化設計，可選用2種倍率物鏡組。
4. 強光自動斷電裝置保護人眼與裝備。
5. 內建光源開啓及弱電警示燈。

KEY FEATURES

1. A helmet mounted or head strap for hand-free operation.
2. The quick-release lever to allow one-handed attach or detach.
3. A modularized design to offer 2 kinds of objective power.
4. Automatic high light cutoff to protect user's eyes and the intensified tube.
5. The build-in indicators to warn user attention.

產品規格

1. 倍率：1X(4X)
2. 視場：40(8.5)度
3. 眼距：55~71公釐
4. 調焦範圍：25公分~∞
5. 屈光範圍：+2~-5D
6. 接目距離：25公釐
7. 操作溫度： -25~+45℃
8. 觀測距離：300(600)公尺
9. 電池操作時間：連續20小時
10. 重量：小於540公克（含電池）
11. 體積：小於160X210X90公厘

Specifications

1. Magnification : 1X(4X)
2. FOV : 40(8.5)degrees
3. Interpupillary Adjustment : 55~71mm
4. Focus Range : 25cm ~ ∞
5. Dioptric Adjustment : +2~-5D
6. Eye Relief : 25mm
7. Operat Temperature : -25 to +45 ℃
8. Oberservation Range : 300(600) meters
9. Battery operation time : 20 hours
10. Weight : less than 540 grams (with battery)
11. Dimension : less than 160X210X90 mm

Materiel Production Center (MPC). 401st Factory. Optoelectronics Business. TS-96 Night Vision Goggle. 2018 Kaohsiung International Maritime and Defense Exhibition. Brochure 2/2.

Materiel Production Center (MPC). 401st Factory. Optoelectronics Business. TS-96 Night Vision Monocular. 2018 Kaohsiung International Maritime and Defense Exhibition. Brochure 1/2.

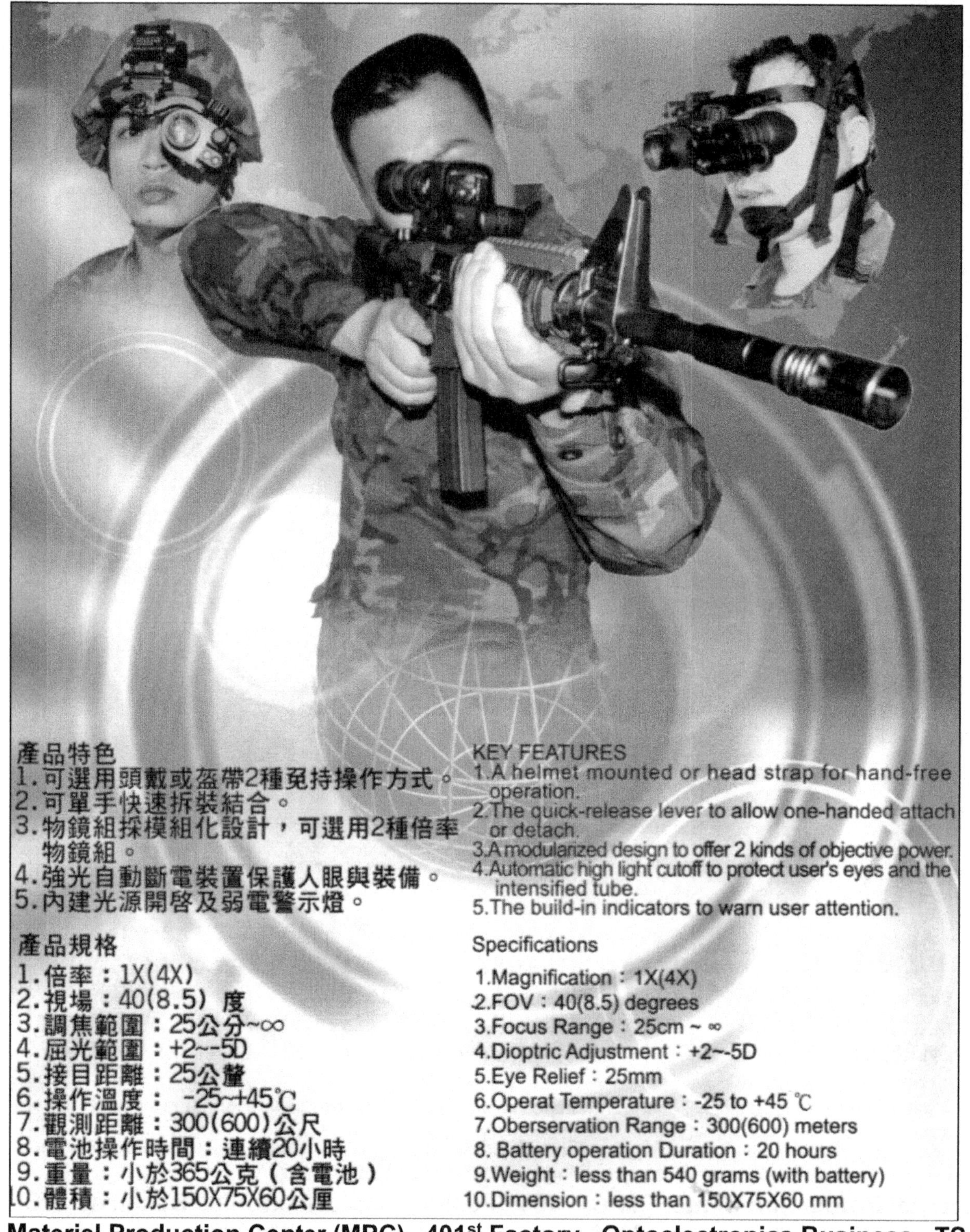
產品特色
1.可選用頭戴或盔帶2種免持操作方式。
2.可單手快速拆裝結合。
3.物鏡組採模組化設計，可選用2種倍率物鏡組。
4.強光自動斷電裝置保護人眼與裝備。
5.內建光源開啓及弱電警示燈。
KEY FEATURES
1.A helmet mounted or head strap for hand-free operation.
2.The quick-release lever to allow one-handed attach or detach.
3.A modularized design to offer 2 kinds of objective power.
4.Automatic high light cutoff to protect user's eyes and the intensified tube.
5.The build-in indicators to warn user attention.
產品規格
1.倍率：1X(4X)
2.視場：40(8.5) 度
3.調焦範圍：25公分~∞
4.屈光範圍：+2~-5D
5.接目距離：25公釐
6.操作溫度： -25~+45℃
7.觀測距離：300(600)公尺
8.電池操作時間：連續20小時
9.重量：小於365公克（含電池）
10.體積：小於150X75X60公厘
Specifications
1.Magnification：1X(4X)
2.FOV：40(8.5) degrees
3.Focus Range：25cm ~ ∞
4.Dioptric Adjustment：+2~-5D
5.Eye Relief：25mm
6.Operat Temperature：-25 to +45 ℃
7.Oberservation Range：300(600) meters
8. Battery operation Duration：20 hours
9.Weight：less than 540 grams (with battery)
10.Dimension：less than 150X75X60 mm

Materiel Production Center (MPC). 401st Factory. Optoelectronics Business. TS-96 Night Vision Monocular. 2018 Kaohsiung International Maritime and Defense Exhibition. Brochure 2/2.

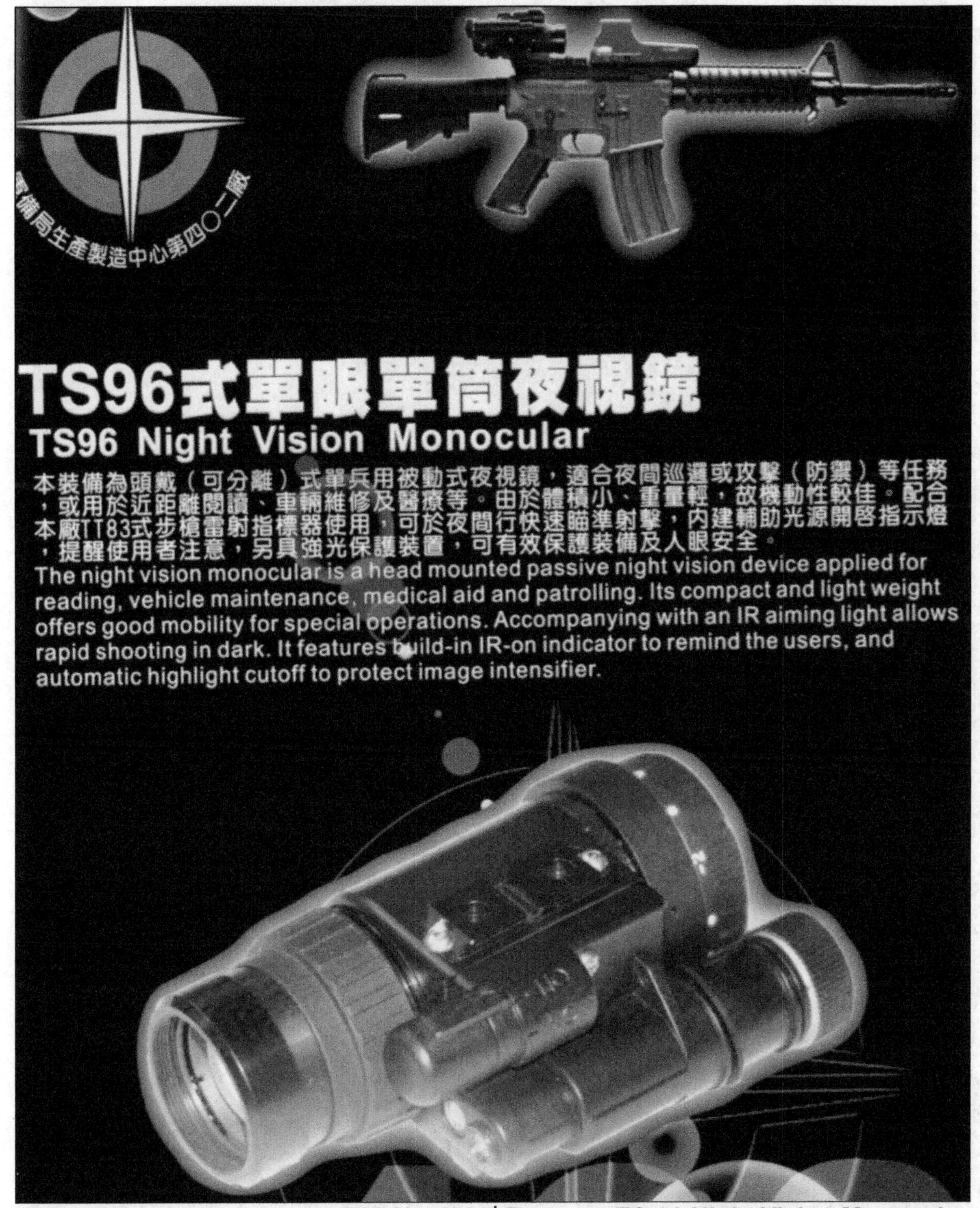

Materiel Production Center (MPC). 402nd Factory. TS-96 Night Vision Monocular. Unidentified Taipei Aerospace and Defense Technology Exhibition (TADTE). Brochure 1/3.

Materiel Production Center (MPC). 402nd Factory. TS-96 Night Vision Monocular. Unidentified Taipei Aerospace and Defense Technology Exhibition (TADTE). Brochure 2/3.

產品特色

1.可選用頭戴或盔帶2種免持操作方式。
2.可單手快速拆裝結合。
3.物鏡組採模組化設計，可選用2種倍率物鏡組。
4.強光自動斷電裝置保護人眼與裝備。
5.內建光源開啟及弱電警示燈。

1.倍率：1X(4X)
2.視場：40(8.5) 度
3.調焦範圍：25公分~∞
4.屈光範圍：+2~-5D
5.接目距離：25公釐
6.操作溫度： -25~+45℃
7.觀測距離：300(600)公尺
8.電池操作時間：連續20小時
9.重量：小於365公克（含電池）
10.體積：小於150X75X60公厘

KEY FEATURES

1. A helmet mounted or head strap for hand-free operatio
2. The quick-release lever to allow one-handed attach or detach.
3. A modularized design to offer 2 kinds of objective pow
4. Automatic high light cutoff to protect user's eyes and t intensified tube.
5. The build-in indicators to warn user attention.

Specifications

1.Magnification：1X(4X)
2.FOV：40(8.5) degrees
3.Focus Range：25cm ~ ∞
4.Dioptric Adjustment：+2~-5D
5.Eye Relief：25mm
6.Operat Temperature：-25 to +45 ℃
7.Oberservation Range：300(600) meters
8. Battery operation Duration：20 hours
9.Weight：less than 540 grams (with battery)
10.Dimension：less than 150X75X60 mm

*Delivery Schedule: 1 Year
*The 402nd factory reserves the right to change specifications.

Materiel Production Center (MPC). 402nd Factory. TS-96 Night Vision Monocular. Unidentified Taipei Aerospace and Defense Technology Exhibition (TADTE). Brochure 3/3 (Close-Up).

NATIONAL CHUNG-SHAN INSTITUTE OF SCIENCE AND TECHNOLOGY (NCSIST)

National Chung-Shan Institute of Science and Technology (NCSIST). Immersive Interaction (i2s3) Shooting Simulation System. 2015 Taipei Aerospace and Defense Technology Exhibition (TADTE). Brochure 1/1.

"Kestrel" series rockets

Introduction

"Kestrel" series rockets are a new generation Shoulder launched weapon. With Optical Sight and Tactical Mount for the Night Vision Goggle, it can meet all the requirements in all-weather combat. The launcher is made of FRP (Fiber Reinforced Plastics), which is lightweight, tough, simple and man-portable. The rocket is water, mud and sand proof, so it can be used in the harsh conditions of battle field. The rockets are suitable for both anti-landing operation and urban environment. The launcher is disposable, and this design can significantly improve the survivability of individual soldier. The rockets are also suitable for mobile or sniping operation. This is truly a next generation HEAT/HESH Rockets with light, robust, and fine design, and will accommodate to asymmetric operation with innovation concept in the future.

Data & Specification

- Sight System : Optical Sight (with Night Vision Goggle)
- Firing Mechanism : Mechanical Trigger
- Kestrel HEAT Rocket :
  Effective Range 400 M
  Penetration Depth 400 mm
- Kestrel HESH Rocket :
  Effective Range 150 M
  Forming a 70~90 cm hole in either 20 cm reinforced concrete walls or 30 cm brick walls

2 rockets for 1 launcher

New

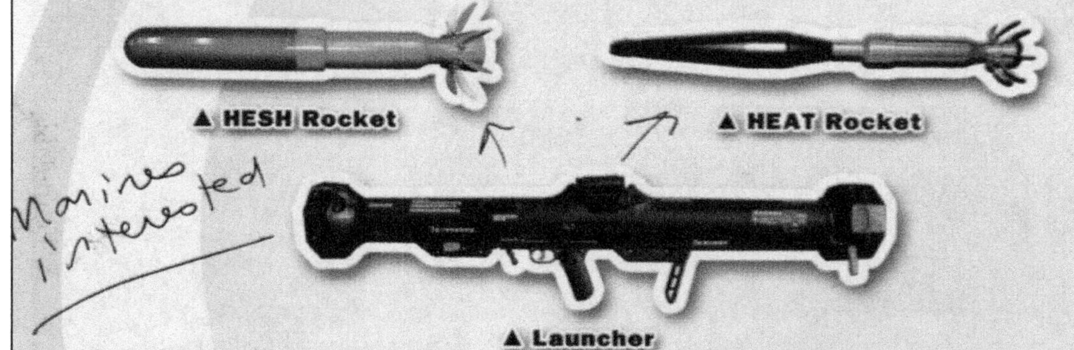

National Chung-Shan Institute of Science and Technology (NCSIST). Kestrel Shoulder Launched Rocket. Unidentified Taipei Aerospace and Defense Technology Exhibition (TADTE). Written note indicates the Taiwan Marine Corps expressed interest in procuring Kestrel. Brochure 1/1.

National Chung-Shan Institute of Science and Technology (NCSIST). Kestrel Shoulder Launched Rocket. 2018 Defence Services Asia (Malaysia). Brochure 1/2.

The Kestrel rocket is one of the lightest shoulder-launched weapon with two types of warheads : high-explosive anti-tank (HEAT) and high-explosive squash head (HESH). Its light feature enable the light troops or supporting forces to assault armored vehicles and fortifications. Its powerful lethality are suitable for an urban or bunker positional warfare.

Features

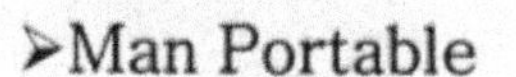

- Man Portable
- Easy to Operate
- Foldable Optical Sight
- Optional Night Vision Goggle
- High Reliability
- High Hit Rate
- Disposable
- Cost Effective
- Deployed in the R.O.C. Marine Corps

Specifications

	HEAT Rocket	HESH Rocket
Effective Range(m)	400	150
Effect	350mm Armor-piercing	70-90cm hole in 30 cm brick wall
Weight(kg)	5	6
Length(mm)	1100	1100
Launcher	Single tube (FRP)	Single tube (FRP)

National Chung-Shan Institute of Science and Technology
Lungtan, Taoyuan, Taiwan (R.O.C.)
Tel : 886-2-2673-9638 ext 351362, 351227
exponshow@ncsist.org.tw
www.ncsist.org.tw

National Chung-Shan Institute of Science and Technology (NCSIST). Kestrel Shoulder Launched Rocket. 2018 Defence Services Asia (Malaysia). Brochure 2/2.

Short-Range Automated Defense Weapon System

Introduction

SMC is developing a short-range automated defense weapon system with features of unmanned, precision, maneuvering and firepower enhancement. By such development, we simultaneously establish technologies and ability of building large-scaled intelligent and electric machinery platforms for conventional weapons, so as to firm the foundation of indigenous defense industry. The system includes three configurations — ground mobile, ship-based and position-fixed models, all been integrated with subsystems such as electric optical imagery identification, target tracking, fire control, fire concentration and accurate servo motors, to ensure the system ability of providing "fast, fierce, aggressive, precision" firepower and the efficiency of decreasing casualty.

National Chung-Shan Institute of Science and Technology (NCSIST). Short-Range Automated Defense Weapon Systems. XTR-101 for the navy platforms or land mobile platforms; XTR-102 for fixed station. 2013 Defence Services Asia (Malaysia). Brochure 1/2.

National Chung-Shan Institute of Science and Technology (NCSIST). Short-Range Automated Defense Weapon Systems.XTR-101 for the navy platforms or land mobile platforms; XTR-102 for fixed station. 2013 Defence Services Asia (Malaysia). Brochure 2/2.

National Chung-Shan Institute of Science and Technology (NCSIST). Antelope Air Defense System. Unidentified Taipei Aerospace Defense Technology Exhibition. Brochure 1/5.

捷羚防空飛彈系統

Antelope Air Defense System

中山科學研究院自1995年開始研發“捷羚防空系統”。在研發初期，中科院致力於發展一高性能射控系統再配以四枚天劍一型飛彈，構成一套先進防空飛彈系統。

捷羚防空飛彈系統是新一代的防空飛彈系統，它整合了目標獲得系統 (TAS)、通訊系統、及操控系統於一身， 完全符合現代防空戰爭的需要。

本系統因為配備有先進的操作及射控系統，故操作非常簡便，僅需兩人（包括一名射手及一名追瞄手）即可完成戰備。可攔截低飛之直昇機、戰鬥機、攻擊機、及轟炸機，它是防守重要戰術設施（如機場及港口）之利器。由於捷羚的高機動性，它也是陸軍部隊或海軍陸戰隊的行軍隨身保鑣。

衆所週知，防空飛彈系統之防空效益完全取決於目標系統的搜索及鎖定能力。因此，中科院在捷羚防空飛彈系統中裝置了一套高性能的目獲系統（TAS）以使其具有很高的目獲攔截成功率，形成滴水不漏的防空網。

CSIST started developing the Antelope in 1995. In the development phase, CSIST focused on developing an agile fire control system armed with four well-proven TC1missiles.

The Antelope is an advanced air defense system, with an integral target acquistion system (TAS), communication system, and operation system. The antelope totally satisfies the needs of modern air defense.

Because of its highly automatic operation and fire control system, the Antelope requires only two persons, a gunner and an observer, to operate. The system can counter all low-flying threats such as choppers, attack fighters, and bombers, so it is a perfect choice for the defense of tactical facilities such as airports and seaports. In addition, it also is a powerful guardian for military troops, since it is a self-propelled air defense system.

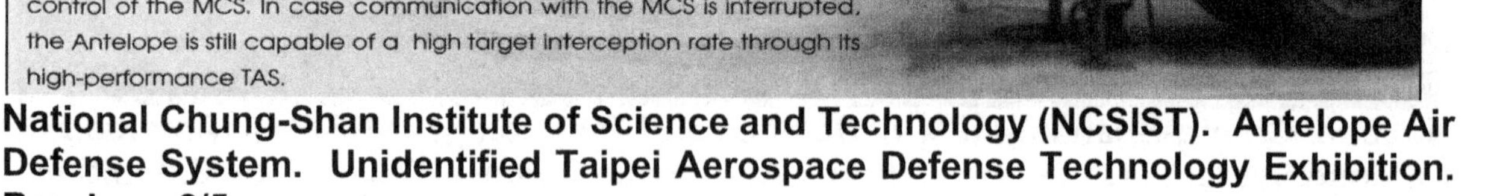

A linkage with a Mission Control System(MCS) is essential for modern air defense. CSIST thus incorporates a long-range communication system to establish a total defense network connected by both the radio and line. The Antelope can obtain early warning information from can be increased through MCS, meaning that the efficiency of air defense can be increased through cooperation between Antelopes under the unified control of the MCS. In case communication with the MCS is interrupted, the Antelope is still capable of a high target interception rate through its high-performance TAS.

National Chung-Shan Institute of Science and Technology (NCSIST). Antelope Air Defense System. Unidentified Taipei Aerospace Defense Technology Exhibition. Brochure 2/5.

系統特性

行進間發射：

捷羚防空飛彈系統裝載於中型貨卡車上可自行移防，並且在車上配備了一部發電機以提供系統操作所須之電力，因此它可以在行進間操作及發射飛彈。

電子反反制能力：

捷羚防空飛彈系統融合了多種目獲系統，其中包括雷達、前視紅外線熱影像機（FLIR）、以及人工目獲追瞄器。此多重目獲系統不僅可以大幅提高目獲成功率， 而且還可以有效地反制敵人的電子反制干擾。

獨立戰備：

捷羚因具有性能優秀之目獲系統，因此它是一個自給自足的防空系統，當缺少指管系統或指管系統已被摧毀時，獨立戰備就是防空系統不可或缺的功能。

Characteristics

Launch-on-move

The self-propelled Antelope is mounted on a medium-size truck and is equipped with a generator to provide electrical power to the system, allowing it to operate while moving. This is an important feature for military troops on the march.

Autonomous operation

The system can easily acquire targets via its own TAS, so it can also serve as an autonomous defense system. This feature is very important if the MCS is absent or destroyed.

Remote operation

To lessen the I threat of being shot by enemy aircraft, the system is equipped with a portable control console which can be operated either in the truck or at any remote position within 70 meters. This feature significantly increases operating reliability when the gunner operates the system from a remote position.

National Chung-Shan Institute of Science and Technology (NCSIST). Antelope Air Defense System. Unidentified Taipei Aerospace Defense Technology Exhibition. Brochure 3/5.

System Characteristics

Night Vision Operation: The Antelope is equipped with radar, FLIR, and a brightness-adjustable console so that the gunner can easily operate the system without any interception degradation during the night.

Easy Operation Power-on: The operating and fire control system in the Antelope is a sophisticated "householder" which handles everything, whether incoming or outgoing. When the system arrives at the battle position, the gunner first turns on the power generator to wake up the system. In the power-up phase, the fire control computer checks system status via built-in-test (BIT). After completing the BIT, the system automatically reports the system status to the gunner via the control console and starts a short warm-up period. Then the gunner can reassign, if necessary, the defense area and preferred operation parameters. After finishing warm-up the system is ready for arming.

Armed Mode: Upon receiving the open firing command from the command center, the system is rapidly placed in the ARM mode simply by pushing the "ARM" button. The system then starts searching for targets in the defense area and locks on to any enemy targets in the defense area that lack a friend identification echo. When the TAS locks the enemy target, the fire control computer automatically slaves the turret and missiles to synchronously lock onto the foes. At this time, the missiles are ready for firing.

SIM Mode: The system incorporates a self-contained simulation mode which provides diverse training programs to make the gunner more familiar with the operating procedures without any other testing or simulation equipment.

National Chung-Shan Institute of Science and Technology (NCSIST). Antelope Air Defense System. Unidentified Taipei Aerospace Defense Technology Exhibition. Brochure 4/5.

Antelope

System Components

Target Acquisition System(TAS)
Operation and control system
Four TC1 missiles Communication system
Communication system
Medium Size of Truck

Flight Tests and Further Application

The Antelope program has finished many successful trials showing the system performance is so extraordinary.

中山科學研究院
Chung-Shan Institute of Science and Technology

桃園龍潭郵政90008-1信箱
電話 : 886-2-2673-9638 · 886-3-471-3022 · 886-080-014723
傳真 : 886-3-471-4183
網址 : http//www.csistdup.org.tw
P.O. Box 90008-1, Lungtan, Taoyuan, Taiwan, R.O.C.
TEL : 886-2-2673-9638 · 886-3-471-3022 · 886-080-014723
FAX : 886-3-471-4183
http//www.csistdup.org.tw

National Chung-Shan Institute of Science and Technology (NCSIST). Antelope Air Defense System. Unidentified Taipei Aerospace Defense Technology Exhibition. Brochure 5/5.

National Chung-Shan Institute of Science and Technology (NCSIST). Mobile Phased Array Radar. 2017 Taipei Aerospace Defense Technology Exhibition. Brochure 1/1.

National Chung-Shan Institute of Science and Technology (NCSIST). Mobile Phased Array Radar. 2018 Defence Services Asia (Malaysia). Brochure 1/2.

Short range air defense system

The phased array radar, regarded as the sensor of a short range air defense system, is capable of providing accurate 3D detecting and tracking data of fighters, UAVs, and cruise missiles. The radar can be mounted on a high mobility 4*4 tactical vehicle and be rapidly deployed to protect valuable assets against air attacks. When linked with the mobile engagement control station, the radar can provide multi-target data for the gun and missile systems to effectively build up multi-layered air defense.

Features

- **Advanced signal processing**
- **Real-time controlling software**
- **Target set : Fixed Wing, Rotary Wing, Cruise Missile and UAV**
- **Identification-friend-or-foe (IFF) integrated**
- **Electronic counter-countermeasures (ECCM)**
- **Clutter suppression**
- **Dual transmitters for higher MTBF**
- **Electronically scanned in elevation and mechanically scanned in azimuth to provide accurate 3D data**

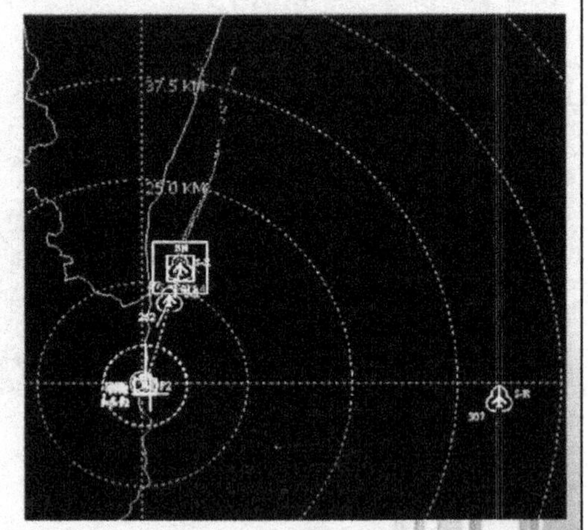

Radar targets on MECS display

Specifications

Crew	2
Instrumented Range	2~72 km
Capacity	64 targets
Scan rate	15 / 30rpm
Mobility	15 min (emplacement) 10 min(march order)
Elevation Coverage	search : 0° ~20° track : -10° ~60°

National Chung-Shan Institute of Science and Technology
Lungtan, Taoyuan, Taiwan (R.O.C.)
Tel : 886-2-2673-9638 ext 351362, 351227
exponshow@ncsist.org.tw
www.ncsist.org.tw

National Chung-Shan Institute of Science and Technology (NCSIST). Mobile Phased Array Radar. 2018 Defence Services Asia (Malaysia). Brochure 2/2.

Vehicle-mounted Mobile Air/Surface Fire Control Radar System

【 Introduction】

The vehicle-mounted mobile air/surface fire control radar system is a multiple-purposes x-band pulse Doppler radar and gun control equipment system which combines searching, tracking , commanding, and controlling. It has high mobility and all weather engagement capability. This fire control radar system and 40mm advanced gun compose a low altitude advanced gun system, which can sequentially complete the searching, identifying, catching, tracking, and engaging to the air target. Going through many times firing practice, it has passed the assessment verification.

Position changing – mobilized transport.

Gun operated by the system.

【Features】

- It has optic-electronic equipment to implement the gun alignment and monitor the target.
- It can be carried by medium tactical wheeler for rapid changing the position.
- It has search radar to acquire targets.
- It has IFF interrogator to identify friend or foe.
- It has TWS function to track 30 targets.
- It has tracking radar to track the target and provide gun control computer with target information.
- It has the gun control computer to calculate data and command the gun to fire.

NATIONAL CHUNG-SHAN INSTITUTE OF SCIENCE & TECHNOLOGY
ELECTRONIC SYSTEMS RESEARCH DIVISION
P.O.Box 90008-22-3 Lung-Tan, Tao-Yuan
32599, Taiwan, R.O.C.
Tel:+886-2-2673-9638 ext.357917
Fax:+886-3-441-1271
http://www.ncsist.org.tw/

National Chung-Shan Institute of Science and Technology (NCSIST). Vehicle-mounted Mobile Air/Surface Fire Control Radar System. 2017 Taipei Aerospace Defense Technology Exhibition. Brochure 1/1.

National Chung-Shan Institute of Science and Technology (NCSIST). C31/C31A1 Point-Initiating, Base Detonating/Self-Destruct (PIBD/SD) Grenade Fuze for 40mm High Velocity-High Explosive Dual Purpose (HV HEDP) Cartridge. 2018 Defence Services Asia (Malaysia). Brochure 1/4.

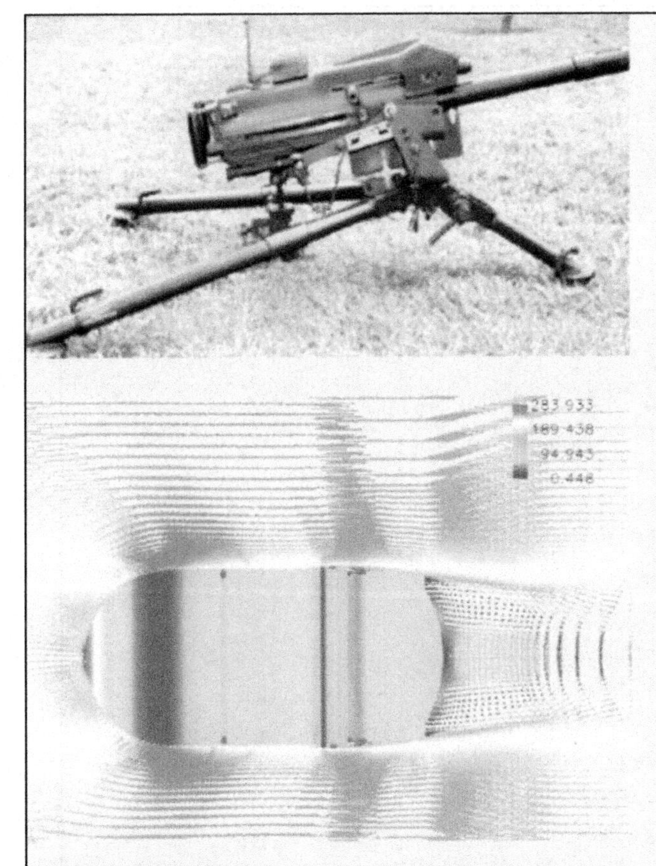

The C31 and C31A1 fuzes are mechanical impact fuzes for 40mm High-Velocity HEDP cartridge, with two independent safety locks and meet the requirements of the MIL-STD-1316. The C31A1 is derived from C31 fuze and has the additional function of self-destruct (SD) mechanism. The SD mechanism enables the projectile to self-destroy in 14 seconds, once the projectile does not work on the soft target or misses the target and flies over the effective range.

Features

- ➢Low Price
- ➢High Reliability
- ➢High Safety
- ➢Easy to Produce

Specifications

Fuze Weight	60±1g
Fuze Length	36.3mm
Fuze Diameter	40.6mm
Setback Safety	15,000G(no arm) 22,500G(armed)
Rotation Safety	1,980rpm(no arm) 6,000rpm(armed)
Arming Distance	18~40m
Muzzle Safety	18m
SD Time	> 14 sec (C31A1)

National Chung-Shan Institute of Science and Technology
Lungtan, Taoyuan, Taiwan (R.O.C.)
Tel : 886-2-2673-9638 ext 351362, 351227

National Chung-Shan Institute of Science and Technology (NCSIST). C31/C31A1 Point-Initiating, Base Detonating/Self-Destruct (PIBD/SD) Grenade Fuze for 40mm High Velocity-High Explosive Dual Purpose (HV HEDP) Cartridge. 2018 Defence Services Asia (Malaysia). Brochure 2/4.

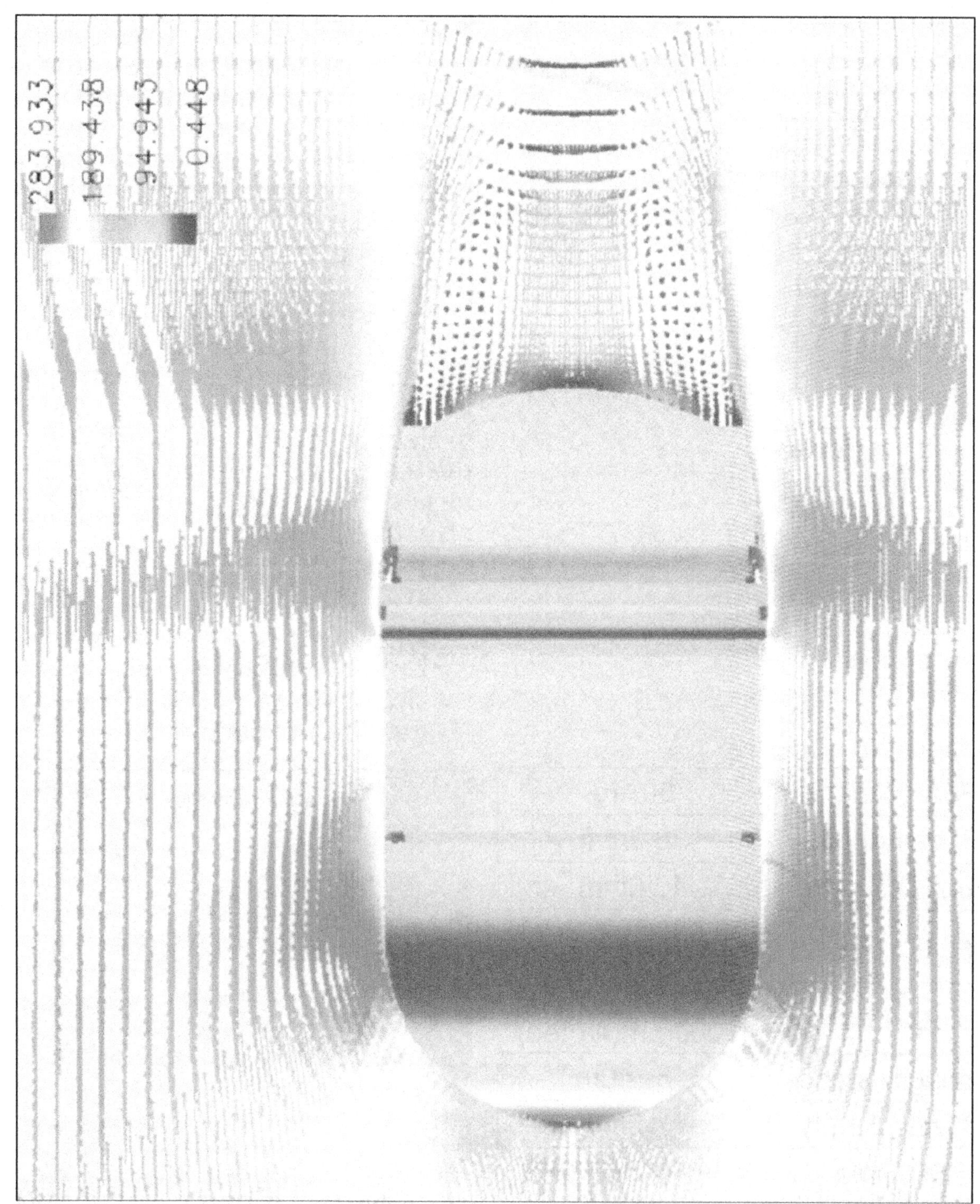

National Chung-Shan Institute of Science and Technology (NCSIST). C31/C31A1 Point-Initiating, Base Detonating/Self-Destruct (PIBD/SD) Grenade Fuze for 40mm High Velocity-High Explosive Dual Purpose (HV HEDP) Cartridge. 2018 Defence Services Asia (Malaysia). Brochure 3/4.

National Chung-Shan Institute of Science and Technology (NCSIST). C31/C31A1 Point-Initiating, Base Detonating/Self-Destruct (PIBD/SD) Grenade Fuze for 40mm High Velocity-High Explosive Dual Purpose (HV HEDP) Cartridge. 2018 Defence Services Asia (Malaysia). Brochure 4/4.

20mm PDSD Fuze

SYSTEM MANUFACTURING CENTER

Description:

The 20mm M505A3 PD fuze is designed for M56A3 HEI cartridge. In order to ensure the safety of user and storage environment, the 20mm PD fuze has to increase the self-destruct function mechanisms. If the projectile miss the target, the 20mm PDSD fuze will self-destroy more than 4 seconds later after the projectile exiting the muzzle. It will not hurt ground men. The 20mm PDSD fuze has no influence upon the trajectory of M56A3 HEI cartridge. Therefore, we can still use the original firing table and will not have any trouble for the fire training.

Features:

◎ Fuze Length—34mm
◎ Fuze Weight—20g approx.
◎ Ballastic Levels
  Velocity：1,030 m/s
  Rotation：124,000 RPM
  Setback：121,800 G
◎ Ammunition—M56A3 HEI
◎ Weapon—T75 20mm Machine Gun
◎ Impact function—The fuze must High order function after impact to 2mm thick 2024-T3 aluminum plate
◎ SD function —The Self-Destruct time must be above 4 seconds

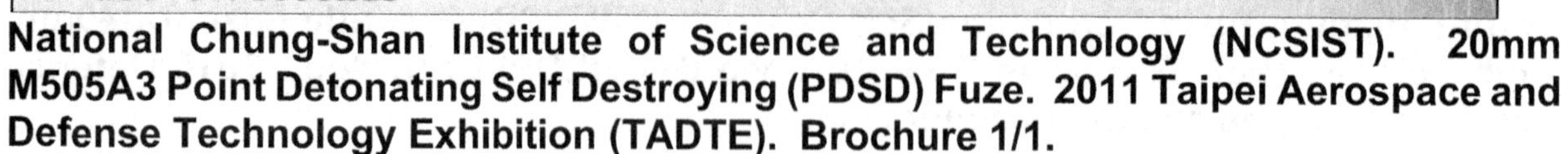

National Chung-Shan Institute of Science and Technology (NCSIST). 20mm M505A3 Point Detonating Self Destroying (PDSD) Fuze. 2011 Taipei Aerospace and Defense Technology Exhibition (TADTE). Brochure 1/1.

40mm HV SD Grenade Fuze

SYSTEM MANUFACTURING CENTER

Description:

This fuze system is 40mm PDSD (Point-Detonated Self-Destruct) fuze for use with 40mm high-velocity HE grenades.It does not only reinforce the safety mechanism but increase self-destruct (SD) mechanism. This SD mechanism enablesthe projectile self-destroy around 17 seconds later after exiting the muzzle when it hits soft targets, i.e. wetland, sands, snow, marsh, tussock…etc., and becomes a dud. This design can reduce the amount of duds and the battlefield recovery time. The 40mm PDSD fuze can be used for active army 40mm high-velocity HE grenades. And do not affect the performanceof trajectory and warhead.

2011

Features:

- ◎ Fuze Length：36.3mm
- ◎ Fuze Weight：61g approx.
- ◎ Ballastic Levels
  - Velocity：240 m/s
  - Rotation：12,000 RPM
  - Setback：65,000 G
- ◎ Ammunition：M430 or M430A1 HEDP
- ◎ Weapon：MK19 MOD3 40mm Machine Gun
- ◎ Impact function：The fuze must High order function after impact to 12.7mm thick ASTM C208 fiberboard plate
- ◎ SD function：The Self-Destruct time must be above 14 seconds

National Chung-Shan Institute of Science and Technology (NCSIST). 40mm High-Velocity/Self-Destruct/Point-Detonated. 2011 Taipei Aerospace and Defense Technology Exhibition (TADTE). Brochure 1/1.

National Chung-Shan Institute of Science and Technology (NCSIST). C4 Point Detonating Fuze for Mortars. 2018 Defence Services Asia (Malaysia). Brochure 1/2.

The C4 Fuze is a SQ and delay Fuze, and can be set Option by the user. The C4 PD Fuze is used on mortar HE projectiles.

The C4 is a ballistic match for DM111 Mechanical PD Fuze, and meets MIL-STD-331C requirements.

Features

- **High Reliability**
- **High Safety**
- **Under Production**

Specifications

Ammunition	**81mm, 120mm mortar rounds**
Type	SQ and delay
Physical Size	Overall length: 88 mm Visible length: 60 mm Diameter: 49 mm Thread: 1.5"-12NF-2A
Weight	200 g
Optional Delay	0.05 sec
First Safety	Pull wire
Second Safety	361 G, all arm
Arming distance	>40 m
Military Standard	MIL-STD-331C

National Chung-Shan Institute of Science and Technology

National Chung-Shan Institute of Science and Technology (NCSIST). C4 Point Detonating Fuze for Mortars. 2018 Defence Services Asia (Malaysia). Brochure 2/2.

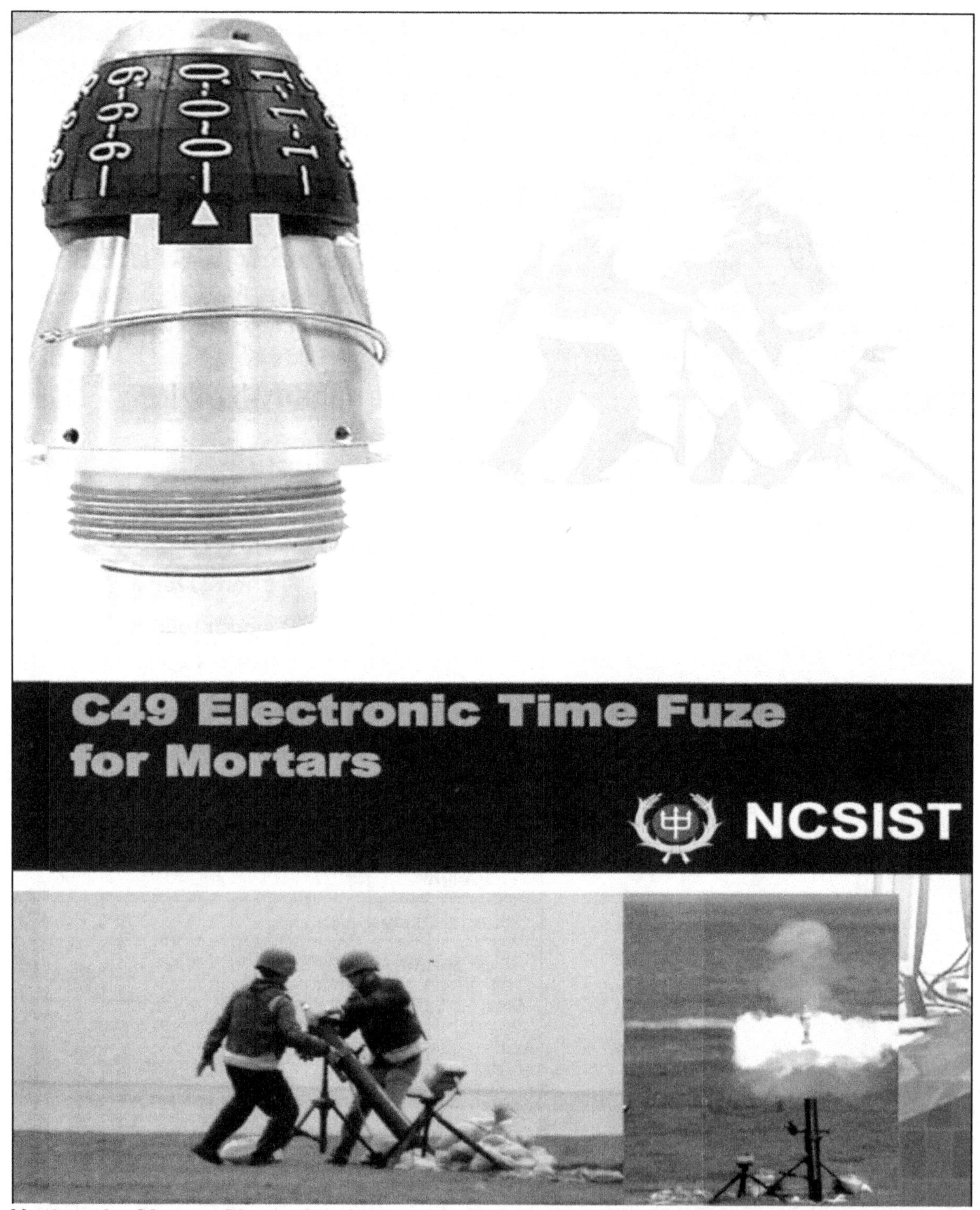

National Chung-Shan Institute of Science and Technology (NCSIST). C49 Electronic Time Fuze for Mortars. 2018 Defence Services Asia (Malaysia). Brochure 1/2.

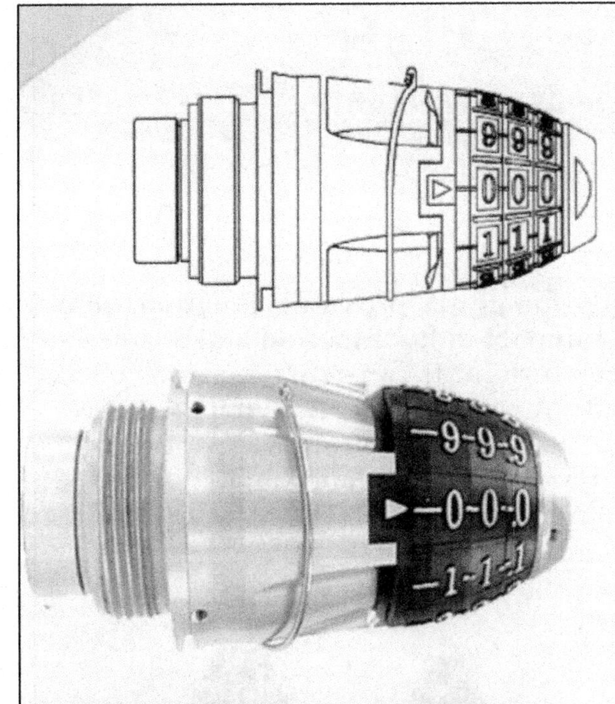

The C49 Electronic Time Fuze is used on mortar smoke, illumination projectiles, and can be set manually by the user. The C49 is a ballistic match for DM93/M776 Mechanical Time Fuze, and meets the MIL-STD-331C and MIL-STD-1316E.

Features

- High Safety
- High Reliability
- High Accuracy

Specifications

Ammunition	60mm, 81mm, 120mm mortar smoke and illuminating rounds, at all charges.
Type	Electronic Time (ET)
Time Mode	Manually setting 5.0 to 99.9 sec with 0.1 sec steps
Accuracy	Better than 0.1 seconds
Power Supply	Turbine Generator
First Safety	Pull wire
Second Safety	500 G, all arm
Overhead Safety	5.0 sec after firing
Military Standard	MIL-STD-331C, MIL-STD-1316E

National Chung-Shan Institute of Science and Technology

National Chung-Shan Institute of Science and Technology (NCSIST). C4 Electronic Time Fuze for Mortars. 2018 Defence Services Asia (Malaysia). Brochure 2/2.

Multi-Option Fuze

SYSTEM MANUFACTURING CENTER

Description:

Multi-function intelligent fuze successfully integrates proximityfunction, electronic time function, impact function, and impact delay function and accomplishes a small electronic ammunition fuze system. It can satisfy various needs for military artillery tactics. It's modular design makes it can match all active army ammunitions from 105mm to 8 inch and offer various options of proximity function, electronic time function, impact function, and impact delay function according to specific operation environment.

Features:

◎ All-weather Operation
◎ Surface Proximity
◎ Electronic Time
◎ Point Detonating
◎ Point Detonating Delay

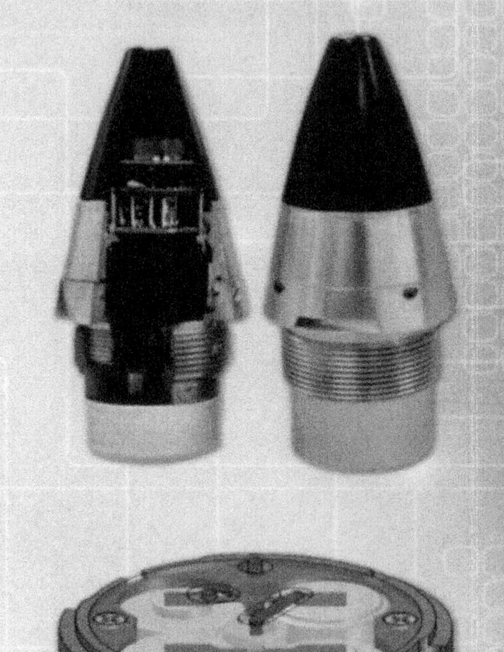

National Chung-Shan Institute of Science and Technology (NCSIST). 105mm-1 Inch Multi-Option Fuze. 2011 Taipei Aerospace and Defense Technology Exhibition (TADTE). Brochure 1/1.

Aeronautical System Research Laboratory. National Chungshan Institute of Science and Technology (NCSIST). Unknow Defense Exhibition. Brochure 1/2.

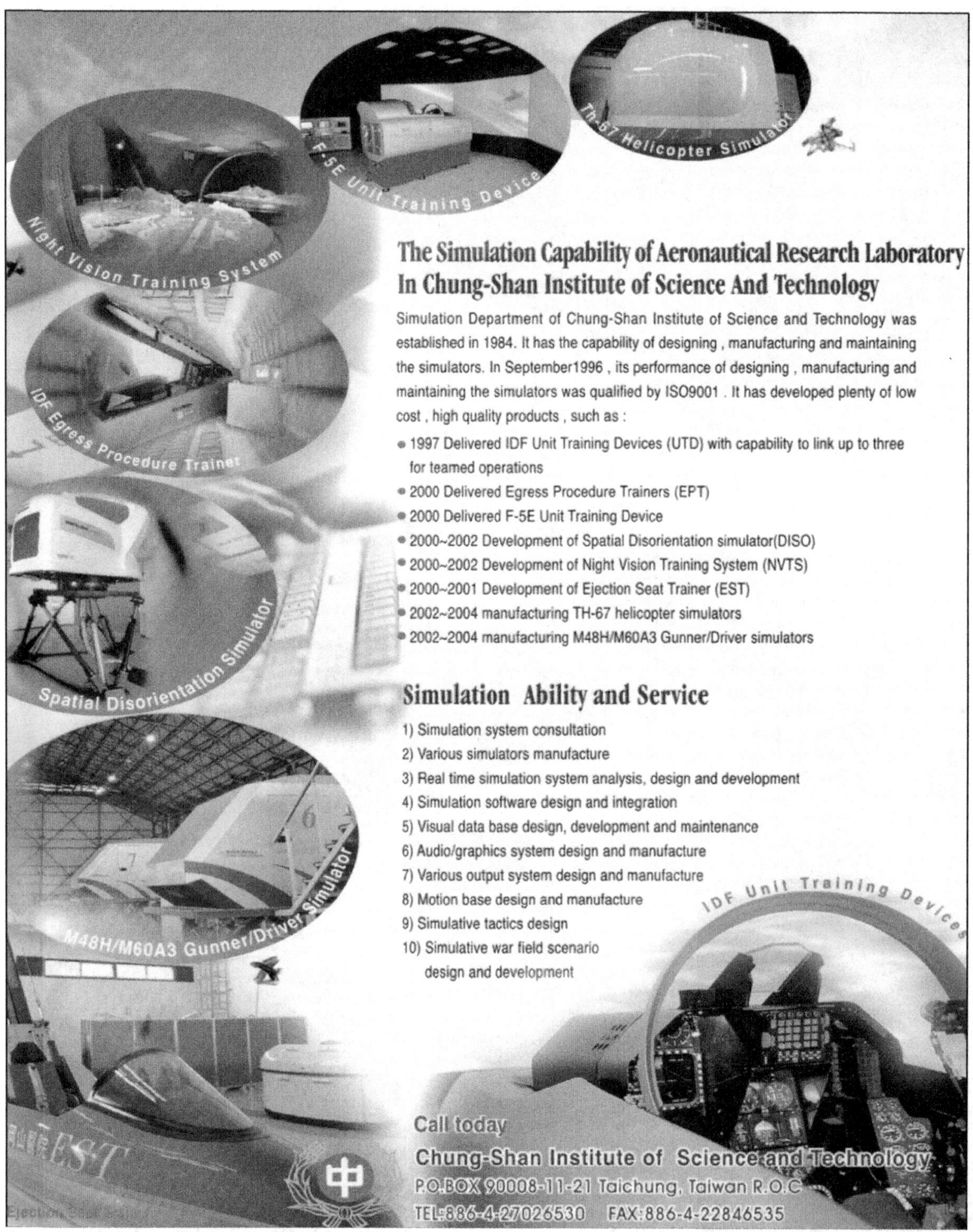

Aeronautical System Research Laboratory. National Chungshan Institute of Science and Technology (NCSIST). Unknow Defense Exhibition. Brochure 2/2.

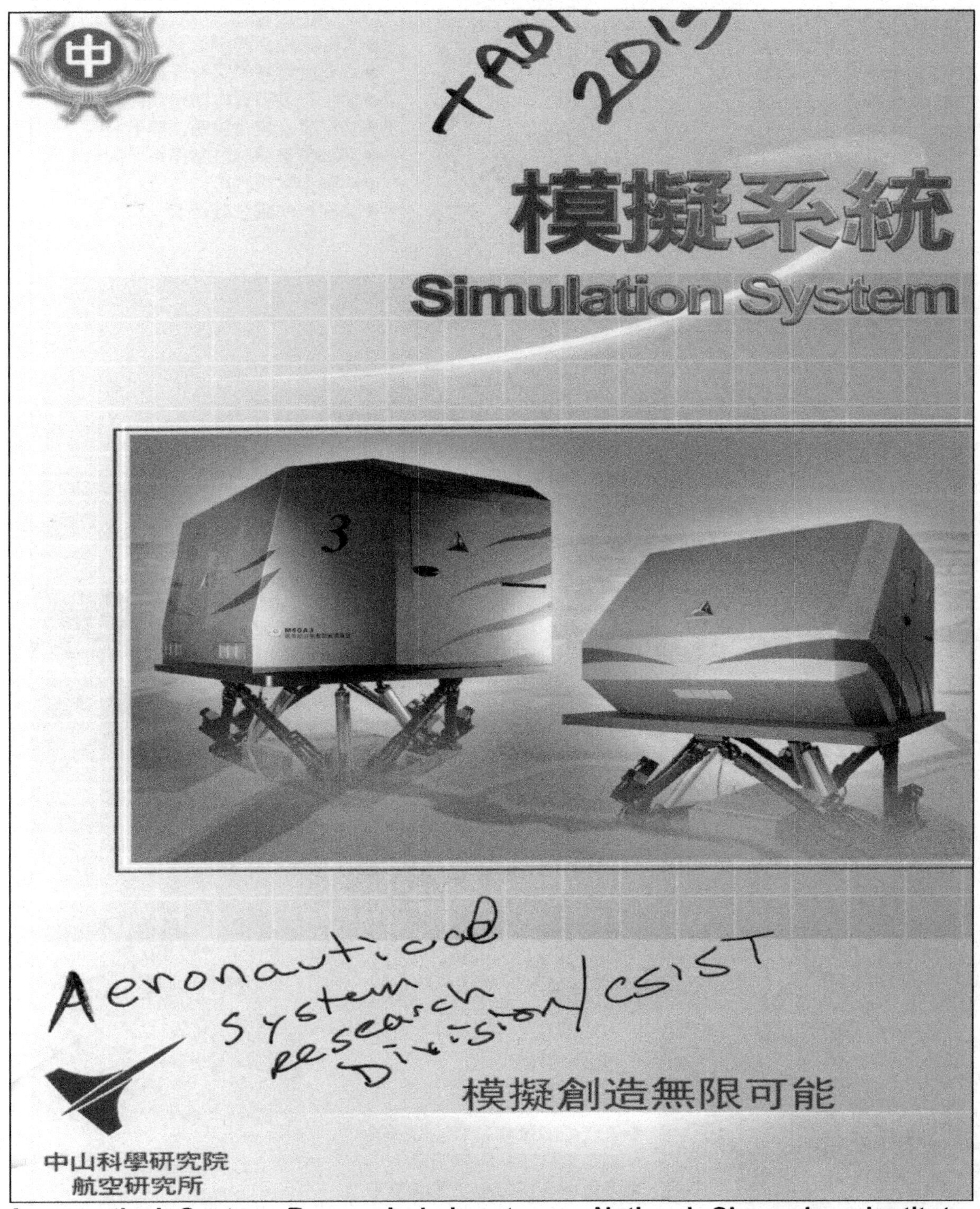

Aeronautical System Research Laboratory. National Chungshan Institute of Science and Technology (NCSIST). Simulation System. Taiwan Aerospace and Defense Technology Exhibition (TADTE). Brochure 1/4.

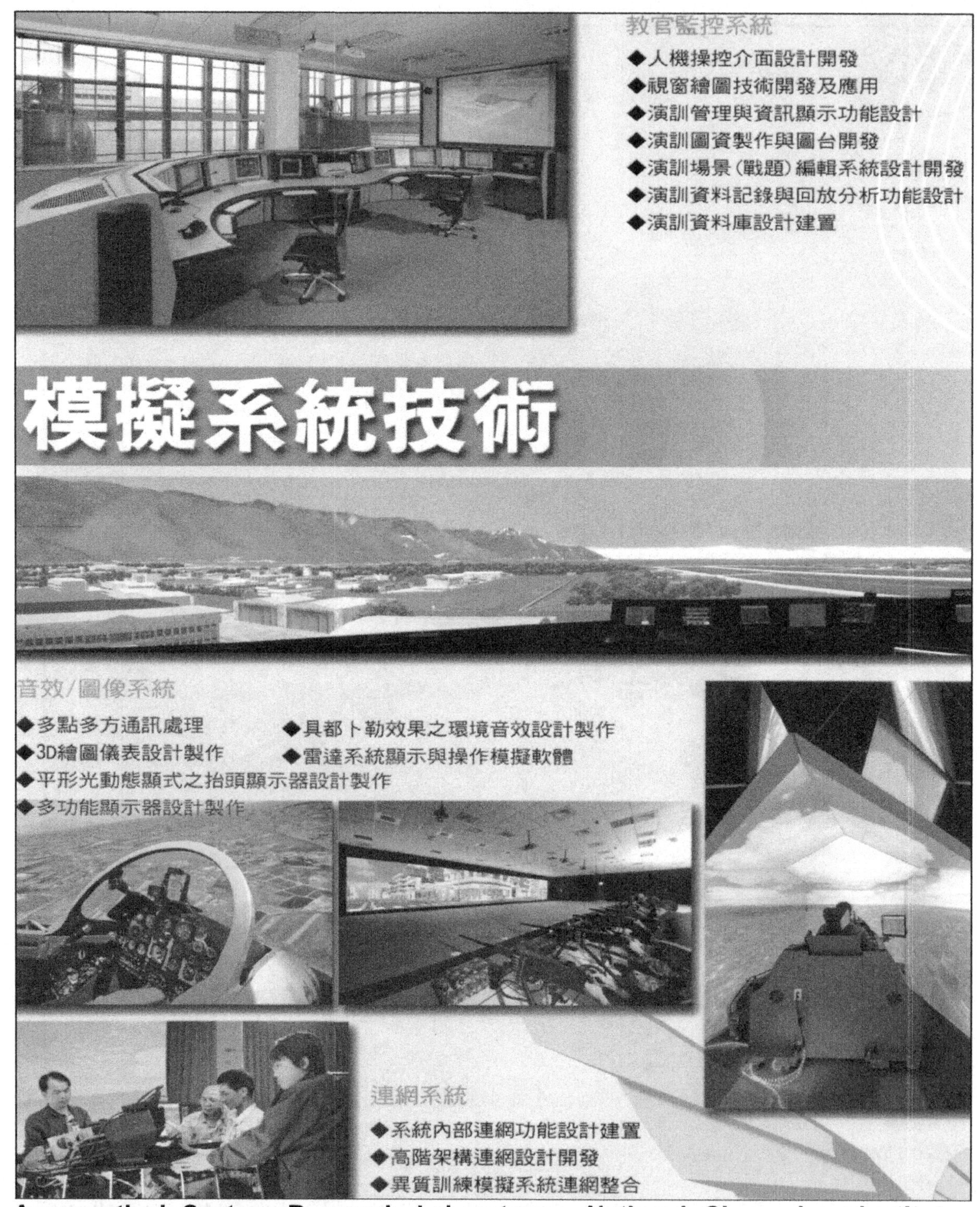

Aeronautical System Research Laboratory. National Chungshan Institute of Science and Technology (NCSIST). Simulation System. Taiwan Aerospace and Defense Technology Exhibition (TADTE). Brochure 2/4.

Aeronautical System Research Laboratory. National Chungshan Institute of Science and Technology (NCSIST). Simulation System. Taiwan Aerospace and Defense Technology Exhibition (TADTE). Brochure 3/4.

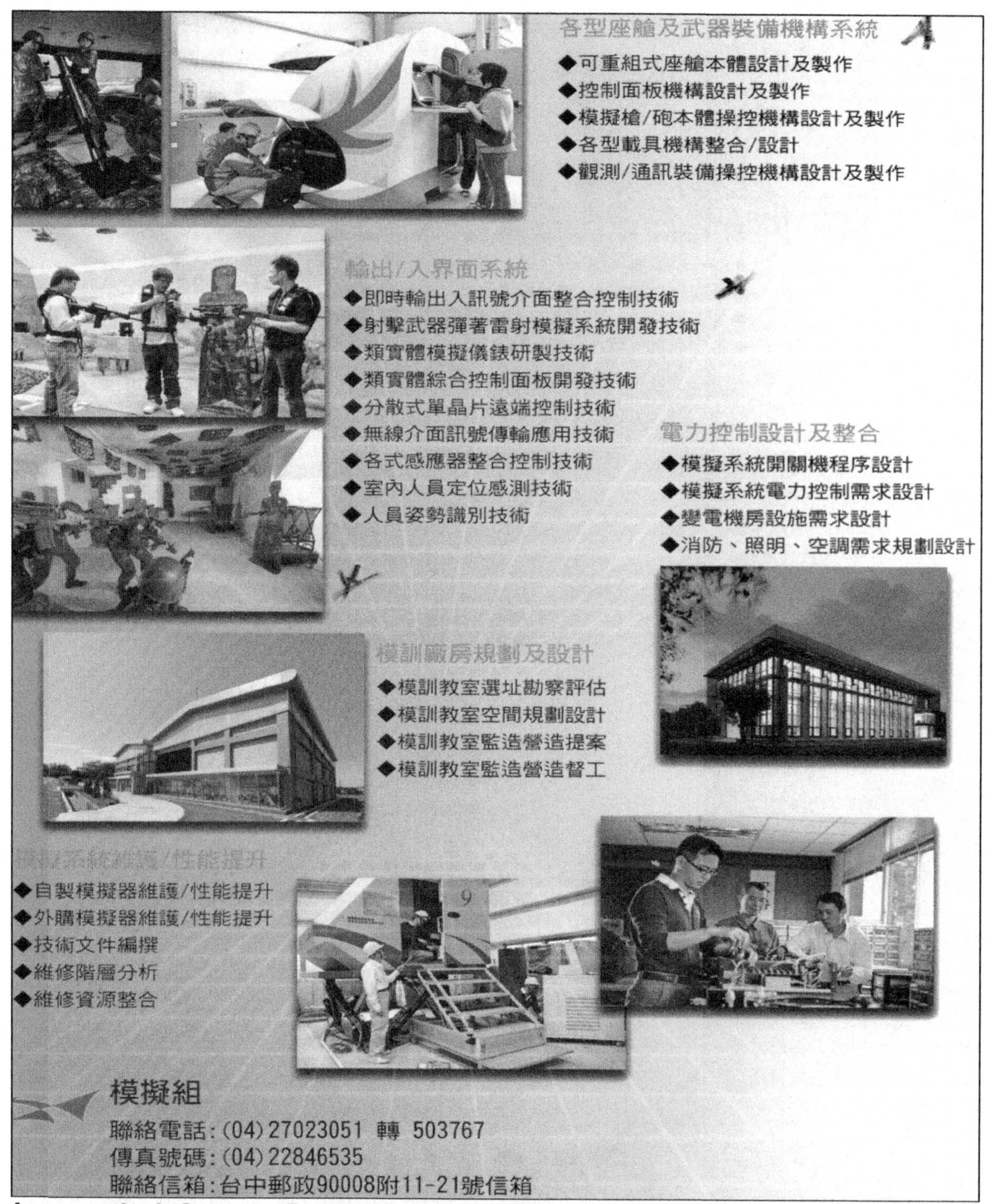

Aeronautical System Research Laboratory. National Chungshan Institute of Science and Technology (NCSIST). Simulation System. Taiwan Aerospace and Defense Technology Exhibition (TADTE). Brochure 4/4.

Radar Scattering Camouflage Net

CHEMICAL SYSTEMS RESEARCH DIVISION

Description:

The radar scattering camouflage net is made of conductive metal fiber and fire retardant polyester, using leaves design to attenuate radar signature. The radar scattering camouflage net is coated a thin film of camouflage patterned anti-near-infrared paint, and shall provide visible 、 near-infrared 、 thermal infrared and radar signature reduction against the background. The radar scattering camouflage net is lightweight, easily constructed in the field and will be providing concealment of mobile equipment.

Features:

Size & Shape: customized
Weight: ca. 350~550 g/m2 (according to exact type)
Flame Resistance: self-extinguishing
Anti-visible & Anti-near-IR Detection
Radar Attenuation > 6 db at 10 ~34 GHz

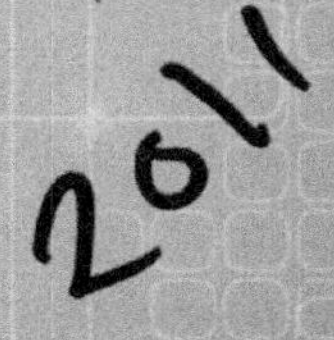

Chemical Systems Research Division. National Chungshan Institute of Science and Technology (NCSIST). Radar Scattering Camouflage Net. 2011 Taiwan Aerospace and Defense Technology Exhibition (TADTE). Brochure 1/1.

Digital camouflage broadband radome

CHEMICAL SYSTEMS RESEARCH DIVISION

Description:

「Digital camouflage broadband radome」 is made by polymer matrix composite material. In this case, the composite material is a foaming honeycomb structural, in demonstration the excellent capability of ultra low insertion loss and mechanical strength. The outside surface of radome is painted of digital camouflage pattern that will be advantage in stealth function on both of visible and near infrared spectra. Molding process is used to fabricate the composite materials radome, the shape of radome is customized available. This kind of radome has been applied to protect broadband detecting radar.

Features:

Size：hight 1.5m、diameter 1.2m、thickness 7mm
Weight：11kg
Materials：composite of glass fiber and honeycomb
Insertion loss ：ultra low insertion loss
Application：Broadband detecting radome

2011

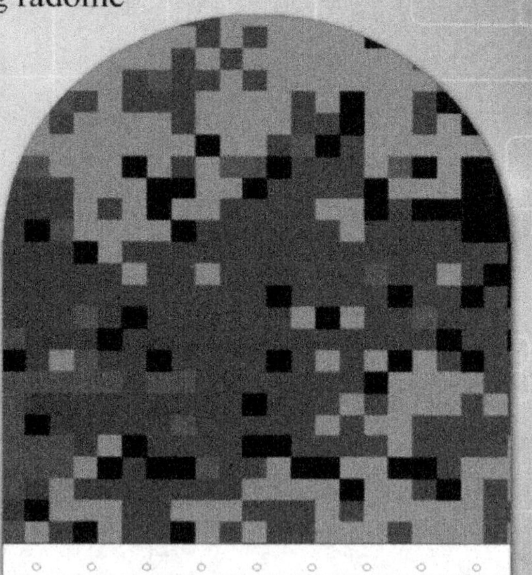

Chemical Systems Research Division. National Chungshan Institute of Science and Technology (NCSIST). Digital Camouflage Broadband Radome. 2011 Taiwan Aerospace and Defense Technology Exhibition (TADTE). Brochure 1/1.

National Chungshan Institute of Science and Technology (NCSIST). Panoramic Vehicle Imaging System. 2018 Defence Services Asia (Malaysia). Brochure 1/3.

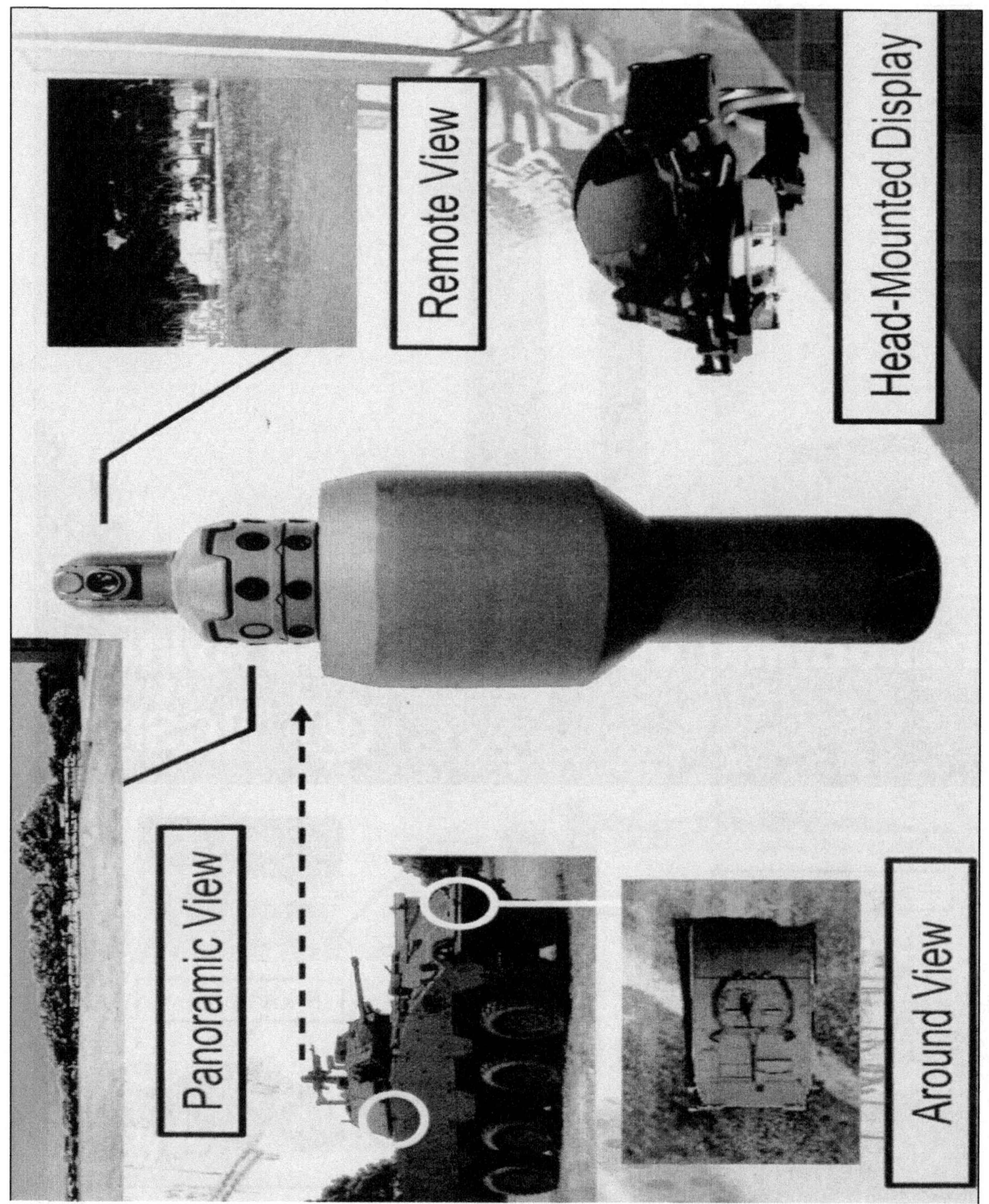

National Chungshan Institute of Science and Technology (NCSIST). Panoramic Vehicle Imaging System. 2018 Defence Services Asia (Malaysia). Brochure 2/3.

The Panoramic Vehicular Imaging System (PaVIS) provides a panoramic view for all confined members inside armored vehicles simultaneously. The system provides the driver with nearby surrounding images and thereby eliminates blind spots. The panoramic cameras offer a 360° view with in-situ video interchangeable among visible, IR and fused images to the dismounted combatants, enabling them to observe the environment outside in detail and to find out the best dismounting time. In addition, the commander can lock on targets according to the surrounding enemy situation and long-distance surveillance control system. The sufficient information of battlefield assists the commander in decision-making and therefore significantly improves fighting efficiency.

Features

- 360° day/night panoramic view of in-situ video
- Visible, thermal and fused image
- Wireless image transfer
- Around view monitoring
- Monitoring remote targets
- Reducing blind spots
- Modular design
- Adaptable to C4I system

Specifications

Display and Control System
- Panorama stitching
- EO payload control
- Communicate with HMD

Around View Monitoring System
- CCD Camera x 4
  Resolution: 720 x 480 pixels
- 360° bird' s eye view

Smart Head-Mounted Display
- Resolution: ≧ 960 x 540 pixels
- I/O:
  Wi-Fi/Bluetooth
- Sensors:
  Geomagnetic sensor
  Accelerometer
  Gyroscopic sensor

EO/IR Payload
- CCD Camera
  Resolution: 1920 x 1080 pixels
  FOV: 2° ~ 50° (h)
- IR Camera
  Resolution: 640 x 512 pixels
  FOV: 32°(h) x 26°(v)
- AZ: 360° / EL: -20° ~ 90°
- Visible / thermal / fused image

Panoramic Camera
- CCD Camera x 8
  Resolution: 720 x 480 pixels
- IR Camera x 4
  Resolution: 640 x 512 pixels
- VFOV: 34°(CCD) x 69°(IR)

National Chung-Shan Institute of Science and Technology
Lungtan, Taoyuan, Taiwan (R.O.C.)
Tel : 886-3-471-2201 ext 351362, 351227
exponshow@ncsist.org.tw
www.ncsist.org.tw

National Chungshan Institute of Science and Technology (NCSIST). Panoramic Vehicle Imaging System. 2018 Defence Services Asia (Malaysia). Brochure 3/3.

National Chungshan Institute of Science and Technology (NCSIST). 2.75-inch Rocket. Helicopters carry two 2.75-inch M260 seven-tube Folding-Fin Aerial Rocket (FFAR) launchers. 2018 Defence Services Asia (Malaysia). Brochure 1/2.

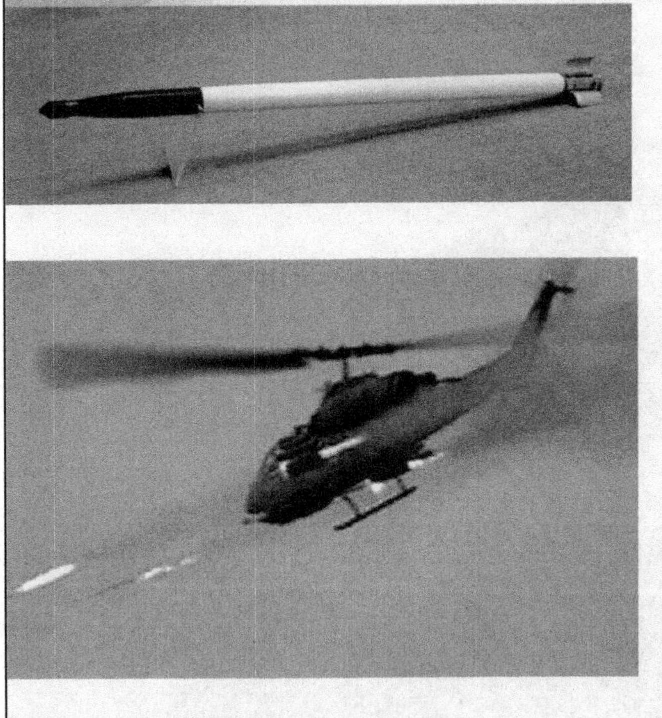

The 2.75-inch rocket is an unguided rocket used primarily in air-to-ground attacks. It fits various launchers and can be equipped with different warheads on high-speed aircraft and low-speed helicopters. NCSIST has experience in producing hundreds of thousands of high performance 2.75-inch rockets with high reliability over the past 20 years. They are currently in service at the ROC Army on AH-64E Apache, AH-1W Cobra and OH-58D Kiowa.

Specifications

Warhead	Model	CM151
	Type	High Explosive
	Weight	4.3 Kg
	Payload	1 Kg Comp. B-4 HE
Motor	Model	CMK66 (rotary wing)
	Propellant Grain	Star-Shaped Center Hole Extruded Double Base
	Weight	6.2 Kg
	Average Thrust	1400 lb
	Effective Range	6000 m
	Compatible Launcher	M260、M261 (rotary wing)

Features

- ➢Easy to operate
- ➢High reliability
- ➢Cost effective
- ➢Salvo firing
- ➢Easy integration
- ➢Easy installation and maintenance
- ➢All-weather operations

National Chung-Shan Institute of Science and Technology
Lungtan, Taoyuan, Taiwan (R.O.C.)
Tel : 886-2-2673-9638 ext 351362, 351227
exponshow@ncsist.org.tw
www.ncsist.org.tw

National Chungshan Institute of Science and Technology (NCSIST). 2.75-inch Rocket. Helicopters carry two 2.75-inch M260 seven-tube Folding-Fin Aerial Rocket (FFAR) launchers. 2018 Defence Services Asia (Malaysia). Brochure 2/2.

Army Aviation and Special Forces Command (陸軍航空特戰指揮部). 601st Aviation Brigade. 1st Attack Group. Longtan, Taoyuan. AH-64E Apache Attack Helicopter. Possibly a fan patch, not official.

Army Aviation and Special Forces Command (陸軍航空特戰指揮部). 601st Aviation Brigade (陸軍航空第 601 旅) – Longtan, Taoyuan. 2nd Attack Group. AH-64E Apache Longbow Attack Helicopter.

Army Aviation and Special Forces Command (陸軍航空特戰指揮部). 601st Aviation Brigade (陸軍航空第 601 旅) – Longtan, Taoyuan. 2nd Attack Group. AH-64E Apache Longbow Attack Helicopter.

Army Aviation and Special Forces Command (陸軍航空特戰指揮部). AH-64E Apache Longbow patch. "Brave, Honor, Precision". Unofficial patch often produced by fans or even individual units for morale purposes.

Army Aviation and Special Forces Command (陸軍航空特戰指揮部). 601st Aviation Brigade (陸軍航空第 601 旅) – Longtan, Taoyuan – 2nd Attack Group. AH-64E Apache Longbow Attack Helicopter. Final Operational Capability (FOC): July 11, 2018.

Army Aviation and Special Forces Command (陸軍航空特戰指揮部). 602nd Aviation Brigade. Hsinshe, Taichung. 1st Attack Group: AH-1W Super Cobra Attack Helicopter.

Army Aviation and Special Forces Command (陸軍航空特戰指揮部). 602nd Aviation Brigade. Hsinshe, Taichung. 1st Attack Group: AH-1W Super Cobra Attack Helicopter.

Army Aviation and Special Forces Command (陸軍航空特戰指揮部). 602R Aviation Brigade. Hsinshe, Taichung. 2nd Attack Group: AH-1W Super Cobra Attack Helicopter. "602R" indicates it was still a regiment before the force modernization realignment converted everything to brigades.

Army Aviation and Special Forces Command (陸軍航空特戰指揮部). 602nd Aviation Brigade. Hsinshe, Taichung. AH-1W Super Cobra Attack Helicopter. TALLY = Target In Sight (radio communication).

Army Aviation and Special Forces Command (陸軍航空特戰指揮部). 602nd Aviation Brigade. Hsinshe, Taichung. AH-1W Super Cobra Attack Helicopter. TALLY = Target In Sight (radio communication).

Army Aviation and Special Forces Command (陸軍航空特戰指揮部). 602nd Aviation Brigade. Commemorative patch for the 35th Han Kuang Exercise (2019) on the No. 1 National Freeway (Sun Yat-sen) in Changhua County's Huatan Township (emergency landing exercises for both fixed wing and rotary have been held on sections of the same freeway in Tainan's Madou and Rende districts and Chiayi County's Minsyong Township). Note to reader: No. 1 National Freeway runs along the west coast of Taiwan facing the Taiwan Strait.

Both commemorative patches for the 35th Han Kuang Exercise (2019) on the No. 1 National Freeway (Sun Yat-sen) in Changhua County's Huatan Township (emergency landing exercises for both fixed wing and rotary have been held on sections of the same freeway in Tainan's Madou and Rende districts and Chiayi County's Minsyong Township). Note to reader: No. 1 National Freeway runs along the west coast of Taiwan facing the Taiwan Strait.

Army Aviation and Special Forces Command (陸軍航空特戰指揮部). 602nd Aviation Brigade. Hsinshe, Taichung. AH-1W Super Cobra Attack Helicopter.

Army Aviation and Special Forces Command (陸軍航空特戰指揮部). OH-58D Kiowa Warrior Reconnaissance Helicopter.

Army Aviation and Special Forces Command (陸軍航空特戰指揮部). 601st Aviation Brigade (陸軍航空第 601 旅) – Longtan, Taoyuan. Reconnaissance Battalion x1 (Kiowa OH-58D Reconnaissance Helicopter).

Army Aviation and Special Forces Command (陸軍航空特戰指揮部). OH-58D Kiowa Warrior Reconnaissance Helicopter. Bravo Flight (10,000 feet altitude) commemorative patch.

Taiwan fan patch for the OH-58D Kiowa Warrior Reconnaissance Helicopter. The OH-58D carries two 2.75-inch M260 seven-tube Folding-Fin Aerial Rocket (FFAR) launchers. The rockets are manufactured at the National Chungshan Institute of Science and Technology (NCSIST).

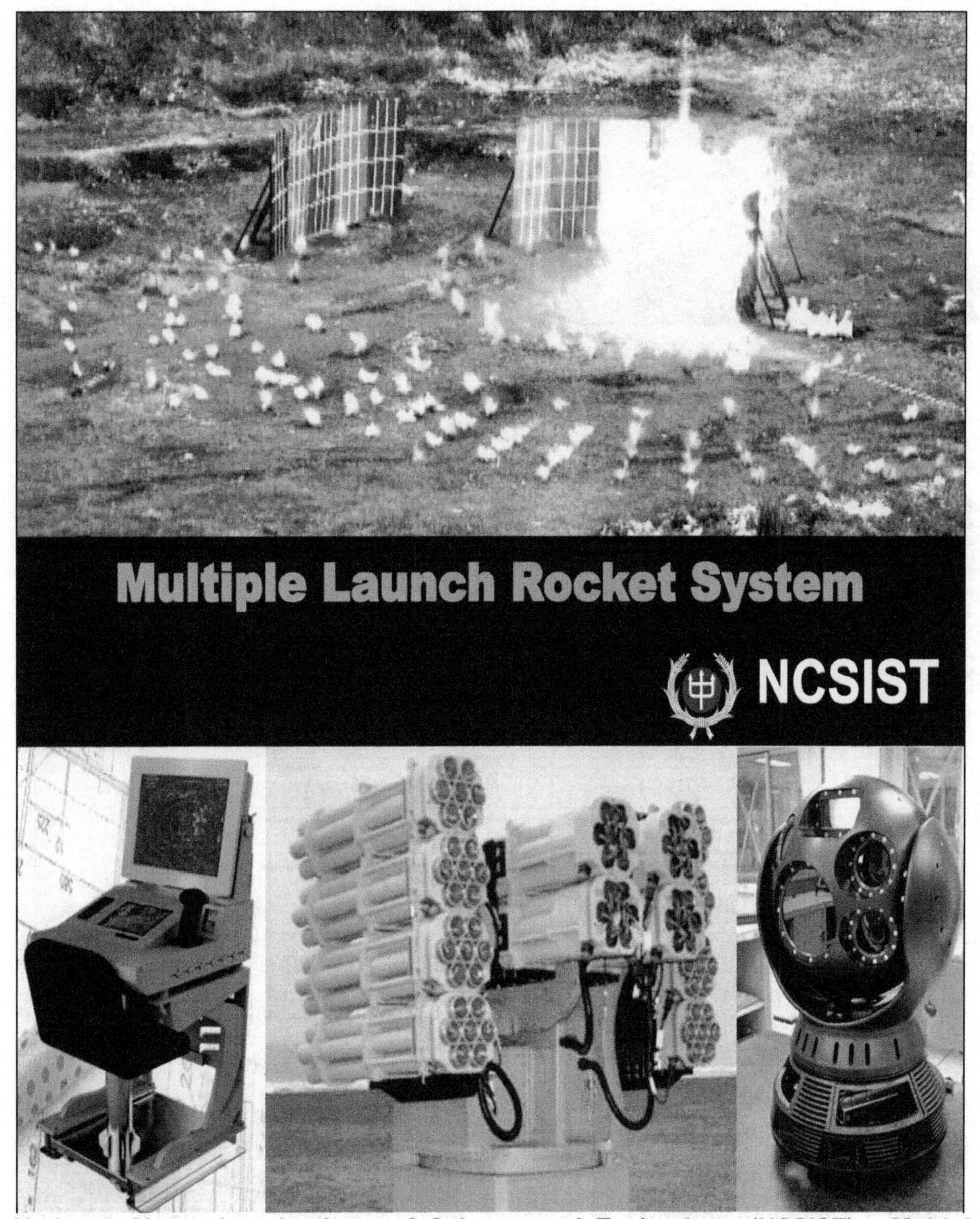

National Chungshan Institute of Science and Technology (NCSIST). Multiple Launch Rocket System. 2018 Defence Services Asia (Malaysia). Brochure 1/2.

The Multiple Launch Rocket System (MLRS) is a remote control weapon system used to detect and destroy incoming enemies and light-armed vehicles. It is composed of electro-optical sensors, mobile consoles, and multiple war head rockets capable of target tracking, image recognition, and ballistics calculation. The MLRS provides short range joint regional firepower for modern warfare.

Features

- Remote controllable
- Graphical user interface
- Easy to reload
- Target-tracking and self-aiming
- Multiple Types of Rockets Compatible

Specifications

	EO sensors	Rocket Turret
Weight	80kg	1,800kg
Azimuth	360°(60°/sec)	330°(30°/sec)
Elevation	-110°~+110°	-15°~+70°
Payload	Color HD Camera Thermal Imager Laser Range-Finder	12 Launchers 84 Rockets Effective Range: 200~10,000m

National Chung-Shan Institute of Science and Technology
Lungtan, Taoyuan, Taiwan (R.O.C.)
Tel : 886-2-2673-9638 ext 351362, 351227
exponshow@ncsist.org.tw
www.ncsist.org.tw

National Chungshan Institute of Science and Technology (NCSIST). Multiple Launch Rocket System. 2018 Defence Services Asia (Malaysia). Brochure 2/2.

電動升降天線桅桿 Electric Drive antenna mast

中山科學研究院
Chung-Shan Institute of Science and Technology

產品簡介

天線升降桅桿用以避開地面通信障礙，增加訊號傳輸距離，發揮雷達系統最佳功能。本項技術與產品具有運作平順、維護容易、抗風性強、任意斜坡上亦能進行高斜角舉升之優點，達到輕、小、可靠、操作方便之特性要求，已實際運用於通訊聯隊之各種射控指管車與中繼車。

Introduction

Electric antenna mast is used to avoid the terrain communication obstacles, increase signal transmission distance, and benefit radar system. This project led by CSIST and cooperated with domestic metal, composite and cybernation industries, has been successfully developed various auto-elevated masts which could be set in limited spaces. Mast's structure composed of carbon fiber composite tubes resulted in lightweight, the gaining of strength and weight-lifting; continuous-motion mechanism desire and strengthened metal components make it possible to elevate steadily, maintain easily, and be able to lift on highly inclined slopes. In communication troops, it practically works on launch-controlled and signal-relayed trunks. This technology had been extensively used on civilian aspects includes wireless communication, storage, transit and lighting, etc.

TADTE 2013

產品特性

- 可伸展至15m及收合至2m。
- 管件特性：採用複合材料，質輕、勁度強、不腐蝕。
- 操作條件：傾斜5度內皆可正常升降操作。
- 最大負載能力：依客戶需求設計。
- 抗風能力：風速20m/sec之下可正常升降，風速35m/sec結構不損壞。
- 電力驅動系統：單一桅桿依序升降或各節桅桿同時升降。
- 具防水設計與無線遙控功能。

Feature

- Can be extended to 15m and retracted to 2m.
- Feature of tube composites, light weight, high stiffness, non-corrosive.
- Operation condition: can be tilted 5 degrees normal lifting operation.
- Max load capacity : custom-made.
- Wind resistance: normal lift with wind velocity 20m/sec, no failure with wind velocity 35m/sec.
- Electric drive system.
- Waterproof and wireless remote control.

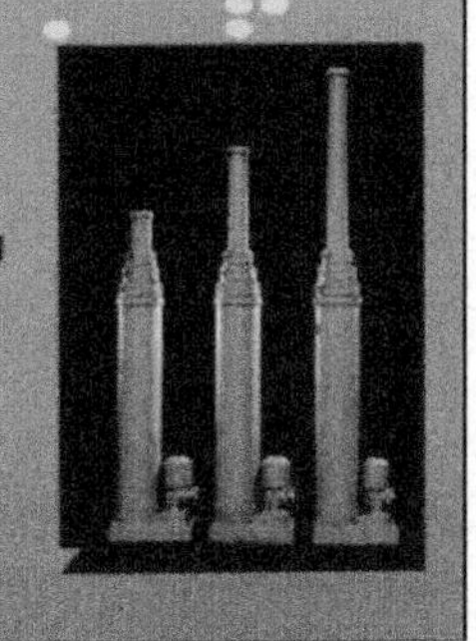

航空研究所 Aeronautical Systems Research Division

Aeronautical System Research Laboratory. National Chungshan Institute of Science and Technology (NCSIST). 2013 Taipei Aerospace Defense Technology Exhibition. Brochure 1/2.

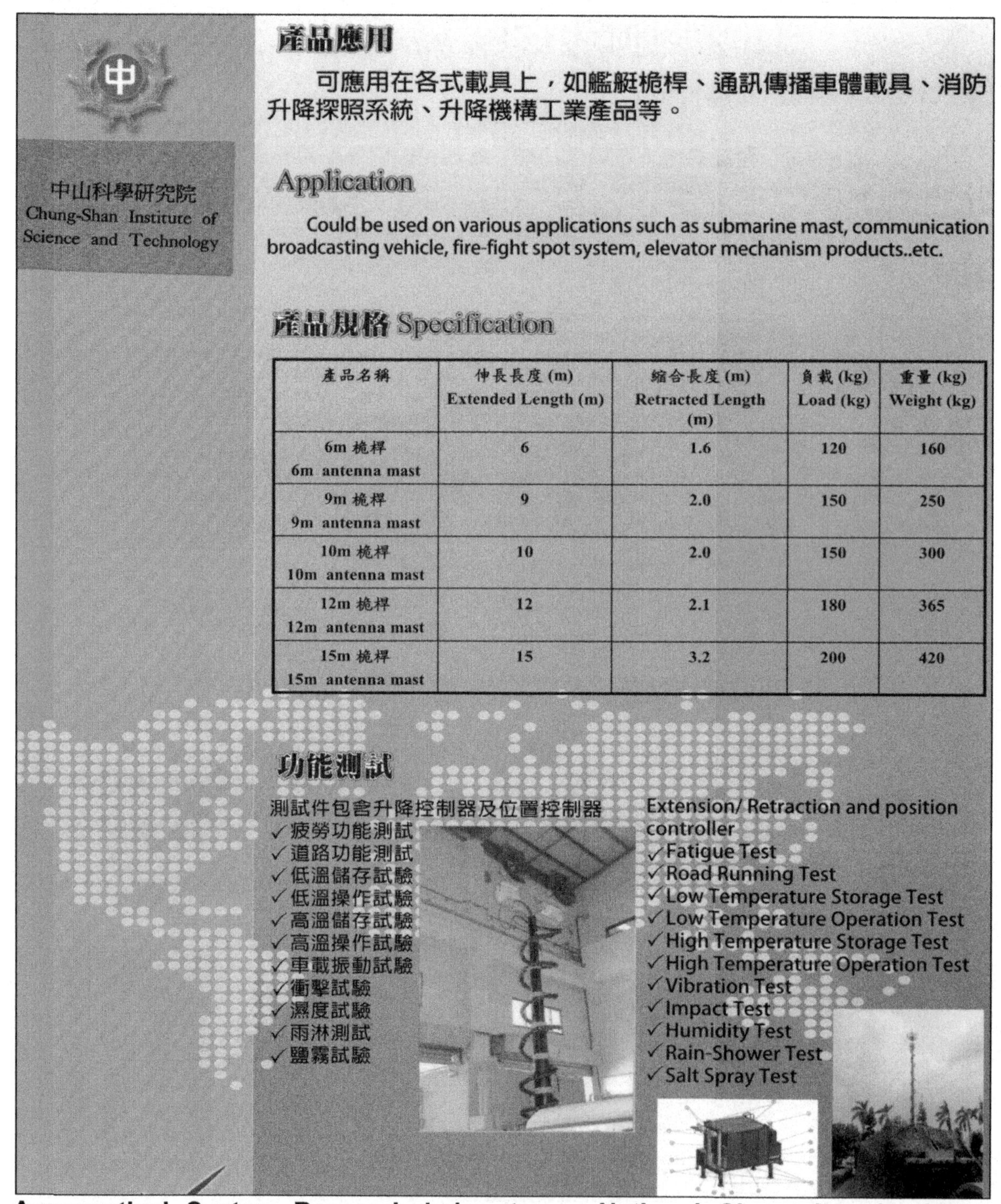

中山科學研究院
Chung-Shan Institute of Science and Technology

產品應用

可應用在各式載具上，如艦艇桅桿、通訊傳播車體載具、消防升降探照系統、升降機構工業產品等。

Application

Could be used on various applications such as submarine mast, communication broadcasting vehicle, fire-fight spot system, elevator mechanism products..etc.

產品規格 Specification

產品名稱	伸長長度 (m) Extended Length (m)	縮合長度 (m) Retracted Length (m)	負載 (kg) Load (kg)	重量 (kg) Weight (kg)
6m 桅桿 6m antenna mast	6	1.6	120	160
9m 桅桿 9m antenna mast	9	2.0	150	250
10m 桅桿 10m antenna mast	10	2.0	150	300
12m 桅桿 12m antenna mast	12	2.1	180	365
15m 桅桿 15m antenna mast	15	3.2	200	420

功能測試

測試件包含升降控制器及位置控制器

- ✓ 疲勞功能測試
- ✓ 道路功能測試
- ✓ 低溫儲存試驗
- ✓ 低溫操作試驗
- ✓ 高溫儲存試驗
- ✓ 高溫操作試驗
- ✓ 車載振動試驗
- ✓ 衝擊試驗
- ✓ 濕度試驗
- ✓ 雨淋測試
- ✓ 鹽霧試驗

Extension/ Retraction and position controller

- ✓ Fatigue Test
- ✓ Road Running Test
- ✓ Low Temperature Storage Test
- ✓ Low Temperature Operation Test
- ✓ High Temperature Storage Test
- ✓ High Temperature Operation Test
- ✓ Vibration Test
- ✓ Impact Test
- ✓ Humidity Test
- ✓ Rain-Shower Test
- ✓ Salt Spray Test

Aeronautical System Research Laboratory. National Chungshan Institute of Science and Technology (NCSIST). 2013 Taipei Aerospace Defense Technology Exhibition. Brochure 2/2.

National Chungshan Institute of Science and Technology (NCSIST). Intelligence Disaster Prevention and Relief Aiding System (iDPRA). 2018 Defence Services Asia (Malaysia). Brochure 1/2.

National Chungshan Institute of Science and Technology (NCSIST). Intelligence Disaster Prevention and Relief Aiding System (iDPRA). 2018 Defence Services Asia (Malaysia). Brochure 2/2.

National Chungshan Institute of Science and Technology (NCSIST). Mobile Point Defense Phased Array Radar System. Unknown Taipei Aerospace Defense Technology Exhibition. Brochure 1/6.

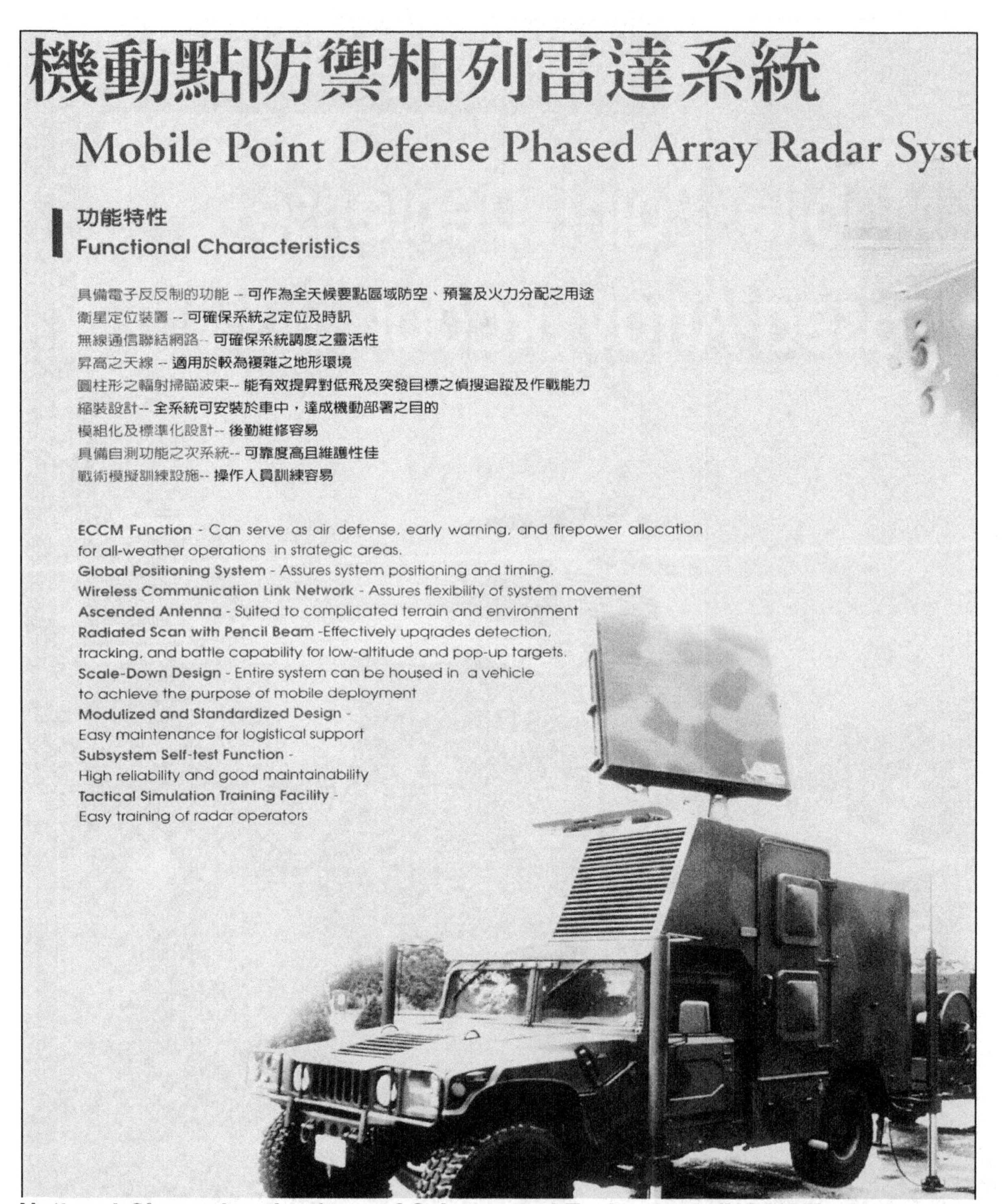

機動點防禦相列雷達系統

Mobile Point Defense Phased Array Radar Syst

功能特性
Functional Characteristics

具備電子反反制的功能 -- 可作為全天候要點區域防空、預警及火力分配之用途
衛星定位裝置 -- 可確保系統之定位及時訊
無線通信聯結網路-- 可確保系統調度之靈活性
昇高之天線 -- 適用於較為複雜之地形環境
圓柱形之輻射掃瞄波束-- 能有效提昇對低飛及突發目標之偵搜追蹤及作戰能力
縮裝設計-- 全系統可安裝於車中，達成機動部署之目的
模組化及標準化設計-- 後勤維修容易
具備自測功能之次系統-- 可靠度高且維護性佳
戰術模擬訓練設施-- 操作人員訓練容易

ECCM Function - Can serve as air defense, early warning, and firepower allocation for all-weather operations in strategic areas.
Global Positioning System - Assures system positioning and timing.
Wireless Communication Link Network - Assures flexibility of system movement
Ascended Antenna - Suited to complicated terrain and environment
Radiated Scan with Pencil Beam -Effectively upgrades detection, tracking, and battle capability for low-altitude and pop-up targets.
Scale-Down Design - Entire system can be housed in a vehicle to achieve the purpose of mobile deployment
Modulized and Standardized Design - Easy maintenance for logistical support
Subsystem Self-test Function - High reliability and good maintainability
Tactical Simulation Training Facility - Easy training of radar operators

National Chungshan Institute of Science and Technology (NCSIST). Mobile Point Defense Phased Array Radar System. Unknown Taipei Aerospace Defense Technology Exhibition. Brochure 2/6.

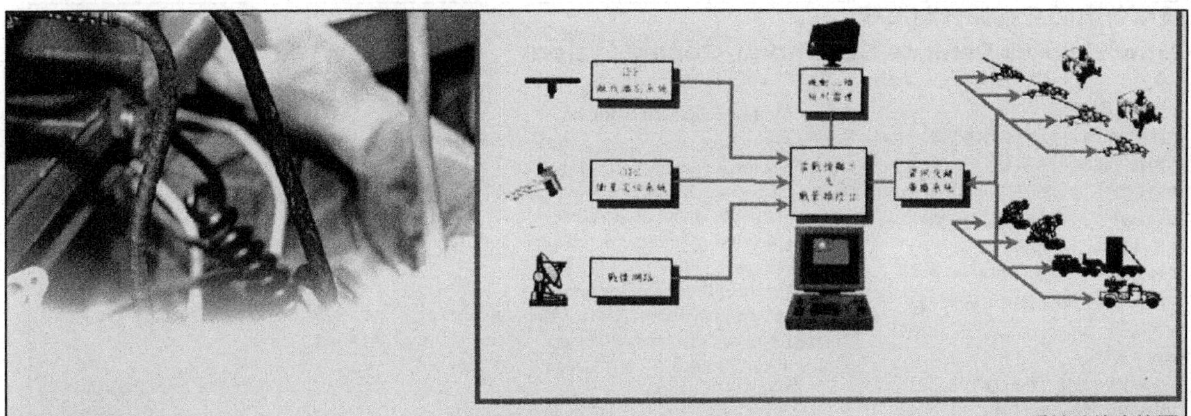

系統功能方塊圖
System Function Block Diagram

系統組成
System Configurations

旋轉式多功能三維相列雷達 -- 執行近空警戒搜索及多目標追蹤之任務
標準彈砲火力分配指管系統 -- 為全系統之火力分配及指揮管制中心
敵友識別系統 -- 執行目標敵我之問詢及識別
通信裝置 -- 提供系統與各火力單元間通信，及本系統與其他指戰管系統間之數據及語音通信
衛星定位裝置 -- 提供系統定位及時訊
電源設施 -- 供給各裝備之主要及緊急之電源需求
載具 -- 提供系統機動能力
空調 -- 提供系統工作環境

Rotational Multii-function 3D Phased Array Radar-
Performs low-altitude alertness/search and multi-target tracking missions
Standard Missile/Artillery Firepower Allocation Direct System -
Serves as firepower allocation and command/control center for overall system
Identification Friend or Foe System - Conducts target interrogation and identification between friend and foe
Communication Facility - Provides internal system communication with fire units, as well as modem/voice communication between the radar system and the external battle command/control system
Global Positioning System- Provides system position and timing
Power Supply Facility - Provides primary and emergency power for each piece of equipment
Vehicle - Provide the system mobility.
Air Conditioning - Provides for the system's operating environment

National Chungshan Institute of Science and Technology (NCSIST). Mobile Point Defense Phased Array Radar System. Unknown Taipei Aerospace Defense Technology Exhibition. Brochure 3/6.

陸軍野戰防空指揮管制系統
Army Field Air Defense Command/Control System

系統規格

工作頻率：X 波段（36 頻道可選）
搜索警戒區域：60 公里
偵追範圍：360 度（方位）；-10 ~ +55 度（俯仰）
偵追精度：15 米（距離）；0.2 度（角度）
偵追能量：64 個目標
天線轉速：15 或 30 RPM
搖控指管火力單元數目：至少12個
敵我識別：MODE A、B、C 及 MODE 4
載具：悍馬車
部署時間：小於 20 分鐘
平均故障時間：300 小時
平均維修時間：0.5 小時

System Specifications

Operating Frequency: X Band (36 channels)
Search (alertness) Range: 60 Km
Detection/Tracking Coverage: 360 degrees (azimuth); -10 ~ +55 degrees (elevation)
Detection/Tracking Accuracy:15 meters (range): 0.2 degrees (angle)
Tracking Capacity: 64 targets
Antenna Rotation Speed: 15 or 30RPM
Remote Command/Control Fire Unit Number: At least 12 units
IFF: Mode A,B,C, and Mode 4
Vehicle: Hummer
Deployment Time: Less than 20 minutes
Mean Time Between Critical Failure (MTBCF): 300 hours
Mean Time to Repair (MTTR): 0.5 hours

空軍近空多目標火力分配系統
Air Force Low-Altitude Multi-Target Firepower Allocation System

系統規格

工作頻率：X 波段（36 頻道可選）
最大搜索警戒區域：75 公里
偵追範圍：360 度（方位）；-10 ~ +55 度（俯仰）
偵追精度：15 米（距離）；0.3 度（角度）
偵追能量：64 個目標
天線轉速：15 或30 RPM
天線可昇高之高度：大於12 公尺
搖控指管火力單元數目：至少12 個
敵我識別：MODE A、B、C 及 MODE 4
抗風能力：最大陣風 20 公尺（秒）
平均故障時間：300 小時
平均維修時間：0.5 小時

System Specifications

Operating Frequency: X Band (36 channels)
Maximum Search (alertness) Range: 75 Km
Detection/Tracking Coverage: 360 degrees (azimuth); -10~+55 degrees (elevation)
Detection/Tracking Accuracy: 15 meters (range); 0.3 degrees (angle)
Tracking Capacity: 64 targets
Antenna Rotation Speed: 15 or 30 RPM
Antenna Raised Height: Above 12 meters
Remote Command/Control Fire Unit Number: At least 12 units
IFF: Mode A,B,C, and Mode 4
Wind Resistance: Up to 20 m/s with maximum wind gust < 10 seconds
Mean Time Between Critical Failure (MTBCF): 300 hours
Mean Time to Repair (MTTR): 0.5 hours

National Chungshan Institute of Science and Technology (NCSIST). Mobile Point Defense Phased Array Radar System. Unknown Taipei Aerospace Defense Technology Exhibition. Brochure 4/6.

旋轉式三維相列雷達
Rotational 3D Phased Array Radar

雷達系統功能

旋轉式三維相列雷達可執行三度空間偵搜追蹤及目標監控，完成要點區域防空空域之警戒預警及威脅目標之掌握。並可彈性聯結各式防空火砲、飛彈或彈砲組合之作戰火力單元，依據最佳火力分配準則，可分派指定多組作戰單元，進行多層縱深及多目標同時接戰，突破敵方飽和攻擊，形成區域局部空優，除確保人員、武器及裝備之存活，並有效達成要點區域防空之作戰任務。

Radar System Functions

Rotational three dimension phased array radar performs three-dimensional search, detection, tracking, and target surveillance. It provides alertness and early warning for air space in strategic areas, and manages threat targets. It can also flexibly is link with a variety of fire units including artillery, missiles, or missile/artillery combinations. Each fire unit is conducted assigned according to the algorithm of optimal firepower allocation and simultaneous engagement is conducted with multi-layer defense depth and multiple targets. This radar can help break through an enemy's saturation assault and establish local-area superiority. In addition, to assure the survival of personnel, weapons, and equipment, it effectively fulfills the battle mission of air defense in a strategic area.

MOBILE POINT DEFENSE
PHASED ARRAY RADAR SYSTEM

National Chungshan Institute of Science and Technology (NCSIST). Mobile Point Defense Phased Array Radar System. Unknown Taipei Aerospace Defense Technology Exhibition. Brochure 5/6.

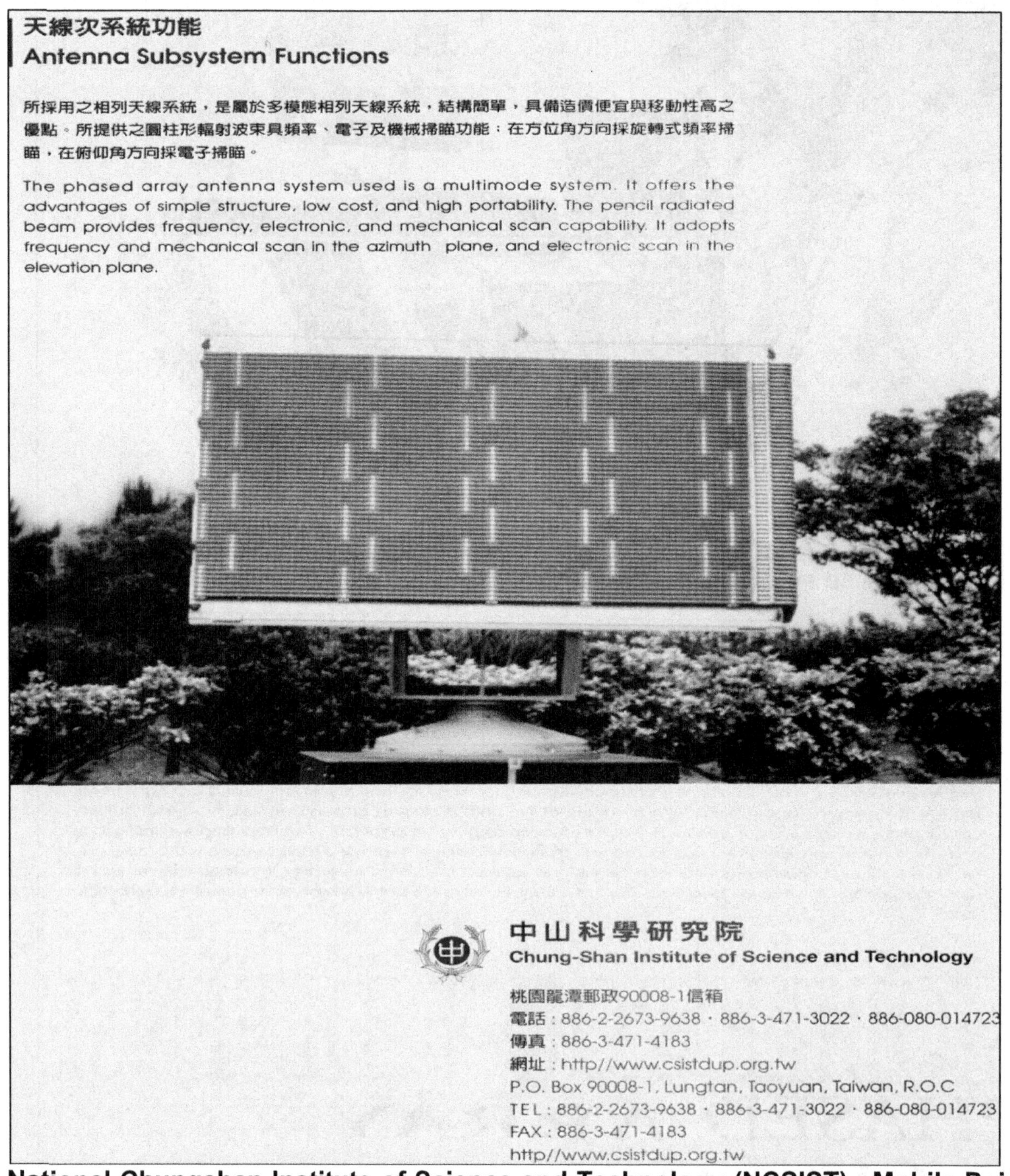

National Chungshan Institute of Science and Technology (NCSIST). Mobile Point Defense Phased Array Radar System. Unknown Taipei Aerospace Defense Technology Exhibition. Brochure 6/6.

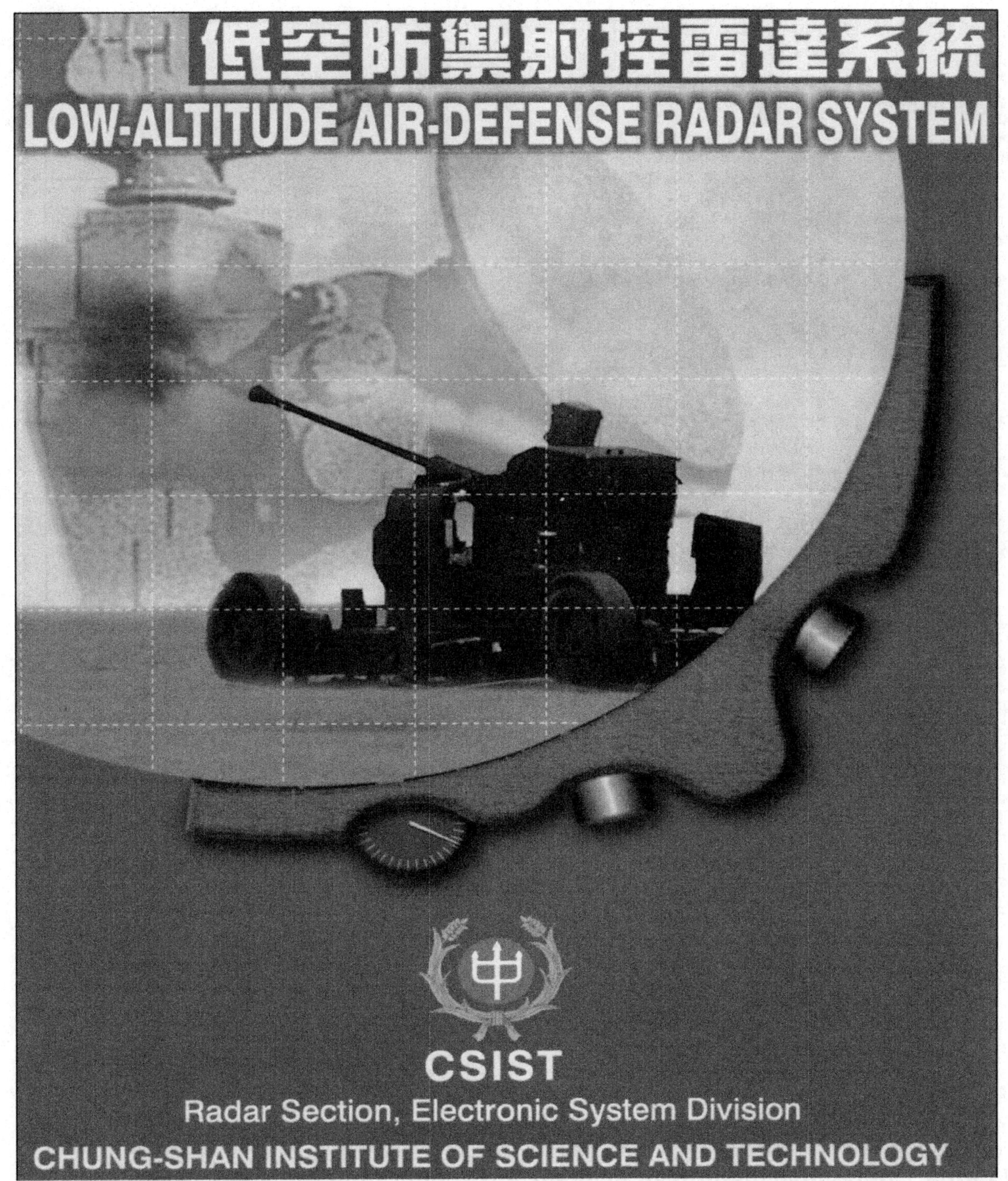

National Chungshan Institute of Science and Technology (NCSIST). Radar Section, Electronic System Division. Low-Altitude Air-Defense Fire Control Radar (LAADFCR). Unknown Taipei Aerospace Defense Technology Exhibition. Note to reader: This was probably an earlier version of the CS-MPQ-78. Brochure 1/6.

低空防禦武器系統
Low-Altitude Air-Defense System, LAADS

低空防禦武器系統是整體防空作戰的最後一道防線，需具備快速接戰及精確命中的能力，以全天候保障機場、港口、軍事基地等固定要點之安全，進而也可機動佈署提供反登陸作戰指揮中心的安全屏障，因此低空防禦武器系統長期均為先進國家投入研發的重點項目之一。中山科學研究院研製之低空防禦武器系統(Low-Altitude Air-Defense System, LAADS)採用彈砲整合之主流架構，以低空防禦射控雷達系統(Low-Altitude Air-Defense Fire Control Radar, LAADFCR)為核心，結合40L70快砲及車載劍一飛彈(TC-1)而成，佈署容易、反應快速、殺傷力強、可有效攔截低空高速之威脅目標。

低空防禦射控雷達系統
Low-Altitude Air-Defense Fire Control Radar, LAADFCR

低空防禦射控雷達系統是國防部中山科學研究院從事國防奠基研究計畫的重要成果之一，其主要性能如下：

- 搜索雷達提供全方位目標情資，指引防空飛彈接戰。
- 追蹤雷達提供即時精確目標指令，導控防空快砲追瞄射擊。
- 光電追蹤儀提供近距作戰能力。
- 整合光學和雷達追蹤系統可增加敵方電子反制之難度。
- 敵友識別儀提供目標辨識情資。
- 具備信號處理及電子防護功能。
- 具備結合早期預警指管系統之通訊界面。

Low-Altitude Air-Defense Fire Control Radar System (LAADFCR) integrated with 40/L70 Guns and TC-1 missile launching systems, which enhances the multi-layer combat performance, upgrades the air-defense capability and achieves the goal of "reducing military personnel increasing the fire power and kill probability." The system can be deployed at land base, battlefield, fortress, harbor or airport to conduct low-altitude air-defense missions in all weather condition. The main system performances are as follows:

- Search radar provides coverage against to the airdefense missiles.
- Track radar provides real-time and precise data of target to control the air-defense guns.
- Electro-optical tracker is an auxiliary device for closerange fighting ability.
- Integration of Electro-optical and radar tracking system increases difficulties of hostile ECM operations Performed by the enemy.
- Identification/Friend-or-Foe allows the discrimination of targets.
- Equipped with signal processor and electronics protection.
- Equipped with interfaces to early warning Command and Decision system.

National Chungshan Institute of Science and Technology (NCSIST). Radar Section, Electronic System Division. Low-Altitude Air-Defense Fire Control Radar (LAADFCR). Unknown Taipei Aerospace Defense Technology Exhibition. Brochure 2/6.

彈砲結合
Guns and Missiles

彈砲結合可提昇對威脅目標的擊殺率，是國軍對射控雷達系統的基本需求。低空防禦射控雷達系統測試期間，先後完成與陸射型TC-1飛彈及40公厘快砲結合測試，充分掌握了目標偵追、精度測校、指令解算、彈著修正、遠距操控等關鍵性技術。本系統亦具備結合其它國軍各型近空防禦飛彈及快砲指引接戰能力。

Guns and Missiles combined in a Low-Altitude Air-Defense Weapon System has been proven to be effective and has higher kill probability to the engaged targets. CSIST has completed the integration of functional tests with 40L70 Gun and TC-1 missile, and has fully capability to apply the core technologies on target tracking, command and decision, gun order computations, remote controls and accuracy calibrations etc.. We also has the capability to extend the integration of weapons to other guns and missiles under service.

※低空防禦射控雷達功能規格※

搜索雷達 SEARCH RADAR		
目標偵搜範圍	30Km,360°	Detection Coverage
多目標追蹤數	20	TWS Tracks
距離追蹤精度	±30m	Range Track Accuracy
方位追蹤精度	±5mil	Az Track Accuracy
追蹤雷達 TRACK RADAR		
目標追蹤距離	30Km	Target Track Range
距離追蹤精度	±10m	Range Track Accuracy
方位追蹤精度	±1mil	Az Track Accuracy
俯仰追蹤精度	±1mil	El Track Accuracy
光電追蹤儀 EO TRACK		
目標追蹤距離	10Km	Target Track Range
雷射測距精度	±5m	Laser Range Accuracy
方位追蹤精度	±0.5mil	Az Track Accuracy
俯仰追蹤精度	±0.5mil	El Track Accuracy
敵友識別儀 IFF		
目標識別範圍	30Km,360°	Identification Coverage
目標識別模式	Mode 1,2,3/A,4	IFF Modes
武器界面 WEAPON INTERFACE		
飛彈結合界面	9	Missile Interface
快砲結合界面	4	Gun Interface

National Chungshan Institute of Science and Technology (NCSIST). Radar Section, Electronic System Division. Low-Altitude Air-Defense Fire Control Radar (LAADFCR). Unknown Taipei Aerospace Defense Technology Exhibition. Brochure 3/6.

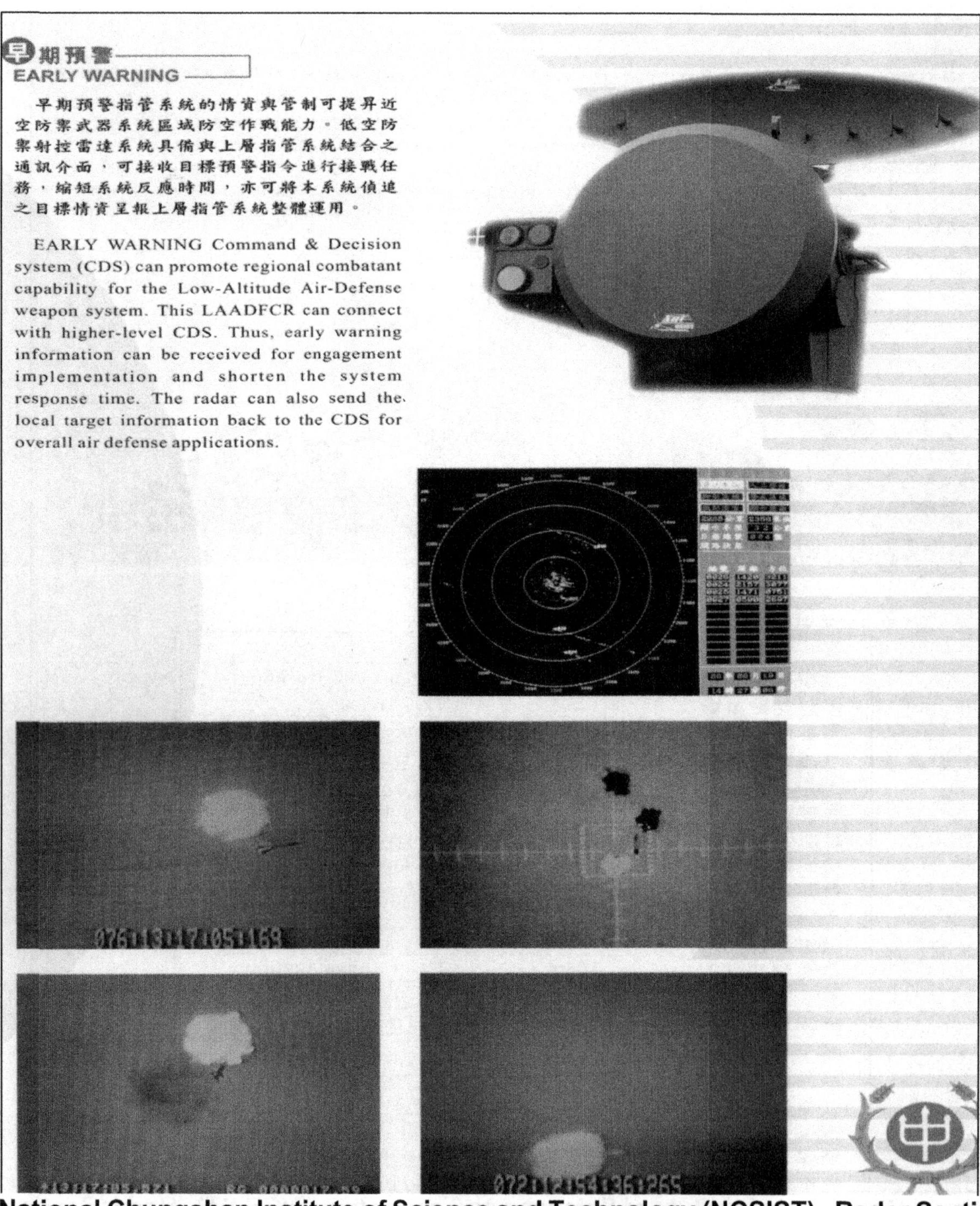

早期預警
EARLY WARNING

早期預警指管系統的情資與管制可提昇近空防禦武器系統區域防空作戰能力。低空防禦射控雷達系統具備與上層指管系統結合之通訊介面，可接收目標預警指令進行接戰任務，縮短系統反應時間，亦可將本系統偵追之目標情資呈報上層指管系統整體運用。

EARLY WARNING Command & Decision system (CDS) can promote regional combatant capability for the Low-Altitude Air-Defense weapon system. This LAADFCR can connect with higher-level CDS. Thus, early warning information can be received for engagement implementation and shorten the system response time. The radar can also send the local target information back to the CDS for overall air defense applications.

National Chungshan Institute of Science and Technology (NCSIST). Radar Section, Electronic System Division. Low-Altitude Air-Defense Fire Control Radar (LAADFCR). Unknown Taipei Aerospace Defense Technology Exhibition. Brochure 4/6.

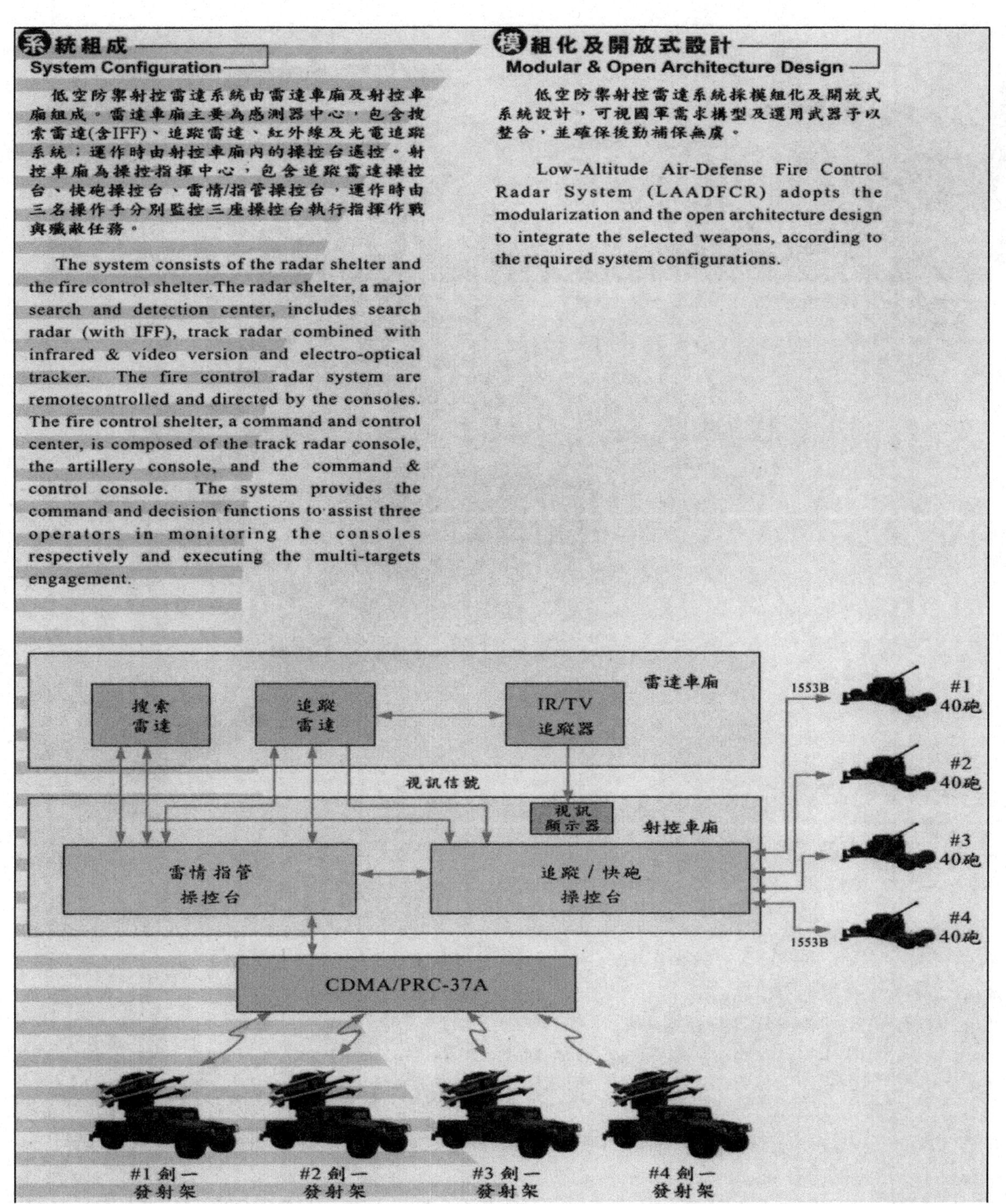

系統組成
System Configuration

低空防禦射控雷達系統由雷達車廂及射控車廂組成。雷達車廂主要為感測器中心，包含搜索雷達(含IFF)、追蹤雷達、紅外線及光電追蹤系統；運作時由射控車廂內的操控台遙控。射控車廂為操控指揮中心，包含追蹤雷達操控台、快砲操控台、雷情/指管操控台，運作時由三名操作手分別監控三座操控台執行指揮作戰與殲敵任務。

The system consists of the radar shelter and the fire control shelter. The radar shelter, a major search and detection center, includes search radar (with IFF), track radar combined with infrared & video version and electro-optical tracker. The fire control radar system are remotecontrolled and directed by the consoles. The fire control shelter, a command and control center, is composed of the track radar console, the artillery console, and the command & control console. The system provides the command and decision functions to assist three operators in monitoring the consoles respectively and executing the multi-targets engagement.

模組化及開放式設計
Modular & Open Architecture Design

低空防禦射控雷達系統採模組化及開放式系統設計，可視國軍需求構型及運用武器予以整合，並確保後勤補保無虞。

Low-Altitude Air-Defense Fire Control Radar System (LAADFCR) adopts the modularization and the open architecture design to integrate the selected weapons, according to the required system configurations.

National Chungshan Institute of Science and Technology (NCSIST). Radar Section, Electronic System Division. Low-Altitude Air-Defense Fire Control Radar (LAADFCR). Unknown Taipei Aerospace Defense Technology Exhibition. Brochure 5/6.

中山科學研究院
Chung-Shan Institute of Science and Technology

桃園龍潭郵政90008-1信箱
電話：886-2-2673-9638・886-3-471-3022・886-0800-014723
傳眞：886-3-471-4183
網址：http://www.csistdup.org.tw
P.O.Box 90008-1,Lungtan,Taoyuan,Taiwan,R.O.C.
TEL：886-2-2673-9638・886-3-471-3022・886-0800-014723
FAX：886-3-471-4183
http：//www.csistdup.org.tw
E-mail：pdduii@www.csistdup.org.tw

National Chungshan Institute of Science and Technology (NCSIST). Radar Section, Electronic System Division. Low-Altitude Air-Defense Fire Control Radar (LAADFCR). Unknown Taipei Aerospace Defense Technology Exhibition. Brochure 6/6.

National Chungshan Institute of Science and Technology (NCSIST). Radar Section, Electronic System Division. CS-MPQ-78 Low-Altitude Air-Defense Fire Control Radar (LAADFCR). Unknown Taipei Aerospace Defense Technology Exhibition. Brochure 1/6.

CS/MPQ-78近空防禦射控雷達

CS/MPQ-78 Low-Altitude Air-Defence Fi

近空防禦 武器系統是防空作戰時最後且必要的一道防線。當來襲敵機突破外層防空網的攔截時，為確保防區的安全，則唯一有效的防禦利器即是性能精良的射控雷達結合防空飛彈及快砲所構成的武器系統，其快速的接戰能力及精確的命中率將為防區提供全方位、全天候的保障。因此，近空防禦武器系統不僅是最經濟、實用、有效的防空武器，且為先進國家長期投入研發之重要國防武器之一。

CS/MPQ-78 近空防禦射控雷達系統主要特性如下：

- 搜索雷達提供360° 全方位多目標情資，指引防空飛彈接戰。
- 追蹤雷達提供即時精確目標指令，導控防空快砲追瞄射擊。
- 光電追蹤儀輔助電戰威脅時近距作戰能力。
- 敵友識別儀提供目標辨識情資。
- 具備信號處理消除雜波及電子反干擾 (ECCM) 防護功能。
- 具備結合早期預警指管系統的通訊界面。

彈砲結合 功能是國軍對射控雷達系統的基本需求。CS/MPQ-78近空防禦射控雷達系統為期兩年的測試期間，先後完成結合陸射型TC-1飛彈、35公厘快砲、40公厘快砲對空目標追瞄、鎖定、射擊等測試項目，充分掌握了目標偵追、精度測校、指令解算、彈著修正、遠距操控等關鍵性技術。本系統亦具備結合各型近空防禦飛彈及快砲指引接戰能力。

The CS/MPQ-78 Low-Altitude Air-Defence Fire Control Radar System is one of the most significant R&D achievements of the Chung Shan Institute of Science and Technology (CSIST).

The main system features of the system are as follows:

- Search radar provides 360° coverage against multiple threats, and provides guidance to air-defence missiles.
- Track radar provides real-time, precise target data to control air-defence guns.
- The electro-optical tracker is an auxiliary device that provides for close-range fighting capability.
- Identification/Friend-or-Foe allows the discrimination of targets.
- Signal processing functions include clutter rejection and ECCM protection.
- Interfaces with the early-warning Command and Decision System.

CS/MPQ-78 近空防禦射控雷達系統
CS/MPQ-78 Low-Altitude Air-Defence Fire Control Radar System

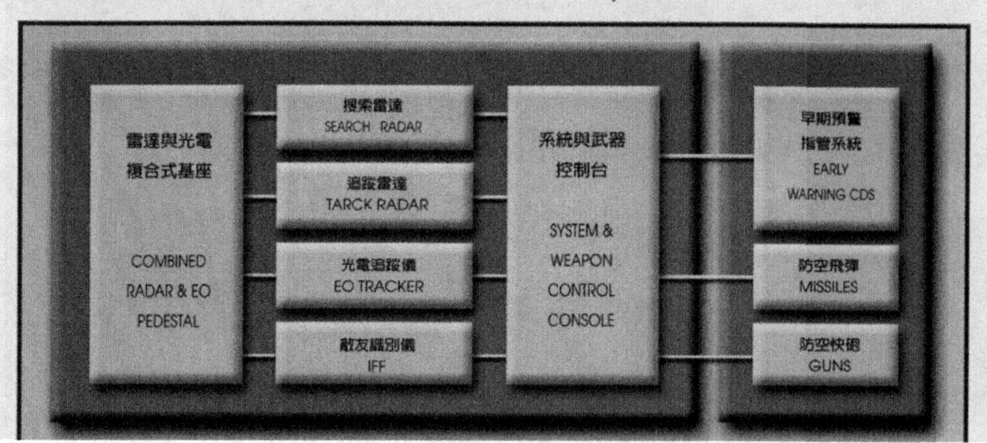

National Chungshan Institute of Science and Technology (NCSIST). Radar Section, Electronic System Division. CS-MPQ-78 Low-Altitude Air-Defense Fire Control Radar (LAADFCR). Unknown Taipei Aerospace Defense Technology Exhibition. Brochure 2/6.

達系統

e Control Radar System

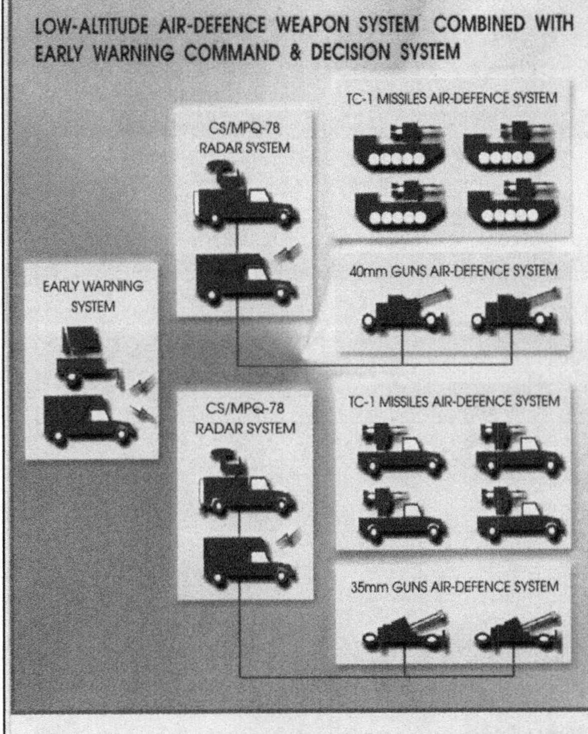

早期預警 指管系統的情資與管制可提升近空防禦武器系統區域防空作戰能力。CS/MPQ-78 近空防禦射控雷達系統具備與上層指管系統結合之通訊介面，可接收目標預警指令進行接戰任務，亦可將本系統偵追目標情資呈報指管系統運用。

EARLY WARNING The Command & Decision System (CDS) can upgrades the regional combatant capability of the Low-Altitude Air-Defence weapon system. The CS/MPQ-78 Low-Altitude Air-Defence Fire Control Radar can connect with higher level CDS. This allows the recept of early-warning information for engagement implementation. The radar can also distribute target information to the CDS.

雷達系統構型：

有分離車廂式與單一車廂式兩種型式。前者由一雷達車廂與一操控車廂組成，分別由悍馬車(HMMWV)承載運行。後者為一拖車，雷達全系統含操控台均整合於此車內，其由拖車頭拖曳運行。

The CS/MPQ-78 Low-Altitude Air-Defence Radar system is available in two versions: a separate-shelter version and a trailer version. The former has two shelters, a radar shelter and an operation shelter. Each of the two shelters is mounted on a Hummers (HMMWV) for transportation. The latter contains the whole entire radar system including the operator console. The trailer is draged by another heavy vehicle such as a tractor march.

National Chungshan Institute of Science and Technology (NCSIST). Radar Section, Electronic System Division. CS-MPQ-78 Low-Altitude Air-Defense Fire Control Radar (LAADFCR). Unknown Taipei Aerospace Defense Technology Exhibition. Brochure 3/6.

CS/MPQ-78 近空防禦射控雷達功能規格:

搜索雷達	
目標偵搜範圍	30Km, 360°
多目標追蹤數	20
距離追蹤精度	±30 m
方位追蹤精度	±5 mil
追蹤雷達	
目標追蹤距離	20 Km
距離追蹤精度	±10 m
方位追蹤精度	±1 mil
俯仰追蹤精度	±1 mil
光電追蹤	
目標追蹤距離	10Km
雷射測距精度	±5 m
方位追蹤精度	±0.5 mil
俯仰追蹤精度	±0.5 mil
敵友識別儀	
目標識別範圍	30Km, 360°
目標識別模式	Mode 1,2,3/A,4
武器界面	
飛彈結合界面	4
快砲結合界面	2

CS/MPQ-78 LOW-ALTITUDE AIR-DEFENCE FIRE CONTROL RADAR SPECIFICATIONS

SEARCH RADAR	
Detection Coverage	30Km, 360°
TWS Tracks	20
Range Track Accuracy	±30 m
Az Track Accuracy	±5 mil
TRACK RADAR	
Target Track Range	20 Km
Range Track Accuracy	±10 m
Az Track Accuracy	±1 mil
El Track Accuracy	±1 mil
EO TRACK	
Target Track Range	10Km
Laser Range Accuracy	±5 m
Az Track Accuracy	±0.5 mil
El Track Accuracy	±0.5 mil
IFF	
Ident. Coverage	30Km, 360°
IFF Modes	Mode 1,2,3/A,4
WEAPON INTERFACE	
Missile Interface	4
Gun Interface	2

National Chungshan Institute of Science and Technology (NCSIST). Radar Section, Electronic System Division. CS-MPQ-78 Low-Altitude Air-Defense Fire Control Radar (LAADFCR). Unknown Taipei Aerospace Defense Technology Exhibition. Brochure 4/6.

實彈射擊 驗證是測評全系統功能的重要指標，民國86年（公元1997年）3月執行一結合快砲實彈射擊驗證任務，準確擊落空中靶標兩枚，圓滿達成計劃目標。

FIRING TESTS on towed targets are an important index for evaluating the overall system. In March 1997, an integration test was con-ducted with guns; two towed targets were destroyed and the project was a complete success.

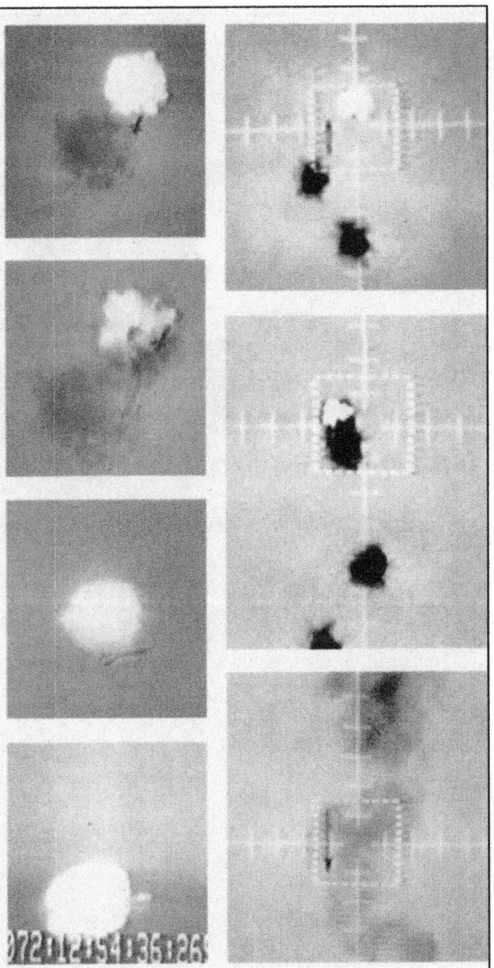

972 12:54:36:26

National Chungshan Institute of Science and Technology (NCSIST). Radar Section, Electronic System Division. CS-MPQ-78 Low-Altitude Air-Defense Fire Control Radar (LAADFCR). Unknown Taipei Aerospace Defense Technology Exhibition. Brochure 5/6.

National Chungshan Institute of Science and Technology (NCSIST). Radar Section, Electronic System Division. CS-MPQ-78 Low-Altitude Air-Defense Fire Control Radar (LAADFCR). Unknown Taipei Aerospace Defense Technology Exhibition. Brochure 6/6.

National Chungshan Institute of Science and Technology (NCSIST). Training and Simulator System Research Division. Simulation and Simulation Technology. TH-67A Creek training helicopter. Full Function Flight Simulator (FFS). 2017 Taipei Aerospace Defense Technology Exhibition. Brochure 1/4.

TH67 全功能飛行模擬機

TH-67 Full

分系統功能簡介 *Description of Sub-system*

主計算機系統

採用SGI O300工作站等級之主計算機，可提供穩定的即時模擬環境；直昇機數學模式則採用ART FlightLab模擬程式，可精確的模擬TH-67直昇機之飛行動態。

Host computer

The SGI O300 workstation computer provides a stable real-time environment for simulation. The mathematic model of helicopter dynamics is generated from ART FlightLab. It provides a precise simulation of the helicopter maneuvering.

投影系統

採用雙曲度之弧形螢幕，具有220度水平視角、70度垂直視角，採用Christine系列投影器，具有5 Foot-Lambert以上之亮度，五個頻道之投影可利用電腦調校方式，確保接處之平順與色彩之一致。

Projection display system

The two-dimension curvature screen has 220° horizontal, 70° vertical view angle. The Christine projectors provide sharp visual scene more than 5 Foot-Lambert brightness. The blending effect between 5 projectors can be adjusted by computer to have a consistence scene.

動感平台

採用FCS ECue 660-8000之六軸電動平台，正常操作之酬載重量為7500公斤，可提供偏航、滾轉、俯仰、上下、左右、及前後六自由度運動之動感法則。

Electrical motion platform

The FCS ECue 660-8000 is consist of six actuators driven by electrical motor. The payload of the system is more than 7500 Kg during normal operation. It provides realistically motion cues during helicopter maneuvering including Yaw, Roll, Pitch, Heave, Sway, Surge motion.

乘員艙

乘員艙使用真實座艙結構，座艙之開關旋鈕均與真實飛機相同，儀錶顯示則採用電腦繪圖方式製作，可有效降低製作與維修成本，並依據訓練需求選取儀器飛行或目視飛行機型。教官監控台可透過座艙內裝設之攝影監視器觀察學員訓練狀況。

Cockpit system

The structure of TH-67 FFS' cockpit is part of the real Helicopter, the control switches and knobs are same as the real. But the instrument display are generated by computer graphic and mask. It will not only reduce the manufacture and maintenance cost, and the cockpit can be changed between IFR and VFR configuration according to the training requirement. The camera inside the cockpit can monitor the pilot training situation from operation control station.

力感系統

採用FCS ECoL 8000力感控制系統，依據飛行員之飛行操作狀態提供力回饋感受，包括迴旋桿之俯仰與滾轉控制、集體桿之起降控制，腳舵板之方向控制則採用摩擦條之設計。

Control loading system

FCS ECoL 8000 Control loading system provides realistically artificial feel of the flight control mechanics according to the flight condition. There are pitch and roll control channels on cyclic stick control and one channel on collective control. The force feedback of rudder pedal is generated by friction grip.

輸出入介面系統

採用Adlink之工業標準輸出入介面控制系統，擷取座艙及教官台之控制信號，透過信號調節、數位/類比之轉換、及資料格式轉換，再將資料透過反射式記憶體送至模擬主計算機，並依據計算結果驅動座艙及教官台相關的指示燈號

I/O interface system

The Adlink I/O interface system with industrial standard will acquire the signals from cockpit and operation station to do the manipulation, D/A conversion, format translation. The output signals will send to the host computer through the Reflective Memory Bus to drive the relative indicator in the cockpit.

National Chungshan Institute of Science and Technology (NCSIST). Training and Simulator System Research Division. Simulation and Simulation Technology. TH-67A Creek training helicopter. Full Function Flight Simulator (FFS). 2017 Taipei Aerospace Defense Technology Exhibition. Brochure 2/4.

Function Flight Simulator(FFS)

音效系統
採用ASTi音效控制系統，可模擬飛行操作之各種音效：如引擎啓動、旋翼轉動、警告音響、摩斯碼聲音、座艙內通話及教官台間的通訊等功能。

Audio system
The ASTi Audio system provide the simulation sound effect during aircraft operation including engine starting, rotor blade rotation, warning tone, Morse code, internal communication, and communication between pilot and operator.

視效產生器
採用Primary Image之視效產生器，根據直昇機之位置與姿態變化，計算視效資料庫中地形、地物之視覺變化，產生適切的視效場景，並透過投影器呈像在弧形銀幕上，使飛行員的視覺產生逼真的臨場感。

Image generator
The image generator will generate the visual image of terrain objects according to the aircraft position and attitude. It will provide realistic visual effect to the pilot through the projectors to display on the screen.

夜視鏡系統
採用nVision之模擬夜視鏡可提供逼真之夜視影像，影像訊號經由夜視處理器之調變，可產生雜訊、陰影、光暈等特殊效果。

NVG system
The nVision simulated NVG will provide realistic night visual scene through the Post Video Processor (PVP). The image signal will provide the special effect of NVG including noise, shadow of light, Halo effect.

教官監控台
教官台採用整合式機櫃的設計型式，可提供教官良好之操作環境。教官台所有操作畫面均以中文化及操作便利性為設計考量。

(1)系統操作設定畫面：包括操作課目選擇、模擬環境參數設定、初始狀態選擇、及錄製與重放操作，以提供訓練課目之選擇與設定、設定、與參數紀錄等功能。

(2)系統狀態顯示畫面：包括系統狀態顯示、飛行軌跡顯示、信號記錄等功能。並於訓練進行期間，提供動感平台位置、加速度、和系統狀態等操作資訊。

(3)儀錶顯示畫面：用以顯示座艙內配置的各類儀錶或指示器之狀態。

(4)視效影像顯示：同步顯示座艙內之視效影像畫面。

Operation control station
The integrated station is designed to provide a friendly operation environment. The graphic user interface is designed in Chinese language and with user convenience

(1) The system operation window includes the selection of training course, parameter of simulation environment, initial condition loading, and record/playback.

(2) The system status display window provides the display of system operating state, flight trajectory, signal recording. It also provides the monitoring data of motion platform.

(3) The instrument display window provides the replicate of the instrument inside the cockpit.

(4) The visual scene display provides the operation scenarios of the helicopter flight.

全功能訓練設施
- 模擬系統、訓練教室、廠區佈置設計及建構的套裝解決方案
- 室內、操作前後、任務歸詢等全訓練程序設計
- 能源與水源等環保要求設計

Full Function Training Facility
- Turnkey system including simulation system, classroom training, facility layout and design, and construction.
- Full training process including classroom training, pre-operation orientation, post-operation orientation, mission case studies, and in mission orientation and lecturing.
- Green building design for energy and water efficiency.

National Chungshan Institute of Science and Technology (NCSIST). Training and Simulator System Research Division. Simulation and Simulation Technology. TH-67A Creek training helicopter. Full Function Flight Simulator (FFS). 2017 Taipei Aerospace Defense Technology Exhibition. Brochure 3/4.

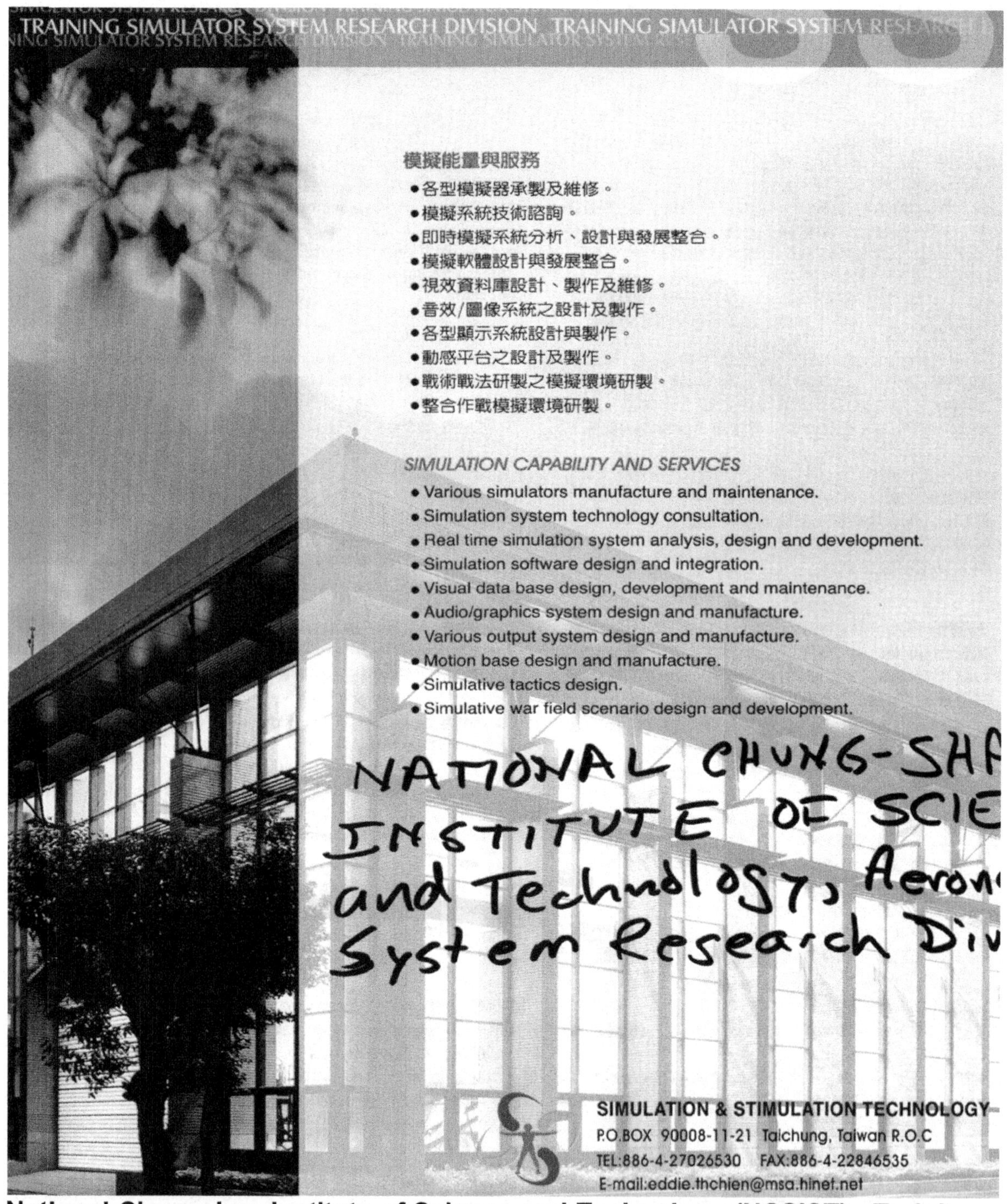

National Chungshan Institute of Science and Technology (NCSIST). Training and Simulator System Research Division. Simulation and Simulation Technology. TH-67A Creek training helicopter. Full Function Flight Simulator (FFS). 2017 Taipei Aerospace Defense Technology Exhibition. Brochure 4/4.

Army Aviation and Special Forces Command (陸軍航空特戰指揮部). TH-67 Creek Pilot Instructor wearing the 603rd and AATC insignia. Identification of instructor cropped to protect identity.

TH-67A Creek Trainer.

National Chungshan Institute of Science and Technology. Simulation and Simulation Technology. M48H/M60A3 Main Battle Tank Driver and Gunner Simulation. Unknown Taiwan Aerospace and Defense Technology Exhibition. Brochure 1/6.

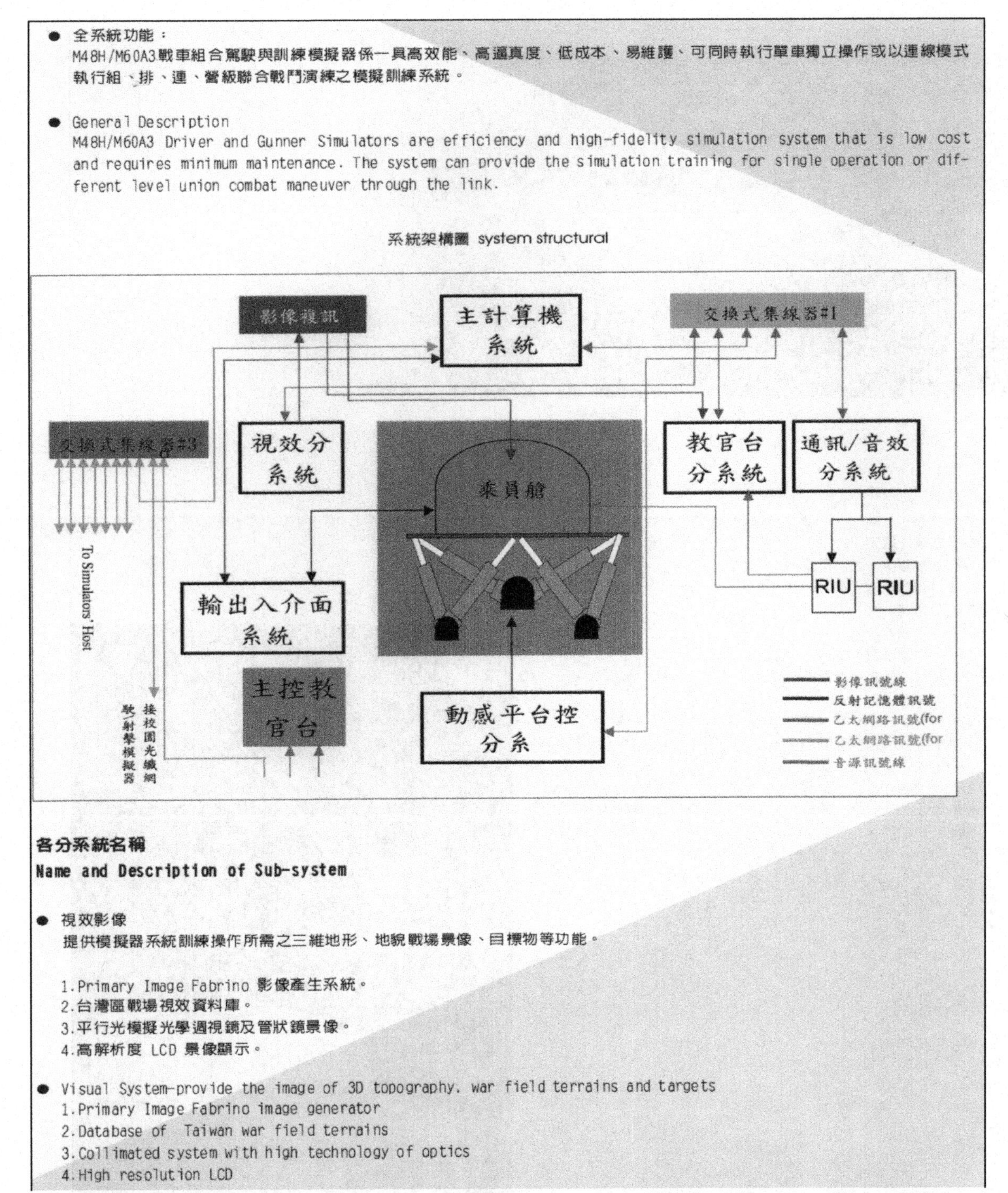

● 全系統功能：
M48H/M60A3戰車組合駕駛與訓練模擬器係一具高效能、高逼真度、低成本、易維護、可同時執行單車獨立操作或以連線模式執行組、排、連、營級聯合戰鬥演練之模擬訓練系統。

● General Description
M48H/M60A3 Driver and Gunner Simulators are efficiency and high-fidelity simulation system that is low cost and requires minimum maintenance. The system can provide the simulation training for single operation or different level union combat maneuver through the link.

系統架構圖 system structural

各分系統名稱
Name and Description of Sub-system

● 視效影像
提供模擬器系統訓練操作所需之三維地形、地貌戰場景像、目標物等功能。

1. Primary Image Fabrino 影像產生系統。
2. 台灣區戰場視效資料庫。
3. 平行光模擬光學週視鏡及管狀鏡景像。
4. 高解析度 LCD 景像顯示。

● Visual System-provide the image of 3D topography, war field terrains and targets
1. Primary Image Fabrino image generator
2. Database of Taiwan war field terrains
3. Collimated system with high technology of optics
4. High resolution LCD

National Chungshan Institute of Science and Technology. Simulation and Simulation Technology. M48H/M60A3 Main Battle Tank Driver and Gunner Simulation. Unknown Taiwan Aerospace and Defense Technology Exhibition. Brochure 2/6.

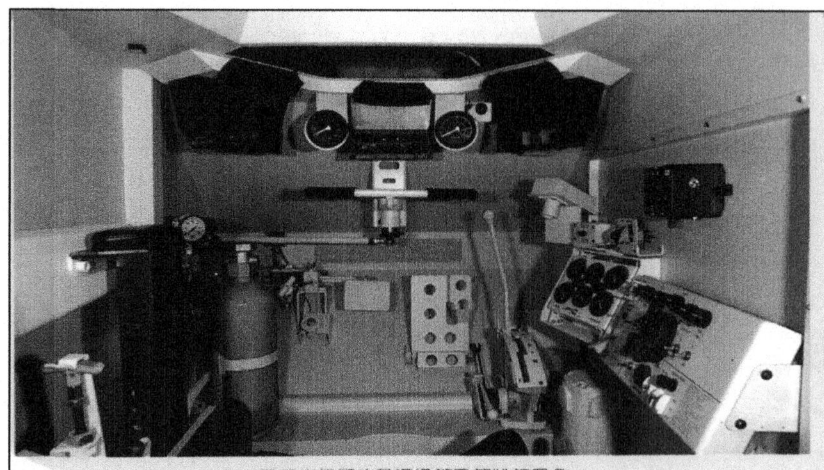

平行光模擬光學週視鏡及管狀鏡景象

Collimatde Sysem with high technology of optics

- 主計算機

  具分時(Time Sharing)、即時(Real Time) 、多人多工(MultiUser.MultiTask)及多處理器(SMP或NUMA)等功能：滿足模擬器系統即時程式管理能力。

  1.2個2GHz 雙CPU 即時電腦。
  2.1 GB 記憶體。
  3.40 GB 硬碟。
  4.Giga Bit網路功能。
  5.Linux 作業系統。
  6.RTAI及REDICE軟硬體即時發展工具。

- Host Computer - With time sharing, real time, multi-user/multi-task functionsand able to manage the real time programs of the simulator systems

  1.Two 2GHz CPU
  2.1GB Memory
  3.40GB Hard Disk
  4.Giga Bit Network founction
  5.Linux operation system
  6.RTAI/REDICE S/W and H/W developing tool

National Chungshan Institute of Science and Technology. Simulation and Simulation Technology. M48H/M60A3 Main Battle Tank Driver and Gunner Simulation. Unknown Taiwan Aerospace and Defense Technology Exhibition. Brochure 3/6.

- 乘員艙

提供射手、車長及裝填手高逼真度之射擊艙空間與艙內元件配置及功能之模擬環境。
1.蜂巢結構之鋁合金板面外艙，滿足平台動態操作之應力需要。
2.楔形切面接合之高科技造形外艙。
3.仿真鑄造之內艙元件，反應真實件之觸感與力感。
4.模擬平行光之光學視鏡。

Cockpit System-provide gunners and drivers the realistic operation environment
1.Aluminum alloy cabin with honey structure
2.High technology with cabin shape
3.High-fidelity instruments in cabinet
4.Compatabls structure collimate system

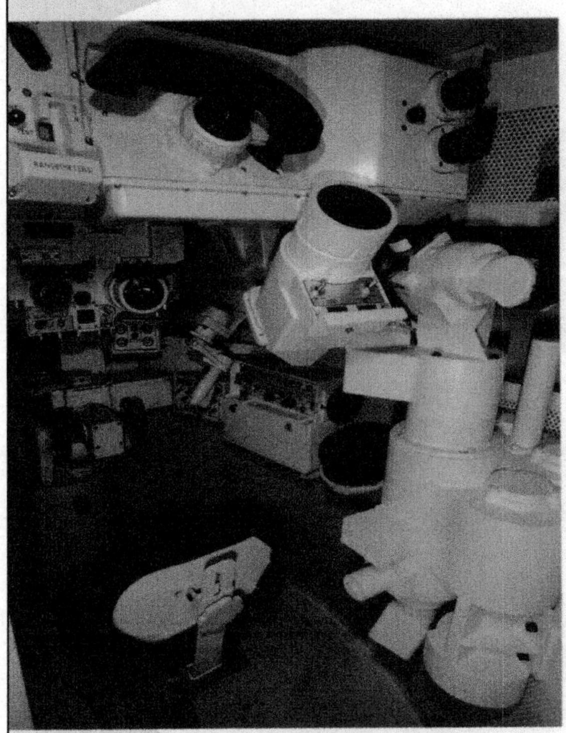

戰車駕駛模擬器乘員艙
Tank Driver Simulator Cockpit System

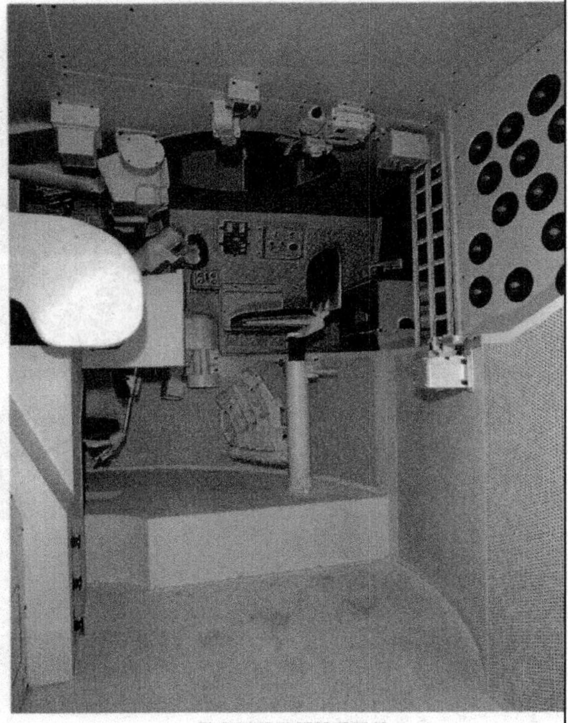

戰車射擊模擬器乘員艙
Tank Gunner Simulator Cockpit System

- 輸出入界面

提供模擬器系統間之類比及數位訊號的傳輸與轉換等功能。
1.工業級界面電腦。
2.Compact PCI 延伸機箱。
3.Compact PCI/ PCI/ DMA光纖界面資料傳輸。

I/O System-Data transfer and communication between analog and digital signals
1.Industrial PC with I/O controllers
2.Compact PCI extend available
3.Compact PCI/PCI/DMA filber interface data tramsmition

National Chungshan Institute of Science and Technology. Simulation and Simulation Technology. M48H/M60A3 Main Battle Tank Driver and Gunner Simulation. Unknown Taiwan Aerospace and Defense Technology Exhibition. Brochure 4/6.

- **動感平台**

提供模擬器系統駕駛運動時之俯仰震動及射擊後座力之動態反應。

1. Fokker Ecue 624-2800動感平台系統。
2. 電動馬達致動設計，噪音小、污染低。
3. 被動式氣源輔助靜態承載，延長電動馬達壽命。
4. 操控電腦機櫃置於平台下方，有效節省裝備配置空間。
5. 斷電時自動回復至零點設計，提高受訓人員安全。
6. 遠端偵錯能力。

- Motion Base-provide the dynamic response of motion

1. Fokker Kcur 624-2800 motion base
2. Low noise and low pollution motion design
3. Pneumatic system to support the static payload. to extend the life of motor.
4. Operation computer is installed under the base to save the space
5. Powerout automatic safety design
6. Remote diagpostic function

動感平台系統
motion base

- **教官台**

提功教官執行模擬器系統功能操控、訓練課目選擇及訓練乘員監控等功能。

1. P4 2.4GHz教官台操作、地理圖資、訊息顯示電腦。
2. 中文操作顯示畫面。
3. 工業級目標物運算電腦。
4. 模擬系統操作、功能設定、乘員監控。
5. 簡易駕駛模組。

- Instructor Operation Station-provide the instructor

1. P4 2.4 GHz computer. providing training control. GIS and system status display
2. Chinese user interface
3. Industrial grade target generating comouter
4. System control. training setup. crew monitoring
5. Simple instructor driving control unit

Simulation Systems Department, ASRD / CSIST
Address : P.O. Box 90008-11-21 Taichung 407, Taiwan, R.O.C.
Tel : +886-4-23756530

National Chungshan Institute of Science and Technology. Simulation and Simulation Technology. M48H/M60A3 Main Battle Tank Driver and Gunner Simulation. Unknown Taiwan Aerospace and Defense Technology Exhibition. Brochure 5/6.

教官台
Instructor Operation Staion

- 音效通訊

提功模擬系統所需之音效及通訊功能。
1. ASTi 音效及通訊電腦。
2. 音效建置模組。

- Sound and Communication System-provide the sound and communication function
  1. ASTi computer
  2. Sound module

像顯易識模之必求其眞
TO DISPLAY THAT OF WHICH SHALL MIMIC THE REALISTIC ORIGIN

形潛莫睹擬之務通其神
TO FUNCTION AS THOUGH IMMERSED IN SENSATIONAL REALISM

中山科學研究院
Chung-Shan Institute of Science and Technology
桃園龍潭郵政90008-1信箱
電話：886-2-2673-9638 · 886-3-471-3022 · 886-0800-014723
傳真：886-3-471-4183
網址：http//www.csistdup.org.tw
P.O. Box 90008-1, Lungtan, Taoyuan, Taiwan, R.O.C.
TEL : 886-2-2673-9638 · 886-3-471-3022 · 886-0800-014723
FAX : 886-3-471-4183
http : //www.csistdup.org.tw
E-mail : pdduii @ www.csistdup.org.tw

National Chungshan Institute of Science and Technology. Simulation and Simulation Technology. M48H/M60A3 Main Battle Tank Driver and Gunner Simulation. Unknown Taiwan Aerospace and Defense Technology Exhibition. Brochure 6/6.

National Chungshan Institute of Science and Technology. Wireless LAN. Unknown Taiwan Aerospace and Defense Technology Exhibition. Brochure 1/6.

CS/PRC-37A 跳頻通信機
Frequency-Hopping Radio

裝備描述

CS/PRC-37A 系列為中科院研發之新一代野戰用 VHF 跳頻無線電機，具有數據和語音抗干擾通信能力，且於定頻通信時可直接與多種現役 VHF/FM 無線電機互通。(諸如：AN/PRC-77、VRC-12 系列等。)
CS/PRC-37A 系列可依需求加裝內嵌式數位式保密器,以提供高度語音及數據通信保密能力：此外並具備多種現代通裝之功能，如波道預置、預置波道掃瞄、電源不足警示、遙控操作及注碼等。

Description

The CS/PRC-37A with is a new-generation tactical VHF fréquency-hopping family of radios,developed by CSIST, with data and voice anti-jamming (ECCM) communication capability and interoperability with most of the fixed-frequency VHF/FM radios currently in operation.
An optional built-in digital cipher provides a high degree of voice and data message security (COMSEC). The functions available in modern communication radios are also included, such as channel preset, preset channel scanning, low battery warning, remote control, late entry, key loading, etc.

裝備組成

CS/PRC-37A 系列之模組化設計可依任何特定之操作需求提供多種組態，以分別適用於短距離及長距離通信。主要包括兩種組態：背負式和車裝型等兩種。
背負式和車裝型均使用相同的無線電主機，背負型可配接短或長天線, 以及使用鎳鎘/鎳氫/鋰等各型電池。車裝型利用載具電瓶為電源，可配賦兩部主機以進行中繼通信，並可加上兩部 50W 功率放大器以進行長距離通信。

Configuration

The modular design of the CS/PRC-37A offers a variety of configurations, which provide short-range and long-range communications, to match any specific operational requirement.
Backpack and vehicular models, which are easily configured with the common receiver-transmitter unit, are the two main configurations.
The back pack radio can be equipped with a short or long whip antenna and Ni-Cd/Ni-H/Li batteries. The vehicular type uses the battery of the carrier as its power supply. It can be equipped with two radios for retransmission, and up to two 50W amplifiers can be engaged for long-range communication.

National Chungshan Institute of Science and Technology. Wireless LAN. Unknown Taiwan Aerospace and Defense Technology Exhibition. Brochure 2/6.

主要功能特性

裝備型式：人員背負式、車裝短距離、車裝長距離、車裝長/短距離、車裝長/長距離

- 頻率範圍 30~88 MHz
- 波道間隔 25 MHz
- 總波道數 2320 預置波道數 8
- 與 AN/PRC-77 定頻語音相容
- 全頻段或部份頻段跳頻抗干擾
- 功率 0.01W~50W 四段選擇
- 內建數據模組 75~16,000 BPS
- 內建保密器模組 (選項配備)
- 內建GPS模組 (選項配備)
- 內建自我測試功能 (BIT)
- 定頻預置波道掃瞄功能
- 定頻呼叫跳頻功能
- 選擇性呼叫功能
- 來電者代碼自動顯示功能
- 明密自動辨識功能
- 中繼轉發功能
- 接收信號強度顯示功能
- 重要參數輸入/查詢密碼管制功能
- 緊急記憶體清除功能
- 無線電注碼功能 (ERF)
- 遙控3.2公里(選項配備)
- 環境規格 MIL-STD-810E
- 電磁防護 MIL-STD-461C

The CS/PRC-37A operates over the 38 to 88MHZ frequency range with 25KHZ channel spacing and is capable of handling voice, analog data, and digital data traffic.

features include:

- 8 preset channels and 2320 channels overall
- Full band/partial band hoppimg to counter jamming
- 0.1W/0.5W/5W/50W power selections
- Integral error correcting and interleving variable data rate adaptop for operation at rates of 75 to 16,000 bps
- Embedded communications security options
- Embedded GPS options
- Comprehensive built-in test (BIT) features with fault diagnosis down to module level, and monitoring of critical functions
- Preset channel scanning
- Hailing
- Selective call
- Caller I.D. auto recognition
- Clear/security mode auto detection
- Automatic retransmit capability
- Receive signal strength indicator
- Password entry controls user access
- Emergency self-destruct memory erase
- Automatic electronic radio-controlled data fill (ERF)
- Ability to add a second RF power amplifier for dual long range and retransmit
- Remote keying and audio operation up to 3km
- Optional full-function remote control at up to 3km
- Environmental characteristics such as rain, fungus,vibration,and shock per MIL-STD-810E
- EMI/EMC per MIL-STD-461C

National Chungshan Institute of Science and Technology. Wireless LAN. Unknown Taiwan Aerospace and Defense Technology Exhibition. Brochure 3/6.

微波組件

Voltage Controlled Oscillator

微波壓控振盪器
Voltage Controlled Oscillator

規 格
本項縮裝SMD微波壓控振盪器研發，規格如下：
(中心頻率：2.083GHz
(可調頻寬：1%
(輸出功率：0dBm(Typ)
(相位雜訊：-100dBc/Hz @ 1.5MHz Offset Carrier
(包裝型式：SMD

性 能
壓控振盪器與鎖相迴路的結合組成了通訊系統的心臟，本項技術符合了省電、體積小及成本低等要求，並且可依不同系統要求調整出不同的頻率範圍。包裝尺寸有300×300×120 mil及270×270×100 mil兩種；包裝方式為SMD，材質為FR4 PCB。

用 途
本產品可應用於無線區域網路(WLAN)、歐規數位行動電話(GSM)、數位無線電話(DECT)、個人通訊服務(PCS)、大哥大(Cellular Phone)、呼叫器(Pager)、儀器設備等。

Specifications
- Center Frequency:2.083GHz
- Tuning Bandwidth:1%
- Output Power:0dBm(Typ.)
- Phase Noise:-100dBc/Hz @ 1.5MHz Offset Carrier
- Package Style:SMD

Features
- Suitable for ISM 2.4 GHz band wireless communications applications
- Low current, low power consumption
- Small volume, light weight
- SMD packaged, readily meeting SMT mass production processing needs

Applications
- Tunable frequency source for microwave applications
- Stable local oscillator wireless communication transmit/receive (T/R) module when combined with phase-lock-loop (PLL)

溫度補償石英晶體振盪器
Temperature-Compensated Crystal Oscillator

規 格
本項被動式溫度補償晶體振盪器研發，規格如下：
- 中心頻率：12.8MHz
- 頻率漂移 (-10°C～+60°C)：±2.5ppm
- 包裝型式：TO

性 能
本產品為一高品質、小型化之TCXO，其頻率穩定度在-10°C到+60°C之溫度範圍可達到±2.5ppm以內，其直流耗電小於 5mA，包裝大小為12mm×18.3mm×5mm，相當於DIP 14 pin 的IC 包裝。

用 途
本產品可用於儀器之頻率源、資料傳送及無線通訊等相關產品上。

Specifications
- Center Frequency:12.8MHz
- Frequency Drift (-10°C～+60°C):±2.5ppm
- Package Style:TO

Features
- Suitable for ultrahigh stability reference oscillator application in a frequency-agile wireless synthesizer
- Low current (5mA), low power consumption
- Small volume (12mm X 18.3mm X 5mm), light weight

Applications
- Ultrahigh stability frequency source for wireless communication systems and high-frequency measurement instruments

National Chungshan Institute of Science and Technology. Wireless LAN. Unknown Taiwan Aerospace and Defense Technology Exhibition. Brochure 4/6.

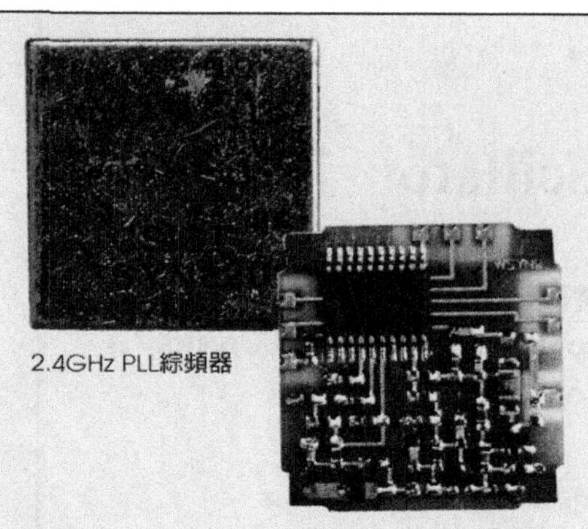

2.4GHz PLL綜頻器

2.4GHz PLL Synthesizer

綜頻器
AGHz PLL

規 格

本項縮裝型 SMD PLL 綜頻器研發，規格如下：

- 頻率範圍：2.12~2.203GHz
- 頻道間隔：1MHz
- 輸出功率：0dBm(Typ.)
- 相位雜訊：-125dBc/Hz @1.5MHz Offset Carrier

性 能

以自行研發之寬頻壓控振盪器為主，使用商用微波鎖相迴路積體電路 (Phase-Lock-Loop IC) 經過程式模擬設計出RC濾波器，在頻道間隔為1MHz時，切換時間可小於224usec，符合IEEE 802.11規範。該成果已成功應用於筆記型電腦之PCMCIA WLAN卡上。

用 途

本產品可做為無線區域網路射頻收發模組中穩定之本地振盪源。

Specifications

- Frequency Range:2.12~2.203GHz
- Frequency Step:1MHz/step
- Output Power:0dBm(Typ.)
- Phase Noise:-125dBc/Hz@1.5MHz Offset Carrier

Features

- Suitable for ISM 2.4 GHz band wireless local area network (WLAN) communication applications
- Low current, low power consumption
- Small volume, light weight
- Fast channel switching speed (≦224usec)

Applications

- Stable local signal source for WLAN transmit/receive (T/R) module

砷化鎵2.4GHz單晶微波積體電路晶片組
2.4GHz GaAs Monolithic Microwave Integrated Circuit (MMIC) Chip Set

規 格

本項2.4GHz單晶微波積體路研發，規格如下：

- 頻率範圍：2.4~2.483GHz
- 低雜訊放大器：增益≧12.5dB，雜訊指數≦2.5dB
- 功率放大器：增益≧25dB，輸出功率≧20dBm
- 升降頻轉換器：昇頻增益≧16dB ，降頻增益≧8dB

性 能

中科院開發完成國內首套砷化鎵2.4GHz單晶微波積體電路晶片組，包含了低雜訊放大器、升降頻轉換器及功率放大器。本套晶片組完全自行設計、製造及封裝，其特性與國外類似產品相當。

用 途

本產品可做為2.4GHz ISM頻段無線通訊用射頻前級。

Specifications

- Frequency Range:2.4~2.483GHz
- Low Noise Amplifier
  Gain:12.5dB(min.)
  Noise Figure：2.5dB(max.)
- Power Amplifier
  Gain:25dB(min.)
  Output Power:20dBm(min.)
- Up/Down Converter
  Upconversion Gain:16dB(min.)
  Downconversion Gain:8dB(min.)

Features

- Suitable for ISM 2.4 GHz band wireless local area network (WLAN) communication applications
- Consists of a low-noise amplifier,power amplifier, and up/down converter chips
- All three chips are packaged in SOIC-8 and SSOP-28 plastic packages

Applications

- ISM 2.4 GHz band WLAN transmit/receive (T/R) module chip sets which meet IEEE 802.11 universal standards

National Chungshan Institute of Science and Technology. Wireless LAN. Unknown Taiwan Aerospace and Defense Technology Exhibition. Brochure 5/6.

無線通訊網路卡
WLAN PCMCIA CARD

規 格

本項 2.4GHz 無線通訊網路卡產品，規格如下：

- 頻率範圍：2.4~2.4835GHz
- Data Rate：1 or 2 Mbps
- Processing Gain： 11 dB
- Interface：PCMCIA
- 輸出功率：100mW(Max.)
- 靈敏度：-87dBm @ 8E-2 FER

性 能

以自行研製之壓控振盪器、綜頻器、低雜訊放大器、功率放大器及升降頻轉換器等 MMIC 晶片組完成 WLAN PCMCIA 卡之雛型研製。本片 WLAN PCMCIA 卡工作在 2.4GHz ISM 頻道，利用 Direct sequence 展頻技術及 CSMA/CA (Collision sense multiple access/Collision avoidance) Protocol，符合 IEEE802.11 標準，其傳輸速率可達 2Mbps. 在開放空間下傳輸距離可達800公尺。

用 途

本片 WLAN PCMCIA 卡工作在 2.4GHz ISM 頻道上，只需修改 Base Band 部份，就可做各種不同用途，如筆記型電腦無線網路系統及區域點對點無線通訊系統。

Specifications

- Frequency Range:2.4~2.4835GHz
- Data Rate:1 or 2 Mbps
- Output Power:100mW(max.)
- Processing Gain:11dB
- Sensitivity:-87dBm @8E-2FER
- Interface:PCMCIA

Features

- Suitable for ISM 2.4 GHz band wireless local area network (WLAN) communication applications
- Utilizes Direct Sequence Spread Spectrum (DSSS) technology and CSMA/CA (Collision Sense Multiple Access/ Collision Avoidance) protocol; meets IEEE 802.11 universal standard
- Outline follows PCMCIA Type II standard

Applications

- ISM 2.4 GHz band WLAN PC adapter card
- Also suitable for wireless bridges and point-to-point data link applications

中山科學研究院
Chung-Shan Institute of Science and Technology

桃園龍潭郵政90008-1信箱
電話：886-2-2673-9638 · 886-3-471-3022 · 886-080-014723
傳真：886-3-471-4183
網址：http//www.csistdup.org.tw
P.O. Box 90008-1, Lungtan, Taoyuan, Taiwan, R.O.C
TEL：886-2-2673-9638 · 886-3-471-3022 · 886-080-014723
FAX：886-3-471-4183
http//www.csistdup.org.tw

National Chungshan Institute of Science and Technology. Wireless LAN. Unknown Taiwan Aerospace and Defense Technology Exhibition. Brochure 6/6.

National Chungshan Institute of Science and Technology. Hand-held IR Camera – CTV/TVS-32. Unknown Taiwan Aerospace Defense Technology Exhibition. Brochure 1/7.

攜式熱像機

Hand-held IR Camera

特徵

- 採高解析、高均勻3-5微米紅外線矽化鉑（PtSi）焦平面陣列感測器，可夜間及白天監控使用。
- 具價格低、重量輕、操作容易之攜式熱像機。
- 採模組化設計維修方便。

規 格

1. 偵測波段：3-5微米
2. 偵檢器：紅外線矽化鉑電荷偶合感測器
3. 像素：256x244
4. 冷卻方式：史特林閉式冷卻器
5. 冷卻時間：<8分鐘
6. 視場角：<4.5°（H)X3.5°
7. 熱靈敏度(NETD)：<0.1°C
8. 偵測距離：人員<1.6公里
9. 顯示方式：內建2.5" LCD Display
10. 輸出訊號：RS-170
11. 電源需求：AC adaptor車用電源或電池(3AH可充電池，使用時間>2.5小時)
12. 重量：<3Kg
13. 尺寸：<30cm X11cm X12cm
14. 工作溫度：-10°C～+50°C

National Chungshan Institute of Science and Technology. Hand-held IR Camera – CTV/TVS-32. Unknown Taiwan Aerospace Defense Technology Exhibition. Brochure 2/7.

Main Features

- Outstanding image resolution (high resolution, superior repose uniformity) with 256 x 244 FPA PtSi detector
- Advanced DDC (detector, dewar, cooler) module with on chip signal processing
- Battery powered
- Low cost, lightweight, user friendly
- Integral miniature LCD video monitor, external standard video output
- Miniature closed-cycle cooler
- Module design for easy maintenance

Specifications

1. Spectral Band: 3 to 5 microns
2. Detector Type: PtSi with CCD readout
3. Array format: 256 x 244 FPA
4. Cooling Method: Stirling closed-cycle cooler
5. Cooling Time: ≦8min.
6. Field of View (FOV): 4.5° (H) x 3.5° (V)
7. Sensitivity (NETD): < 0.1°C @300 K background
8. Detection Range: < 1.6KM for Asian people
9. Display: Built-in 2.5" LCD monitor
10. Video output: RS170, B&W
11. Input Power: 110VAC (AC adapter) or Ni-MH battery (3AH, available time 2.5 hours)
12. Weight: < 3Kg
13. Size: < 30cm x 11cm x 12cm
14. Operating Temperature: -10°C~ +50°C

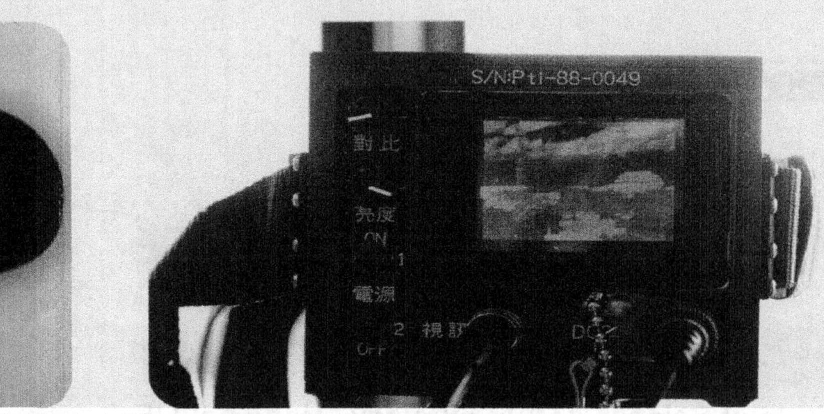

National Chungshan Institute of Science and Technology. Hand-held IR Camera – CTV/TVS-32. Unknown Taiwan Aerospace Defense Technology Exhibition. Brochure 3/7.

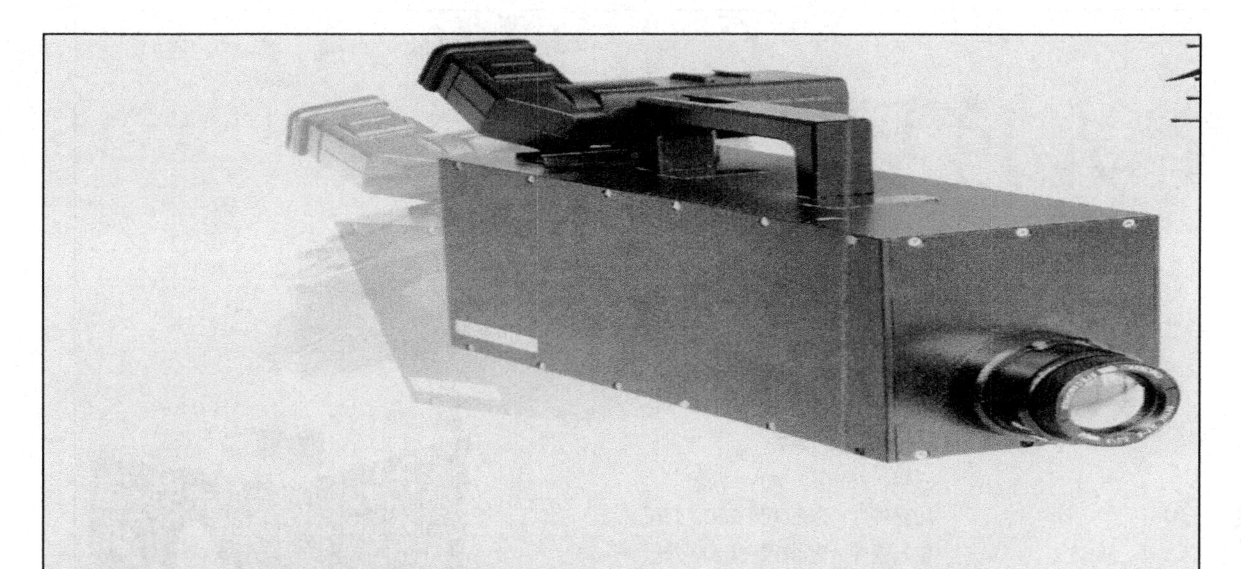

用途

海岸監控、緝私、夜間安全警戒、製程監控、非破壞性工業檢測、設備預防保修、醫療輔助診斷、環保監控、火災監控、地球資源探勘、生態保育、法律搜證、汽車夜間輔助等應用。

Applications

Lightweight surveillance, reconnaissance, and targeting systems
Infantry, scouts, and special forces
Night sight for small and medium weapon systems
Paramilitary forces
Coast guard surveillance and patrol boats

National Chungshan Institute of Science and Technology. Hand-held IR Camera – CTV/TVS-32. Unknown Taiwan Aerospace Defense Technology Exhibition. Brochure 4/7.

CS/TVS-32
全光度監控攝影機

壹、規格

動態範圍：10-5lux~105lux (無星暗夜至正午強光)
視　角：水平7.2，垂直5.4度
有效偵測距離 (以人像靶為準)
日間：25公尺~ 4000 公尺
月光條件：25公尺~ 2500 公尺
星光條件：25公尺~ 1500 公尺
掃瞄系統：RS-170，2:1交錯式
感應器：日間：1/2 吋 FRAME TRANSFER CCD,748H×494V
夜間：25mm 影像增強 CCD，靈敏度 600Ma/lm 以上
訊噪比：40dB
解析力：日　間：580TV lines
月光條件：450 TV lines
星光條件：350 TV lines
控光系統：CCD 及 sensor 雙迴路偵測，內建 CPU，配合外界光度，自動偵測並調整系統至最佳主控模式。
GAMMA：0.45fixed
電　源：AC110V 或 DC24V 或 DC12V 選用
自動警報功能：內建於監視器中，含四個可調位置之監視區域，以配合各監控地點之特殊需求於四個監控區域中有任何移動目標即發出警報。
監視器：NTSC9 吋單色螢幕。
工作溫度(配合防熱罩)：-20°C~+50°C
尺　寸：攝影機本體：長510mm×寬255mm×高155mm
監視器：長 355mm × 寬 220mm × 高 270mm
重　量：11.5 公斤
環境要求：參考 MIL-STD-810D 水密、防震、防腐蝕、耐鹽霧
選配功能：光纖傳輸或無線傳輸
自動錄影裝備
中控警報系統
全方位旋轉平台
大尺寸監視器

National Chungshan Institute of Science and Technology. Hand-held IR Camera – CTV/TVS-32. Unknown Taiwan Aerospace Defense Technology Exhibition. Brochure 5/7.

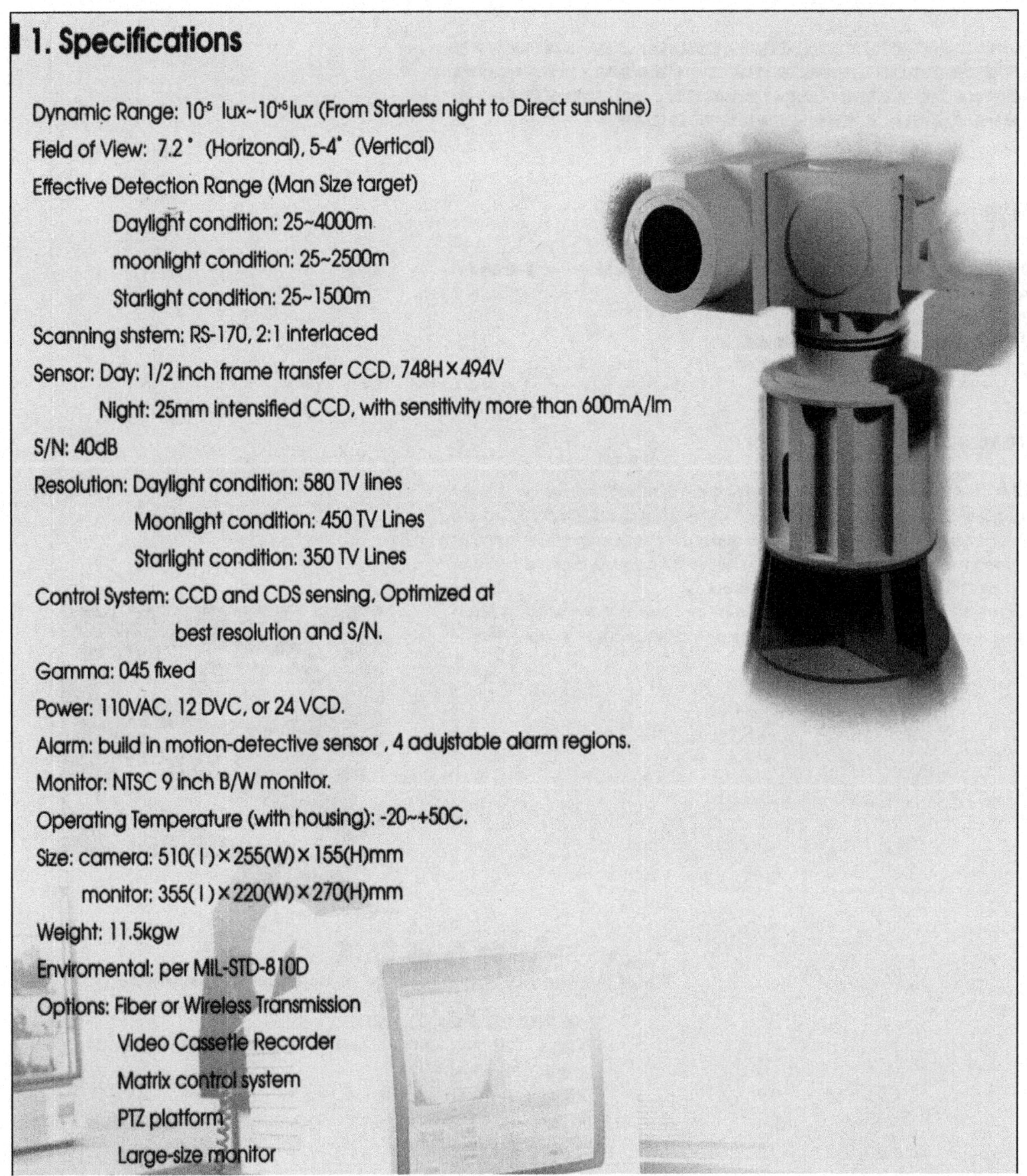

1. Specifications

Dynamic Range: 10^{-5} lux~10^{+5}lux (From Starless night to Direct sunshine)
Field of View: 7.2° (Horizonal), 5-4° (Vertical)
Effective Detection Range (Man Size target)
Daylight condition: 25~4000m
moonlight condition: 25~2500m
Starlight condition: 25~1500m
Scanning shstem: RS-170, 2:1 interlaced
Sensor: Day: 1/2 inch frame transfer CCD, 748H×494V
Night: 25mm intensified CCD, with sensitivity more than 600mA/lm
S/N: 40dB
Resolution: Daylight condition: 580 TV lines
Moonlight condition: 450 TV Lines
Starlight condition: 350 TV Lines
Control System: CCD and CDS sensing, Optimized at best resolution and S/N.
Gamma: 045 fixed
Power: 110VAC, 12 DVC, or 24 VCD.
Alarm: build in motion-detective sensor, 4 adujstable alarm regions.
Monitor: NTSC 9 inch B/W monitor.
Operating Temperature (with housing): -20~+50C.
Size: camera: 510(l)×255(W)×155(H)mm
monitor: 355(l)×220(W)×270(H)mm
Weight: 11.5kgw
Enviromental: per MIL-STD-810D
Options: Fiber or Wireless Transmission
Video Cassette Recorder
Matrix control system
PTZ platform
Large-size monitor

National Chungshan Institute of Science and Technology. Hand-held IR Camera – CTV/TVS-32. Unknown Taiwan Aerospace Defense Technology Exhibition. Brochure 6/7.

性能

CS/TVS-32為軍規設計氣密之日夜兩用監控攝影機，涵蓋之光度範圍自無星暗夜至日光直射正午。在系統監控螢幕內建自動警報區域，可對監控目標區異動加以偵測並發出警報。系統採完全被動設計，本身不發出任何光線，故不會被敵人偵測到。亦不受強光干擾而使性能降低或受損。

用途

替代現有人力，進行營區、要塞、岸站、油彈庫之自動監控。可移動安裝，無人為監控疏失。
日夜之一般偵搜巡防。
配合自動錄影系統，進行治安蒐證錄影。
山林、河川保育，山難、海難救助工作。

Description

CS/TVS-32, a state of the art day and night camera, is fully ruggalized, waterproof, and fully military designed. CS/TVS-32 can be operated from starless night to direct sunshine conditions. With its build-in motion detection system, CS/TVS-32 can be used to monitor unusual morement in the specific area.
CS/TVS-32 is a passive camera system, no active illuminating source is required, and it is impossible to be damage under high-light-level.

中山科學研究院
Chung-Shan Institute of Science and Technology

桃園龍潭郵政90008-1信箱
電話：886-2-2673-9638・886-3-471-3022・886-080-014723
傳真：886-3-471-4183
網址：http//www.csistdup.org.tw
P.O. Box 90008-1, Lungtan, Taoyuan, Taiwan, R.O.C.
TEL : 886-2-2673-9638・886-3-471-3022・886-080-014723
FAX : 886-3-471-4183
http//www.csistdup.org.tw

National Chungshan Institute of Science and Technology. Hand-held IR Camera – CTV/TVS-32. Unknown Taiwan Aerospace Defense Technology Exhibition. Brochure 7/7.

Chung Shyang II (Albatross) Unmanned Aerial Vehicles

Under the Aviation and Special Force Command (ASFC) there is one unmanned aerial vehicle (UAV) group: Tactical Reconnaissance Group x 3 squadrons: 1st Tactical Reconnaissance Squadron, 6th Army (North) at Taiping, Yangmei; 2nd Tactical Reconnaissance Squadron, 10th Army (Central); 3rd Tactical Reconnaissance Squadron, 8th Army (South). The group is outfitted in total with 32 Chung Shyang II (Albatross) UAVs.

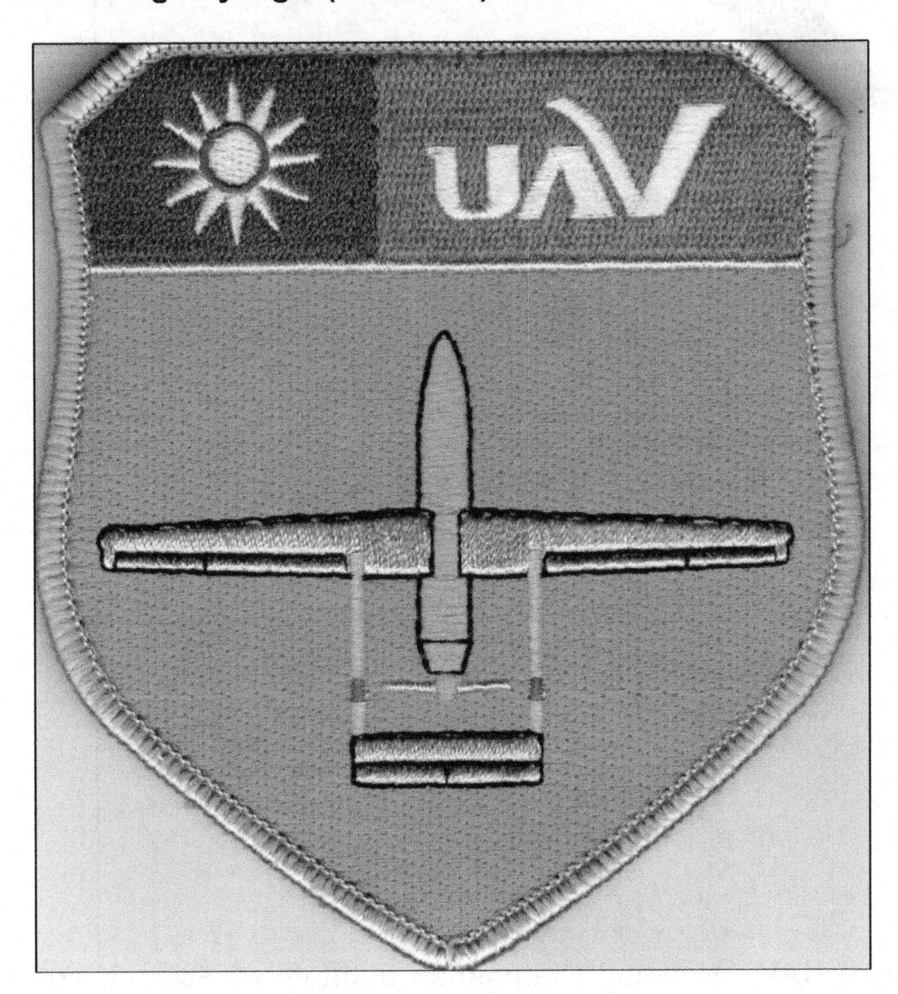

The Albatross Tactical UAS were designed in composite material structures and modular systems. The UAS were equipped with optical EO/IR payloads and characterized by long endurance flight, GPS navigation systems, autopilot ability, real time data and video transmission and communications relay.

In military applications, the UAS are used for day and night surveillance and reconnaissance, target acquisition and designation and battlefield damage assessment, etc. All the real time imagery and data can be transmitted to ground control station for the relay transmission system to assist to transmit information to the C4ISR, which plays preemptive and combined operations effects.

Additionally, the systems can be widely used for civil applications such as agriculture, fisheries, animal husbandry, disaster monitoring, environmental protection, traffic control, target searching, position recognition, coastal patrol, communications relay and hazardous terrain survey.

Albatross (Chung Shyang) UAV. National Chungshan Institute of Science and Technology (NCSIST)/Aeronautical Systems Research Division (ASRD). 2018 Defense Services Asia (Malaysia). Brochure 1/1.

Chung Shyang (Albatross) UAV. National Chungshan Institute of Science and Technology (NCSIST)/Aeronautical Systems Research Division (ASRD). 2011 Taipei Aerospace and Defense Technology Exhibition (TADTE). Brochure 1/2.

Chung Shyang (Albatross) UAV. National Chungshan Institute of Science and Technology (NCSIST)/Aeronautical Systems Research Division (ASRD). 2011 Taipei Aerospace and Defense Technology Exhibition (TADTE). Brochure 2/2.

Aeronautical Structure and Materials. National Chungshan Institute of Science and Technology (NCSIST). 2013 Taipei Aerospace and Defense Technology Exhibition (TADTE). Brochure 1/7.

結構與材料

複材載具結構

- ◆複材機體結構具備質輕、耐久等優點，可提昇載具性能、降低燃油使用量、增加航程及節省成本。
- ◆複材結構具高強度、高勁度、耐高溫、耐腐蝕、高導熱等特性，並可導入吸波、透波、導波及健康診測等設計。
- ◆可依客戶需求，進行客製化研製，包括產品規格、設計、分析、製造及測試驗證。
- ◆可應用於飛機、船艦、車體等各式載具結構及國防武器。

飛機起落架

- ◆具有飛機起降時地面方向控制、落地時吸收衝擊能量及減速作用等功能。
- ◆可搭配使用油氣式或葉片式減震器。
- ◆材料選用包括玻纖、碳纖複合材料、鋁合金與不鏽鋼。
- ◆使用電力線性收放致動器系統。
- ◆以分析及落錘試驗進行驗證。

Aeronautical Structure and Materials. National Chungshan Institute of Science and Technology (NCSIST). 2013 Taipei Aerospace and Defense Technology Exhibition (TADTE). Brochure 2/7.

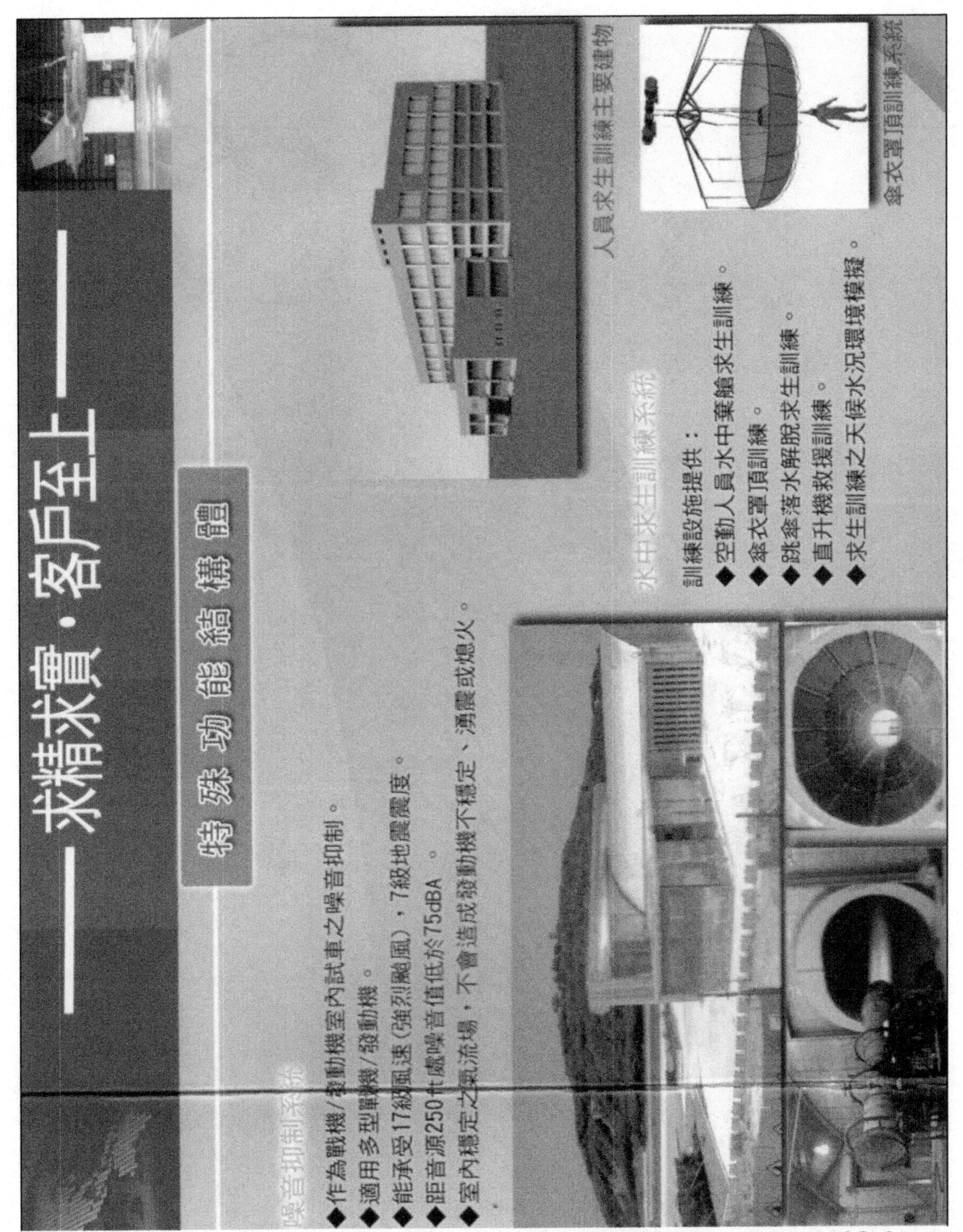

Aeronautical Structure and Materials. National Chungshan Institute of Science and Technology (NCSIST). 2013 Taipei Aerospace and Defense Technology Exhibition (TADTE). Brochure 3/7.

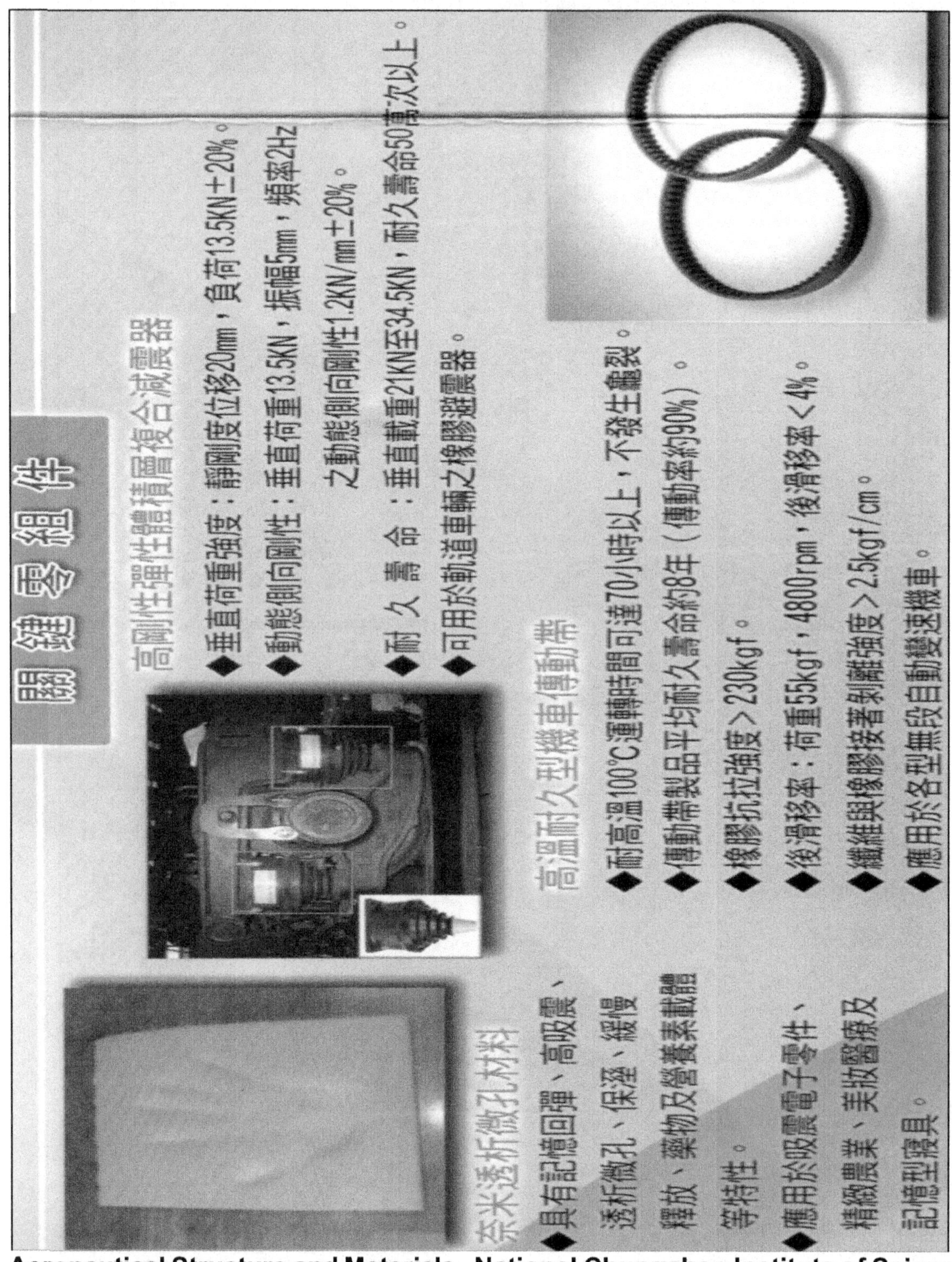

Aeronautical Structure and Materials. National Chungshan Institute of Science and Technology (NCSIST). 2013 Taipei Aerospace and Defense Technology Exhibition (TADTE). Brochure 4/7.

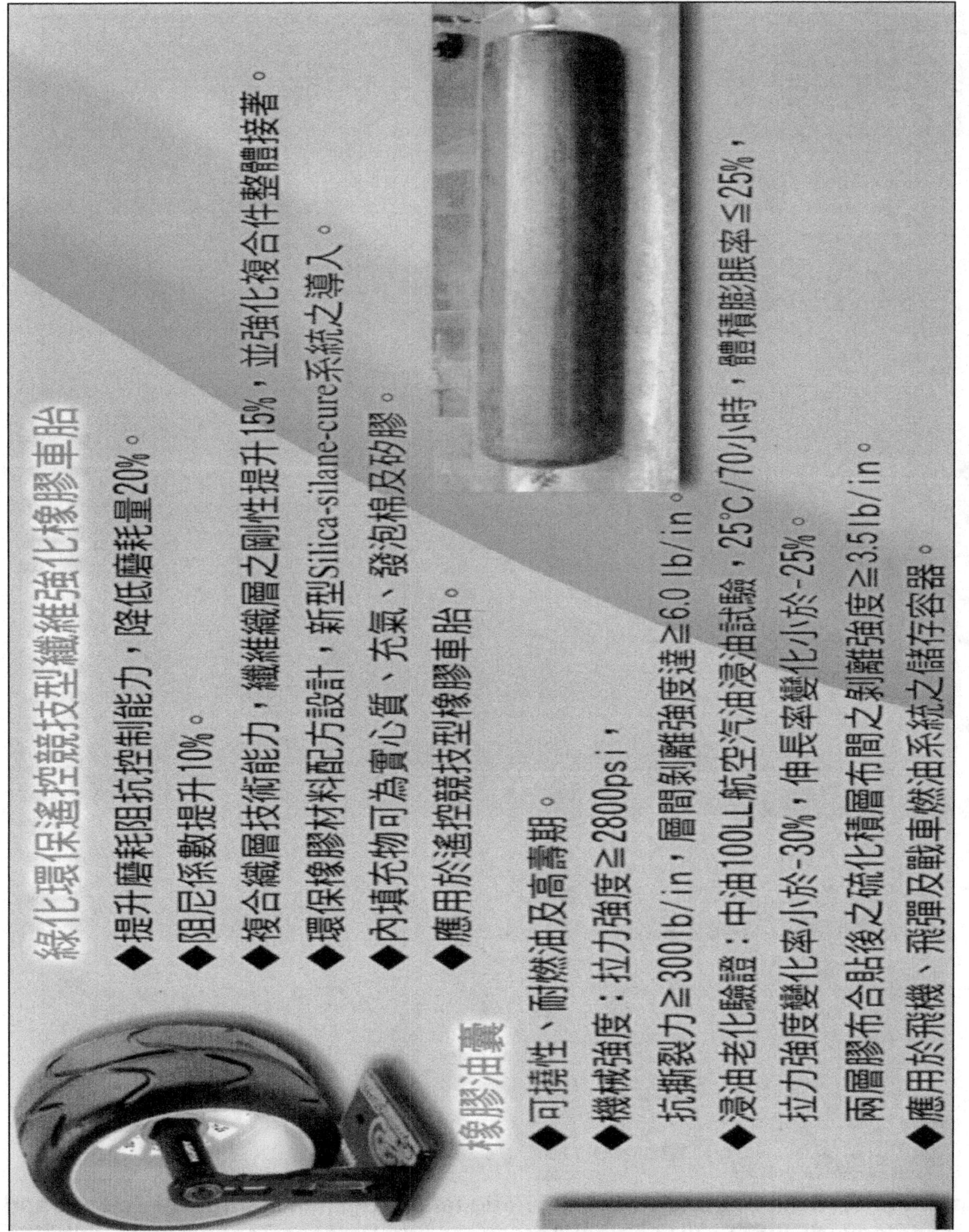

Aeronautical Structure and Materials. National Chungshan Institute of Science and Technology (NCSIST). 2013 Taipei Aerospace and Defense Technology Exhibition (TADTE). Brochure 5/7.

Aeronautical Structure and Materials. National Chungshan Institute of Science and Technology (NCSIST). 2013 Taipei Aerospace and Defense Technology Exhibition (TADTE). Brochure 6/7.

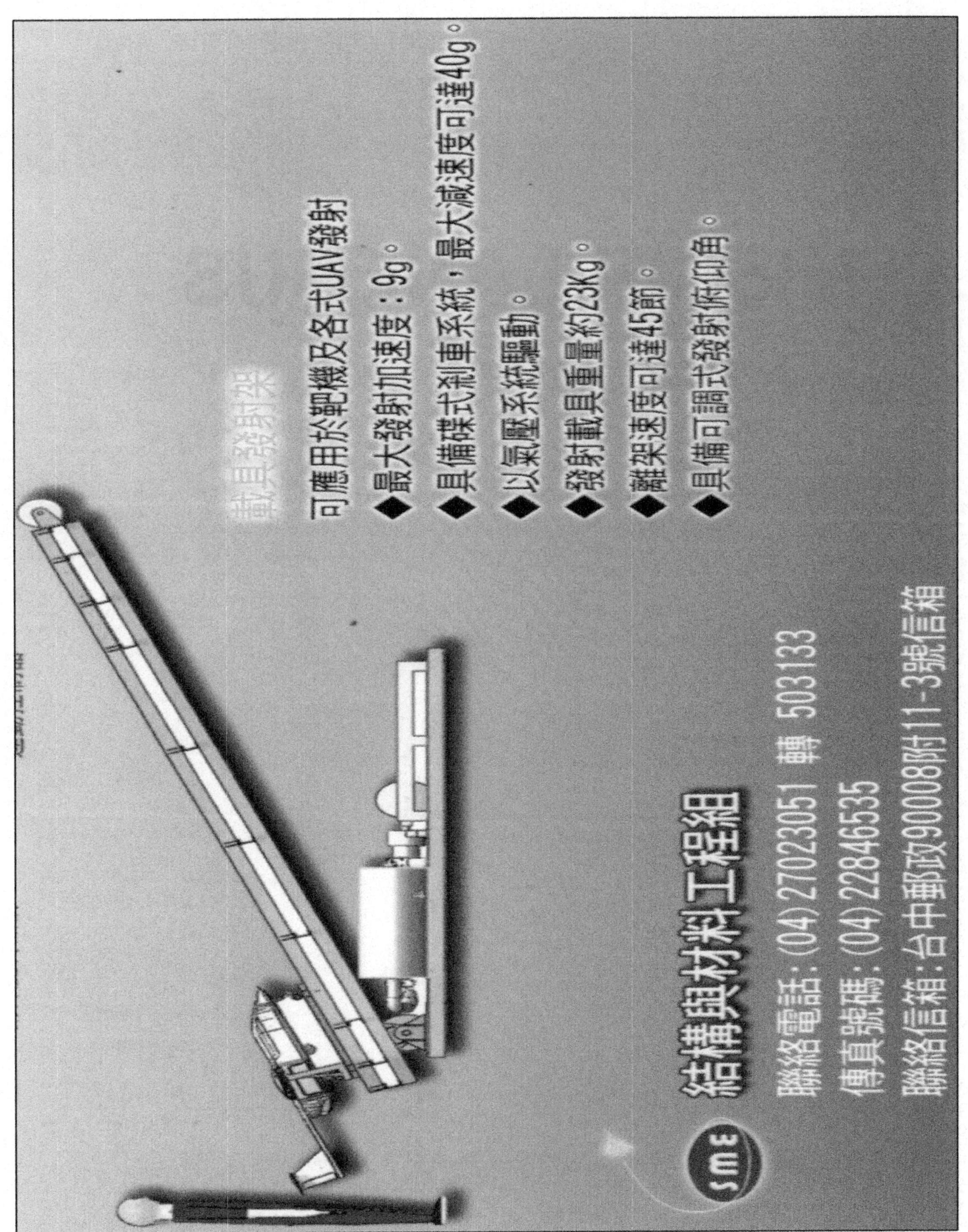

Aeronautical Structure and Materials. National Chungshan Institute of Science and Technology (NCSIST). 2013 Taipei Aerospace and Defense Technology Exhibition (TADTE). Brochure 7/7.

MISCELLANEOUS

BigxReality. Virtual Reality Military/Police Battle Field Training. 2019 Taipei Aerospace and Defense Technology Exhibition (TADTE). Brochure 1/2.

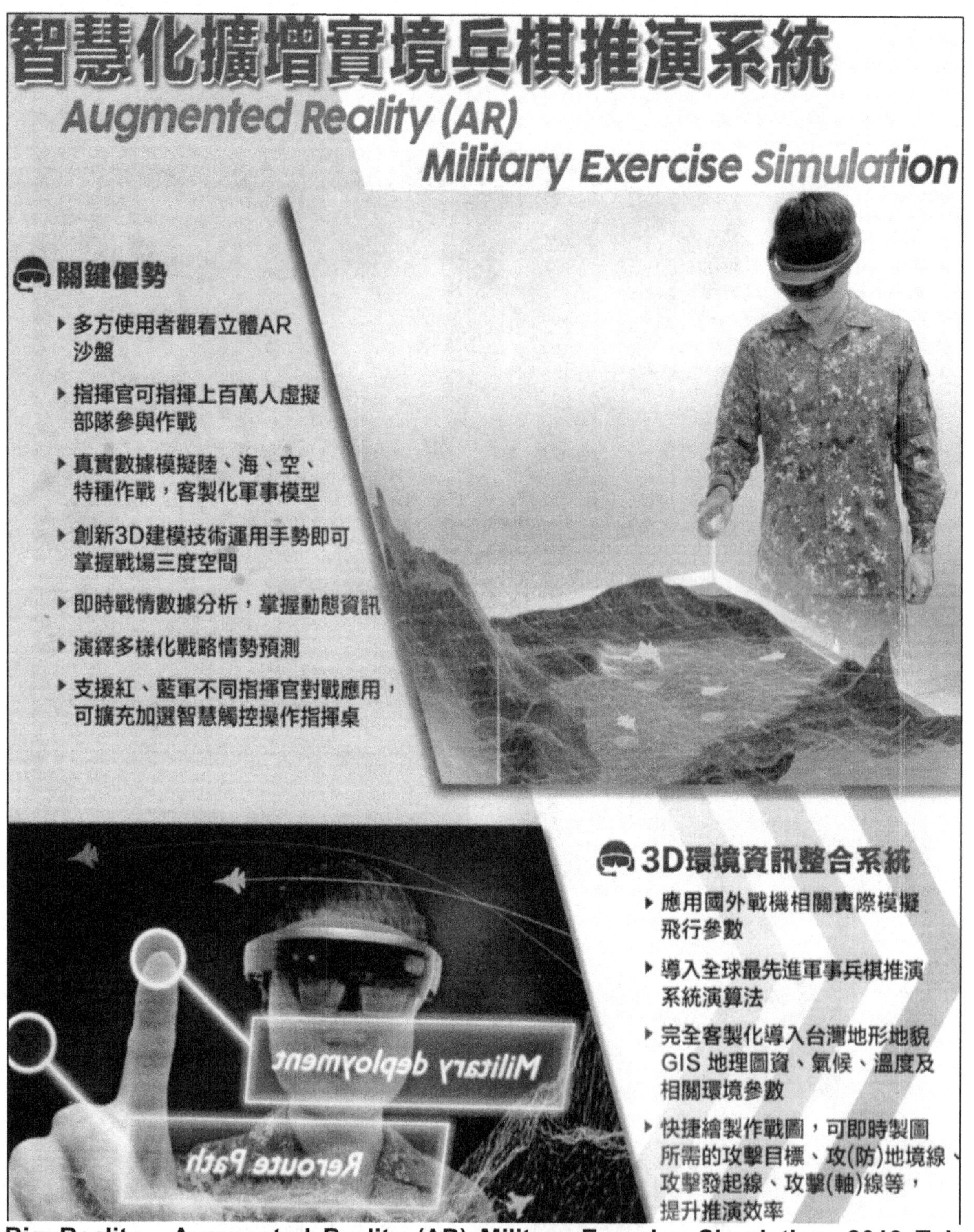

BigxReality. Augmented Reality (AR) Military Exercise Simulation. 2019 Taipei Aerospace and Defense Technology Exhibition (TADTE). Brochure 2/2.

Type \ Property	Weight (g)	V50 (Fps)	NIJ Level
HE-3	1250 ~ 1300	2000	III A
HE-5	1550 ~ 1600	1750	

Unit:mm

Shape \ Dimension	Length	Width	Thick	Curve Radius
Flat	50	50	3 ~ 10	
	50	25	3 ~ 10	
Curved	50	50	3 ~ 10	400
	50	25	3 ~ 10	400
Hexagonal	104		8 ~ 25	

Type	Size	Weight (g)	Ammunition	NIJ Level
HC-47	25X30cm	2400	7.62x39mm	
HC-N3	25X30cm	2850	7.62x51mm NATO	III
HC-N4	25X30cm	3500	7.62x51mm AP	IV

Hocheng Corporation. Advanced Ballistic Products. 2011 Taipei Aerospace and Defense Technology Exhibition (TADTE). 1/2.

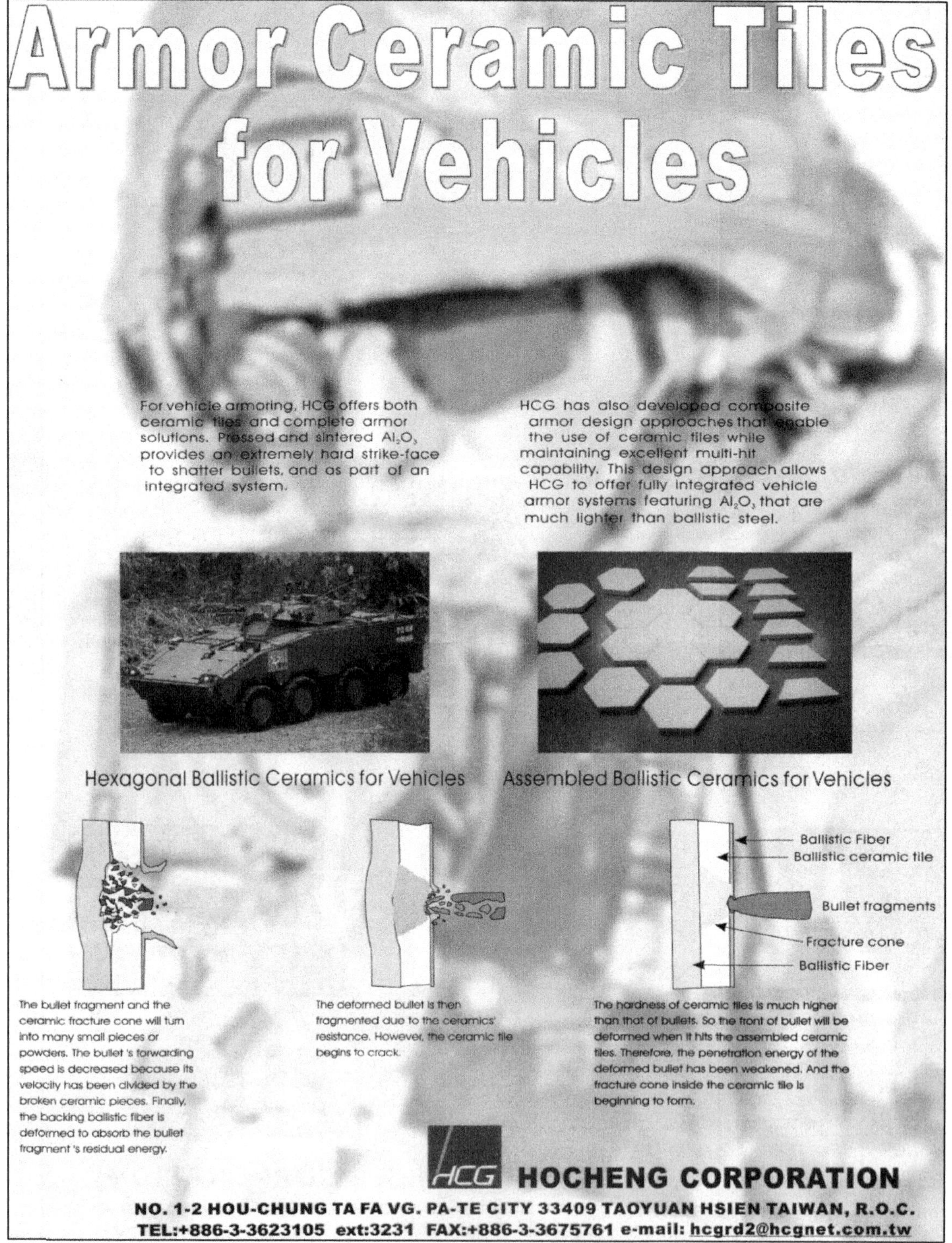

Hocheng Corporation. Armor Ceramic Tiles for Vehicles. 2011 Taipei Aerospace and Defense Technology Exhibition (TADTE). 2/2.

Asia Sun (Taiwan) Inc. AAA Armor Force. 2019 Taipei Aerospace and Defense Technology Exhibition (TADTE). Brochure 1/6.

Asia Sun (Taiwan) Inc. AAA Armor Force. 2019 Taipei Aerospace and Defense Technology Exhibition (TADTE). Brochure 2/6.

Asia Sun (Taiwan) Inc. AAA Armor Force. 2019 Taipei Aerospace and Defense Technology Exhibition (TADTE). Brochure 3/6.

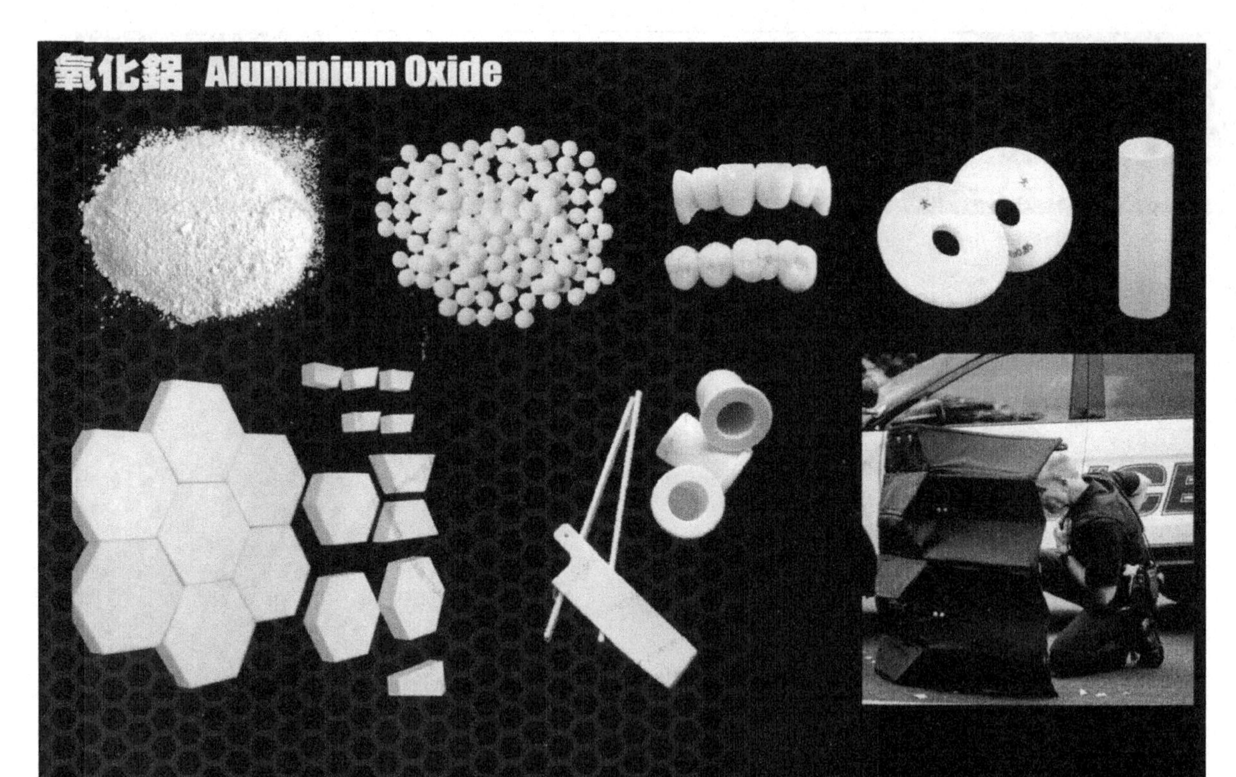

「戰力保存」是小國在強國侵略下，最後創造戰場致勝關鍵條件。「戰力保存」之真諦即為戰力發揮，無法達成戰力發揮之戰力保存，亦將喪失致勝因素。

「台灣亞旭股份有限公司」有鑑於此，特將國軍在確保個人及主戰裝備戰力保存之防彈防護之弱點加以研發精進。進而創造適合國軍官兵最佳戰力保存之防護裝備，並已具高規格、輕質化、智慧化之條件，確保戰鬥員在戰場上發揮戰力，戰勝敵人。

"Preserving combat power" is the element that tips the scales in a battlefield at the endgame for small countries fending off the invasion from stronger ones. The essence of "preservation of combat power" is to exert combat power. If the preservation fails to achieve the purpose of exertion of combat power, it might cost the country the chance to win.

In response to this, "ASIA SUN (TAIWAN) INC." has probed into the weakness of bulletproof performance in our army's personal armors and ameliorated it. We have achieved building protective gears that are high profile, light, and smart and most suitable for preserving our army's combat power to make sure our combatants are able to overcome their enemies in a battlefield with their combat power.

台灣亞旭科技 ASIA SUN (TAIWAN) INC.

聯絡人/Contact：劉欣庭 先生/Edward Liu
電話/Tel：+886-3-4837777 傳真/Fax：+886-3-4837791 郵件/Mail：lab2@asiasun.com.tw
地址：32853桃園市觀音區觀音工業區經建一路49號
Address：No.49, Jingjian 1st Rd., Guanyin Dist, Taoyuan City 32853 Taiwan

Asia Sun (Taiwan) Inc. AAA Armor Force. 2019 Taipei Aerospace and Defense Technology Exhibition (TADTE). Brochure 4/6.

Asia Sun (Taiwan) Inc. AAA Armor Force. 2019 Taipei Aerospace and Defense Technology Exhibition (TADTE). Brochure 5/6.

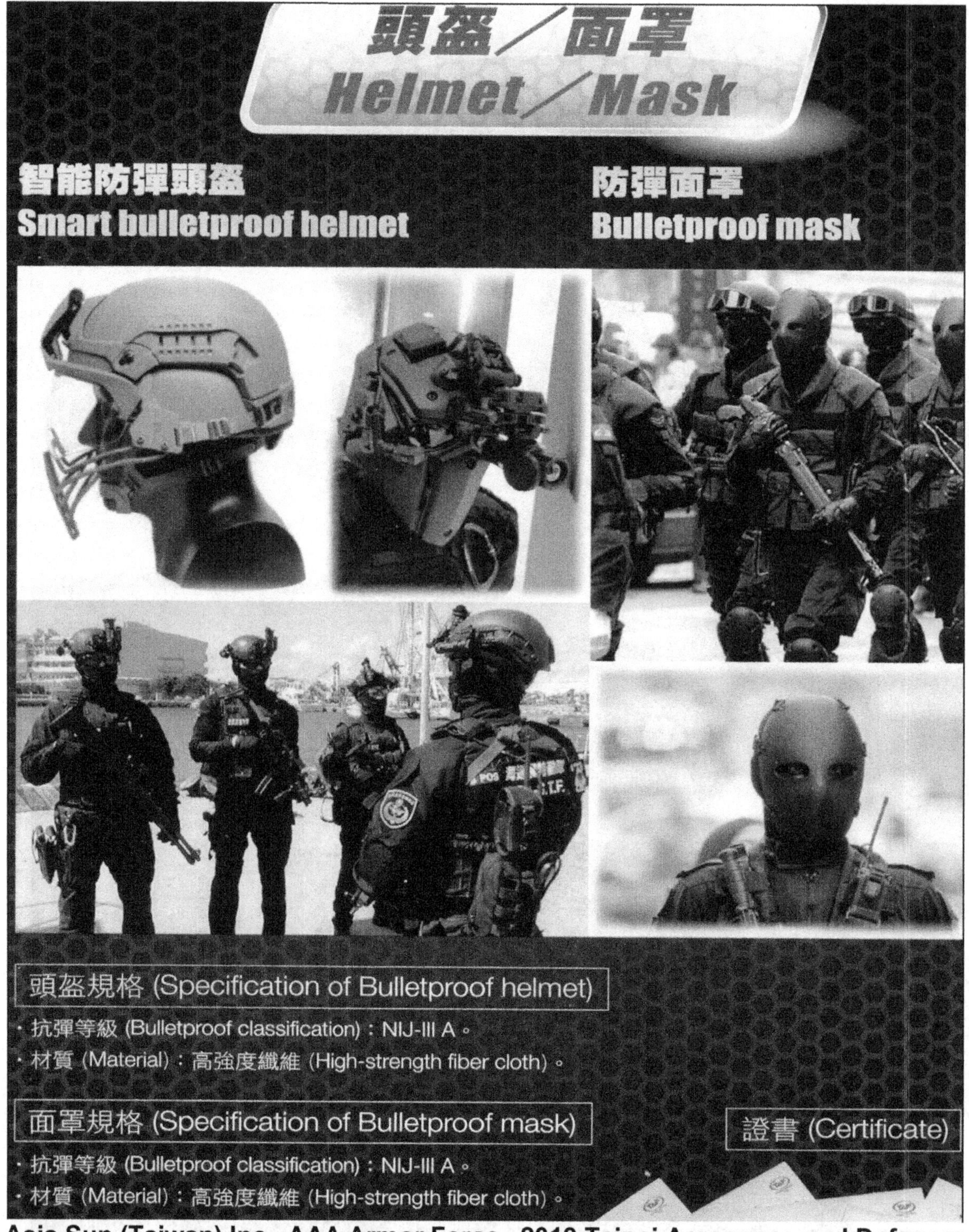

Asia Sun (Taiwan) Inc. AAA Armor Force. 2019 Taipei Aerospace and Defense Technology Exhibition (TADTE). Brochure 6/6.

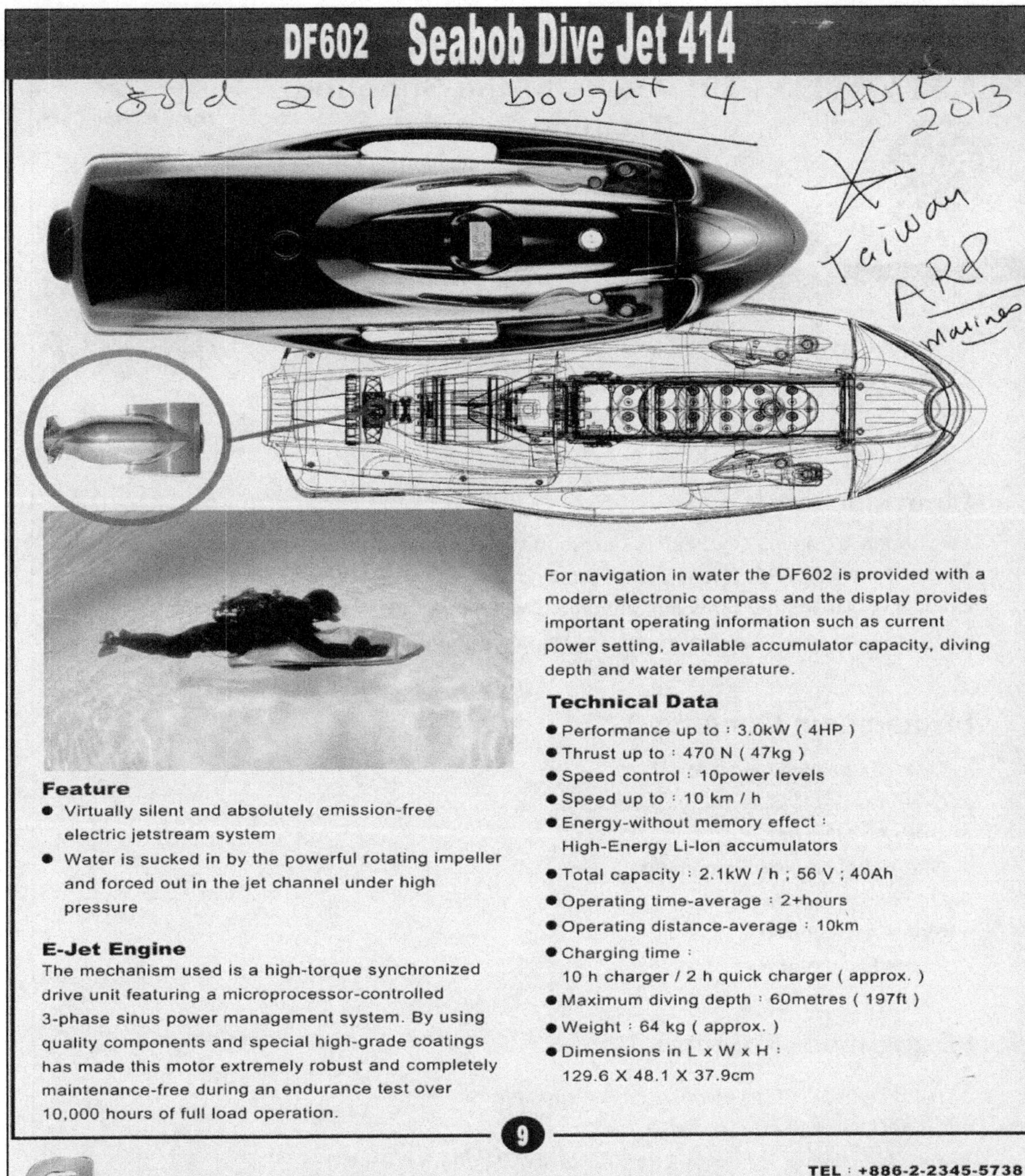

DAFAR INTERNATIONAL INC.

TEL : +886-2-2345-5738
FAX : +886-2-2345-5743
www.dafar.com.tw

Dafar International Inc. DF602 Seabob Dive Jet 414. 2013 Taiwan Aerospace Defense Technology Exhibition (TADTE). Representative said the DF602 had sold four (x4) to the Taiwan Marine Corps' Amphibious Reconnaissance Patrol in 2011. Note to reader: Dafar International appears to have been purchased by Cayago Luxury Seatoys, shuttered the Taiwan office, and moved to in China. Brochure 1/1.

【Introduction】

This shooting training simulator is designed to provide the military and police sectors with shooting training availabilities at their training stages to better training effect, beyond limitations of insufficient training equipment, equipment wear, weather (rain and typhoon), and environmental conditions (high temperature).

【Equipment Contents】

1. One T91 training simulator
2. One In-Cell 22" LCD Touch Screen
3. One rifle bracket
4. One signal processing computer
5. One carrying case
6. Signal cable group
7. Instruction manual

【Equipment Features】

1. High reliability, long lifespan, and easy maintenance
2. Portability due to compact size
3. Low cost suitable for use of basic military units and trainings
4. Improvement of shooting training efficiency and reduction of training cost
5. Utilization of the LCD Touch Screen and simulation system will ultimately increase the overall effectiveness of this military training platform.

IMON Interactive Simulator. Injoy Motion Corporation. T91 Assault Rifle Shooting Training Simulator. 2011 Taipei Aerospace and Defense Technology Exhibition (TADTE). Brochure 1/3.

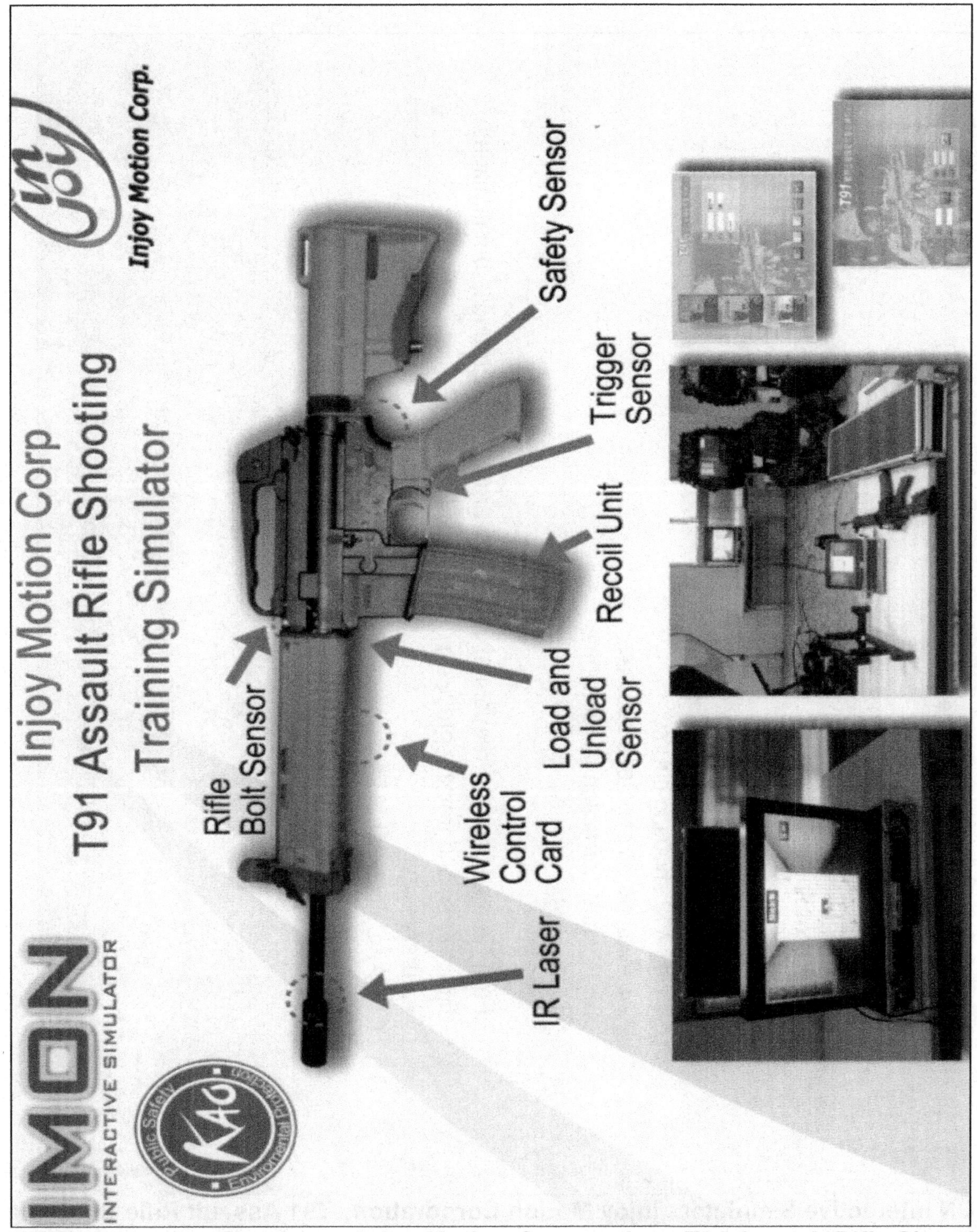

IMON Interactive Simulator. Injoy Motion Corporation. T91 Assault Rifle Shooting Training Simulator. 2011 Taipei Aerospace and Defense Technology Exhibition (TADTE). Brochure 2/3.

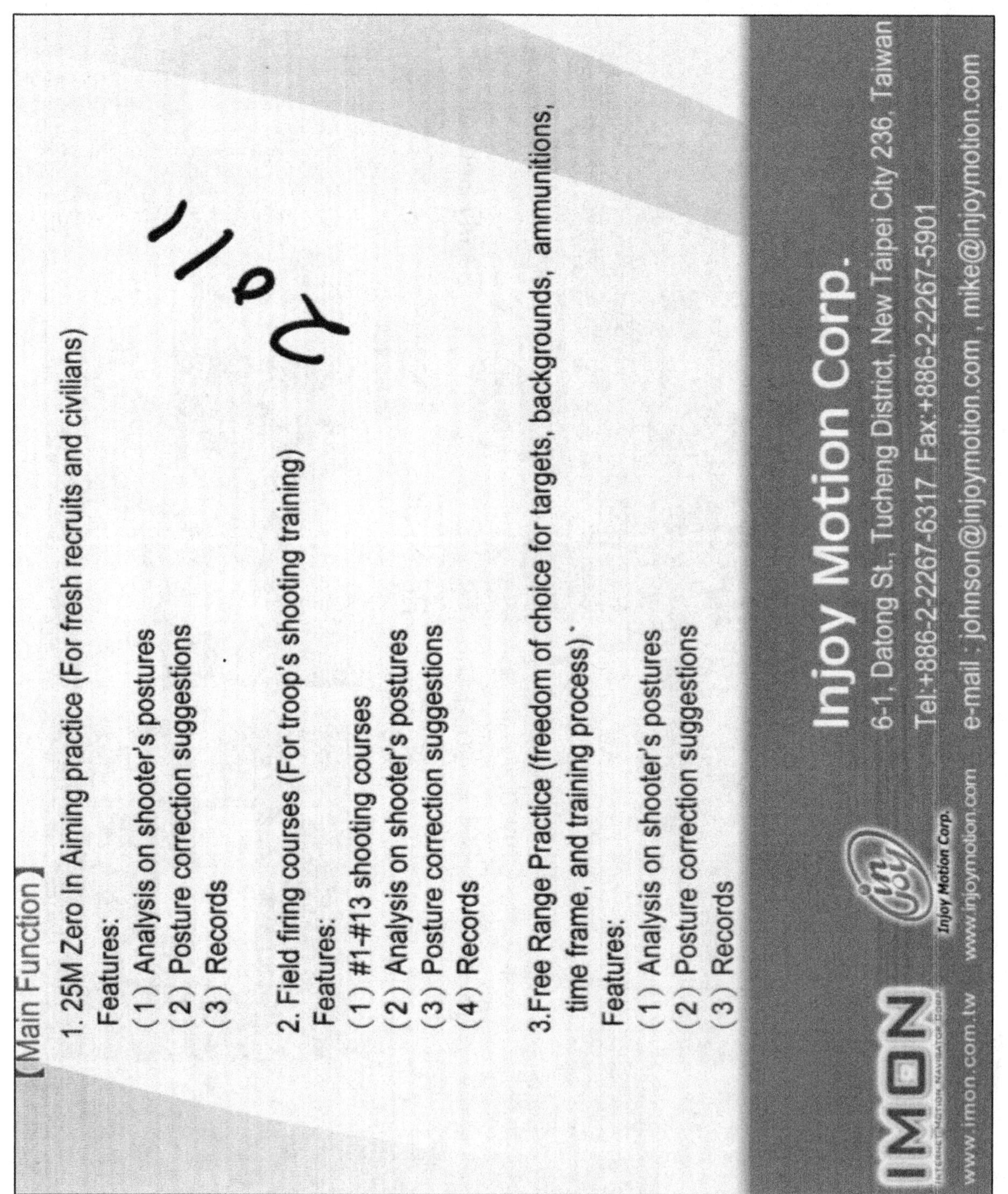

IMON Interactive Simulator. Injoy Motion Corporation. T91 Assault Rifle Shooting Training Simulator. 2011 Taipei Aerospace and Defense Technology Exhibition (TADTE). Brochure 3/3.

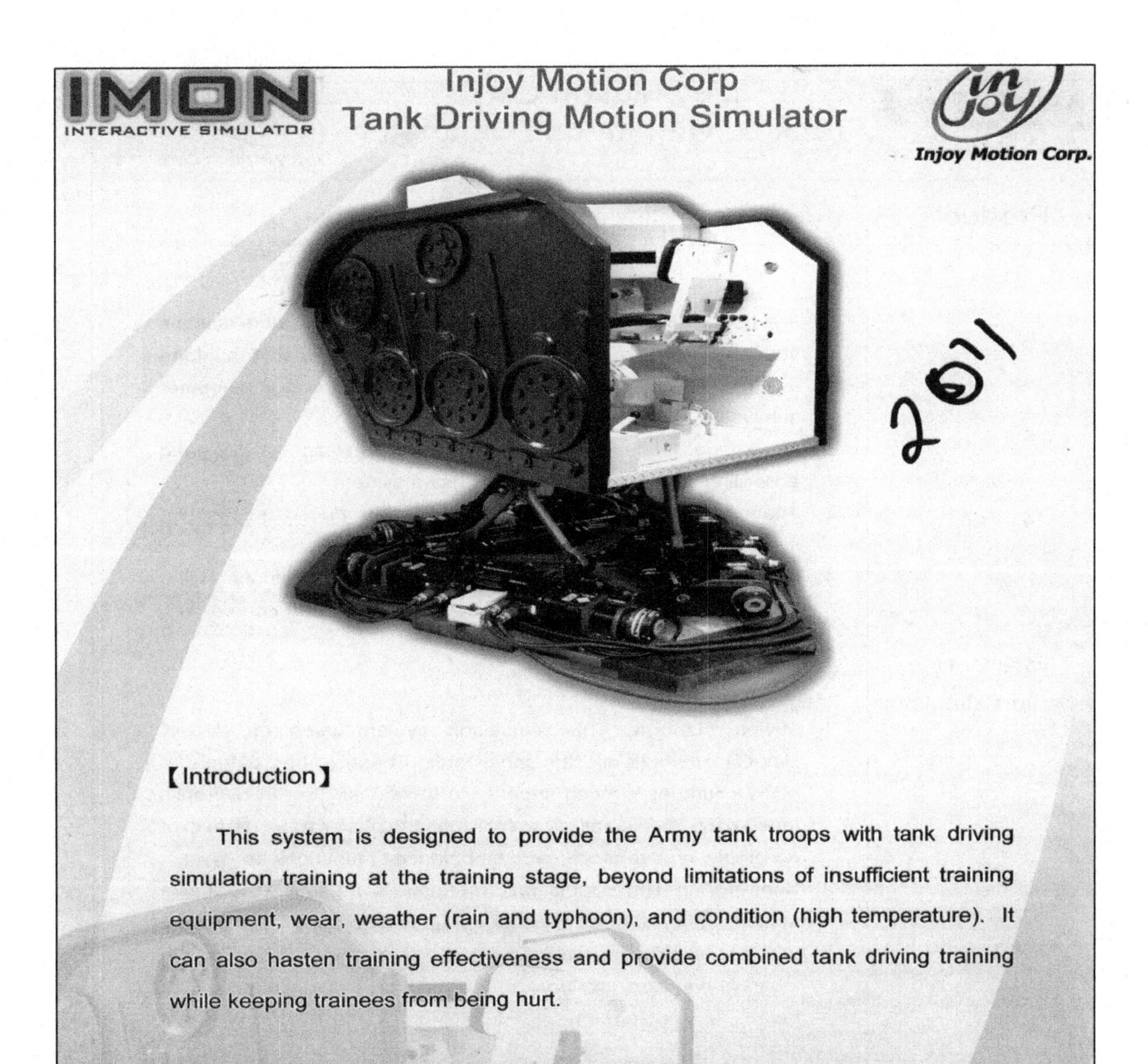

【Introduction】

This system is designed to provide the Army tank troops with tank driving simulation training at the training stage, beyond limitations of insufficient training equipment, wear, weather (rain and typhoon), and condition (high temperature). It can also hasten training effectiveness and provide combined tank driving training while keeping trainees from being hurt.

The tank driving simulator uses an IMON® Hex-Glider 6-axis electric-servo motion base and IMON® Adaptive Motion Cuing Control System to convert a static driving simulator into a medium-sized motion tank driving simulator. It provides customers with a cost-effective total solution of the motion simulation system. In addition, it can be fine tuned to customers' needs, and adaptive to dynamic characteristics of various vehicles such as tanks, wheeled vehicles, aircraft, and boats etc. for best simulation functions.

IMON Interactive Simulator. Injoy Motion Corporation. Tank Driving Motion Simulator. 2011 Taipei Aerospace and Defense Technology Exhibition (TADTE). Brochure 1/2.

Injoy Motion Corp
Tank Driving Motion Simulator

Product	Description
Tank Driving Motion Simulator	1. A simple version of the tank driving simulator, the system is a box structure of the tank driver's seat, and serves as a student stand including vision system, I/O interface system and a computer subsystem. 2. The images appear full colored. The vision system is designed according to the full size of a real vision system. 3. The mechanical components are exactly reproduced and function in the same as those in a real tank. 4. Training capabilities: The system can provide preliminary driving course, basic driving course, special terrain driving course, and combat driving course. Major features: 1. Motion platform: This simulation system uses an IMON® Hex-Glider 6-axis electric-servo motion base (it has patents of many countries.), much superior to traditional Stewart hydraulic /pneumatic or linear actuator motion platform in terms of assembly, maintenance, and the balance of functions and cost. 2. Control unit: This simulation system uses an IMON® Adaptive Motion Cuing Control System on an IMON® Hex-Glider, and provides customers with a cost-effective total solution of the motion simulation system. The simulation system can be fine tuned to customers' needs, and adaptive to dynamic characteristics of various vehicles for best simulation functions.

IMON
Injoy Motion Corp.
www.imon.com.tw www.injoymotion.com

Injoy Motion Corp.
6-1, Datong St., Tucheng District, New Taipei City 236, Taiwan
Tel:+886-2-2267-6317 Fax:+886-2-2267-5901
e-mail : johnson@injoymotion.com , mike@injoymotion.com

IMON Interactive Simulator. Injoy Motion Corporation. Tank Driving Motion Simulator. 2011 Taipei Aerospace and Defense Technology Exhibition (TADTE). Brochure 2/2.

■ ABOUT TW-DIDA

Taiwan Defense Industry Development Association (hereinafter referred to as TW-DIDA) is a non-profit social organization, established according to the laws of the ROC (Taiwan). TW-DIDA began in February, 2017, and was formally founded in July, 2017. With its mission to promote the development of Taiwan's defense industry. TW-DIDA is dedicated to facilitating interactions between the industry with government officials and academia, such as organizing international meetings, conferences, and seminars. TW-DIDA also conducts strategic researches for providing policy advices relating to the defense industry.

The missions of the TW-DIDA include the following：

1. Conducting defense industry development strategy research, and providing relevant policy recommendations.
2. Organizing national and international defense industry collaboration activities.
3. Hosting international defense industry related meetings, conferences, and seminars.
4. Facilitating communication with government agencies, and organizing relevant events.
5. Assisting exchange between the industry and academia by conducting professional trainings and educational activities.
6. Accepting engagements with government agencies and nongovernmental organizations to perform relevant researches and provide consulting and accreditation services.
7. Assisting the government in preparing proposals concerning the implementation of technology projects in three sectors: legal person, industry, and academia.

■ The reasons for establishing the Association

TW-DIDA includes three sections: Policy (Aviation), Industry (Shipbuilding), and Information Security. There are 41 domestic enterprises and legal persons (as attachment) registered as current members, including Aerospace Industrial Development Corporation, CSBC Corporation, Taiwan, and National Chung-Shan Institute of Science and Technology etc. All of the members participated the US-Taiwan Defense Industry Conference in 2015 and 2016.

1. To facilitate the overall economic development of the country, the government has drawn up the bluebook for the defense industry policy pivoting on the principal axes of shipbuilding, aerospace and information security.
2. In response to the government's policy for developing the economics, three key sectors of Taiwan's private industries for national defense, shipbuilding, aerospace and information security, voluntarily formed a team to go to the US to consecutively attend the US-Taiwan Defense Industry Conference in 2015 and 2016, and in the meeting, the team expressed their resolution and readiness to support our Government in developing the national defense industry to the US.
3. In the past two years, the representatives from our industry attending the US-Taiwan Defense Industry Conference stayed in close contact both during the meeting and after returning to Taiwan, and are willing to support the government in seizing the opportunity for upgrading our defense industry once again.
4. The attitude of the above-mentioned industry representatives has been channeled to the government, and the government responded positively in the hope of having three sectors of defense industry to continually to support the government and to make recommendations.
5. To sum up, the association was planned and organized under the Civil Associations Act to effectively rally the industry's strength and support the government in implementing the policy so that the influence of the recommendations from the industry will be strengthened and the resource of the association will utilized to bring the studies of the think tank into full play. It is hoped that the associatio.n will be the point of contact for the communication and coordination between the government and the industry for developing the defense industry policy and for carrying out the development of the defense industry.

ADDRESS：Rm. 412, No.566, Ln. 134, Longyuan Rd., Longtan Dist., Taoyuan City 32544, Taiwan (R.O.C.)
TEL：03-4116573 · FAX：03-4116583 · EMAIL：associate@twdida.org.tw · https://www.twdida.org.tw

Taiwan Defense Industry Development Association (TW-DIDA). 2018 Kaohsiung International Maritime and Defense Exhibition. Brochure 1/1.

2021 ROC (Taiwan) National Defense Report
Ministry of National Defense.

Selected "Defense Sketches" from 2021 Report

2021 ROC (Taiwan) National Defense Report, Editorial Committee. Ministry of National Defense: ISBN: 978-986-5446-94-9. (1/7).

Keeping Communications Operational at Remote Locations

For ages, communications are a decisive element to win a war, and can be best explained by the saying that "Operations are directed by orders while orders are sent by communications." Therefore, it is not only a mandate, but also an irreplaceable thinking and value to keep communications operational for all members of the Kaohsiung and Pingtung Operational Team, Southern Area Command and Control Squadron, Command and Control Protection Group, Information, Communications and Electronic Force Command.

The team is responsible for maintaining operations and providing technical support to military command and control systems under the jurisdiction of the 4th Theater Command, and shall, support the military information and communications system for disaster prevention and relief on demand, so as to keep it uninterrupted during operations.

Even though its members are shrinking and scattered in various posts from the mountaintop to the beachhead with their respective areas of responsibility, the team has been trying to maintain a good working atmosphere and environment for them. They have well executed all information and communications missions to support annual major exercises, and acquired good results in skill qualification exams and competitions. Consequently, the team was selected as one of the ROC Armed Forces exemplary units in 2020. Major Hong Rui-huang, Leader of the team, said, "Despite of having difficulty and hardship during the processes of the missions, the team has ultimately accomplished successes with all its members working as one." Therefore, the honor and glory belong to them.

Working in posts inconveniently located and supported, The team, based on its belief of "keeping communications working even over the clouds," remains undeterred to keep communications operational beyond all barriers. Moreover, Major Hong has not only promised himself to create a friendly working environment and improve his own expertise, but also led his fellow members to attend charitable events on holidays, like a service member of a new generation to dedicate to the service on duty and pay back to the society off duty.

2021 ROC (Taiwan) National Defense Report, Editorial Committee. Ministry of National Defense: ISBN: 978-986-5446-94-9. (2/7).

The Defense Sketch

The 542nd Armor Brigade, ROC Army

Grooming Rigorously as a Rock-solid Force

The 542nd Armor Brigade of the ROC Army is garrisoned in the north of Taiwan. To improve its independent operational capabilities and flexibility of applying tactics, it began to inaugurate "combined arms battalions" in 2020, and now has 3 combined arms battalions, 1 artillery battalion, and multiple companies at its direct command. According to Lieutenant Colonel Hsu Hong-lin, Commander of the 2nd Combined Arms Battalion, 542nd Armor Brigade, the concept of combined arms involves incorporation of multiple service branches, and has become a challenge to its the commander of the reorganized battalion, who needs to integrate anti-armor and mechanized infantry tactics altogether, and leads it for training at a joint base soon after its inauguration. After going through a year of training to establish the bond, his battalion has been verified of having substantial capability improvements by becoming a distinguished unit during a recent Lian-yong drill.

Captain Chen Kai, Commander of the 2nd Combined Arms Company, 2nd Combined Arms Battalion, 542nd Armor Brigade, pointed out that his fellow soldiers are all enlisted voluntarily, and that it is his responsibility to help them acquire their military specialties, realize purposes, and set goals for their career track. He has set up a range of evaluations and physical competitions to build a sense of honor and boost morale for his fellows, who would excel themselves in a joint base training. Then the bond of his company will be much stronger. Captain Chen is very proud of establishing this virtuous circle for his troop.

Sergeant supervisor Liao Jun-wu of the 2nd Combined Arms Company mentioned that enhancing physical fitness is necessary to build the strength of a unit. Currently, their fitness training regime is scientifically designed with meal plans to enhance the fitness of their fellow soldiers in a most efficient way, turning them into both "sportsmen and warriors." He will lead non-commissioned officers (NCOs) to assist his commanding officer in a professional manner so as to maximize the potential for his unit.

2021 ROC (Taiwan) National Defense Report, Editorial Committee. Ministry of National Defense: ISBN: 978-986-5446-94-9. (3/7).

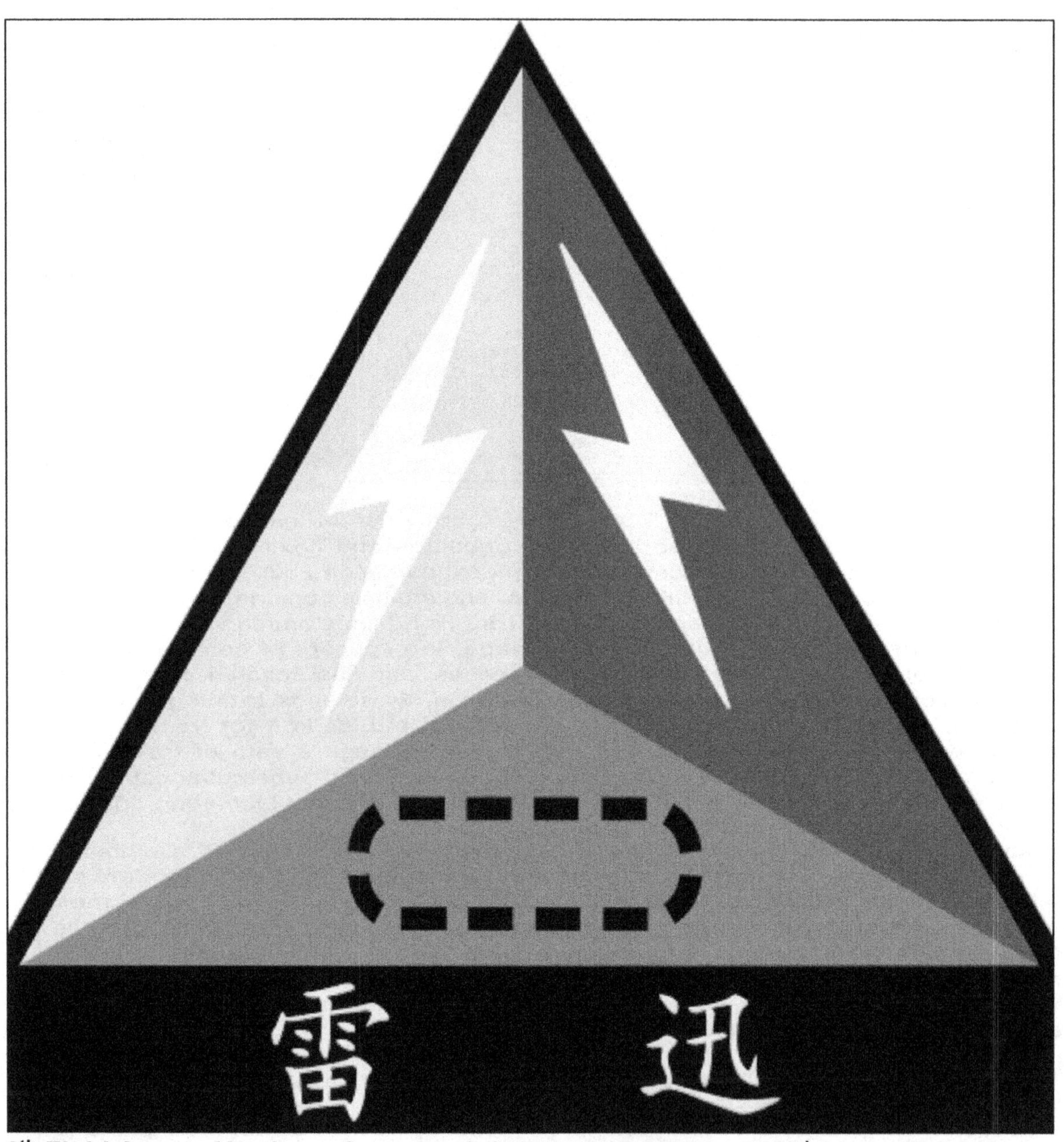

6th Field Army - Northern Command (陸軍第六軍團指揮部). 542nd Armor Brigade (陸軍裝甲第五四二旅). Hukou Township, Hsinchu County; 湖口鄉, 新竹縣.

The Defense Sketch

The 333rd Mechanized Infantry Brigade, ROC Army

Being Independent and Resilient; Exerting Strikes Swiftly

To deal with current enemy threats and fulfill the demands for future homeland defense operations, the ROC Army has inaugurated a series of combined arms battalions since 2020. The goal of these ground striking units is to bring "coordination of branches" into play and strengthen their "independent operational capabilities" so as to swiftly react to contingencies and conduct strike operations at all dimensions.

The 2nd combined arms battalion, 333rd Mechanized Infantry Brigade, ROC Army, was established on 1st June 2020. Its commander, Lieutenant Colonel Zhao Zhi-wei expressed that his battalion was assigned to an anti-landing mission during a joint landing operational drill of the 36th Han Kuang exercise held in Jialu'tang of Pingtung, and they had to coordinate with aviation units to verify their joint command and control capabilities and the effectiveness of marshalling manpower and firepower. He mentioned that the capabilities and battlefield management skills of his battalion were verified through settings of force-on-force, full gears, actual terrains, and realistic operations in the 4th season force-on-force combat readiness drill of the same year. In 2021, they will receive a joint operational training in a joint training base, and then go through a test.

Facing missions and tasks that come one after another, Lt. Col. Zhao said "To bring safety to the people is the core value of the ROC Armed Forces." The morale of his fellow soldiers is high, and they all recognized that a rock-solid defense to protect the people can only be built by their full devotion to combat readiness and training.

The most distinctive feature of a combined arms battalion is that it digitizes the battlefield. It contains a mix of infantry, armor, and artillery units; liaison officers from navy, air force, and army aviation units; specialists, such as UAV and Stinger missile operators and snipers; and Clouded Leopard eight-wheeled armored vehicles for increasing its mobility.

Company Commander Captain You Shi-hong and Private Chen Jia-yan said, "Fellow soldiers of the 1st Infantry Company, 2nd Combined Arms Battalion are a band of brothers and sisters that have the strongest combat power." They mentioned that the perseverance, endurance, and physical fitness of service members can be strengthened by going through realistic battlefield scenarios and rigorous military training and patriotic education. They may consequently forge a robust combat power to exert their maximum potential to safeguard the nation.

2021 ROC (Taiwan) National Defense Report, Editorial Committee. Ministry of National Defense: ISBN: 978-986-5446-94-9. (4/7).

8th Field Army - Southern Command (陸軍第八軍團指揮部). 333 Mechanized Infantry Brigade (陸軍機械化步兵第三三三旅). Wanluan, Wanjin Camp, Pingtung; 屏東萬巒，萬金營區.

The Defense Sketch

The 209th Arsenal, Production and Manufacturing Center, Armament Bureau

Full Dedication to R&D with Major Progress in Indigenous Vehicle-making

The 209th Arsenal of the Armament Bureau is one of the armament production centers with distinctive contribution to our self-reliant defense, such as the multi-year development of an 8-wheeled armored vehicle with various derivatives. The development of a weapon shall aim to meet the operational requirements of its operating services or branches. Taking the CM34 IFV with 30mm chain gun in production as an example, it not only has a fire control system that targets accurately and initiates strikes effectively, but also a body with better performance. The 209th Arsenal and its critical technologies in domestic vehicle-making are credited for realizing major progress in our defense capabilities.

Given Taiwan's terrain, the focus of homeland defense will be urban warfare, said Colonel Su Ren-bao, program head for the 8-wheeled armored vehicle. The program's primary objective is to develop a vehicle that fits in such an operational environment. Without foreign technical inputs, the program teamed up with Taiwan's industrial and research sectors, such as Industrial Technology Research Institute (ITRI) and China Steel Corporation (CSC), in a concerted defense R&D effort and successfully developed the chassis for the 8-wheeled armored vehicle. It paved the way for the buoyant debut of the CM34.

As Master Sergeant Huang Wei-zhe and Specialist Shi Jian-zhi in charge of technological R&D for the program pointed out, R&D works require immersion and passion, and their team had a very clear goal to overcome all difficulties and reach milestones as scheduled. They successfully completed the performance test for the vehicle on the dried riverbed of the Zhuoshui River in 2017 and then the design of a six-wheel steering system in the following year. When the powerful façade of the armored vehicle emerged from the gate of its hangar and was unveiled, people at the scene were all carried away. This successful launch not only represents a great achievement of their career, but also great reward for their strenuous dedication.

2021 ROC (Taiwan) National Defense Report, Editorial Committee. Ministry of National Defense: ISBN: 978-986-5446-94-9. (5/7).

The Defense Sketch

National Chung-Shan Institute of Science and Technology

Innovative Museum with Enthusiastic Volunteers to Deepen the Thinking of Self-reliant Defense

The Science and Technology Exhibition Center is the most visitor-friendly locale in the compound of the National Chung-Shan Institute of Science and Technology (NCSIST). The Exhibition Center offers multiple simulated interactive experiences, such as cockpit firing runs, touchpad-controlled interaction with missiles, and an operational command and control center. The visitors can quickly immerse themselves in this fascinating domain of defense technologies. Since its opening in 2017, it has received more than 40,000 visitors.

In order to expand public participation in defense-related activities, the NCSIST assembled the first volunteer group the Chung Shan Volunteer Team by hiring its retirees and those who have an interest in promoting R&D for self-reliant defense technologies to serve as tour guides.

Chang Chung-shing, founding leader of the Team after serving 25 years in the NCSIST, said that there were 30 members in the beginning, and now the Team has 50 members, who are all enthusiastic to lead visitors to know the chronicles of the R&D for our defense technologies by sharing their experiences and anecdotes. He was glad to make our strength in indigenous defense technologies known to the public after retirement.

His patented design of thrust-stopping charge for missile engines is also displayed in the Exhibition Center. Mr. Chang still feels the excitement and pride of this achievement whenever he introduces it to the visitors.

As Jian Ding-hua, Director of the Civilian-Military Industrial Development Center of the NCSIST, pointed out, the NCSIST has developed technologies in 10 areas from the bottom of the sea to outer space, including aerospace, mechanical system, information and communications, chemical engineering, material and optoelectronics, electronic system, system engineering, system manufacturing, system maintenance, and information management. The Center showcases our technological prowess through exhibits ranging from original R&D design to final production, from weapon systems to critical components, from military to civilian applications, and space technologies. It not only helps to recruit talented people in defense-related technologies, but also demonstrates our R&D capacities and achievements towards a self-reliant defense, establish an awareness of national security for the people, and strengthen the effectiveness of all-out defense education.

2021 ROC (Taiwan) National Defense Report, Editorial Committee. Ministry of National Defense: ISBN: 978-986-5446-94-9. (6/7).

The Defense Sketch

Quick Reaction Company, Military Police Command

Forging Honor with Perseverance; Train Hard to Boost Strength with Hard Training

The Military Police Motorcycle motorcade would demonstrate a remarkable march-past with formation changes during the on national day parades, and the people are always impressed by its low-speed handling skills, imposing posture, and formidable looking. The motorcade was formed by the Quick Reaction Company of the 202nd Area Military Police Command. The Company is not only responsible for the security of the installations of the Military Police Command, but also can augments the defense capacity for our national capital by its maneuverability with heavy motorcycles.

The Quick Reaction Company is formerly known as the prestigious Military Police Motorcycle Company, and in peacetime it provides motorcade escorts for important national celebrations, delegations of foreign dignitaries or diplomatic missions, and Spring and Autumn Memorial Services in peace time, while in wartime it will shall be assigned to scout for reconnoitre hostilities the enemy and assist in support anti-airborne and urban warfare operations. Facing threats of PLA's unrestricted warfare and enhanced long-range power projection, the Company has realigned its strength to dramatically increase its striking power. It can swiftly change the mix of motorcycles and surveillance vehicles to become a critical force at its garrisoning area.

To firmly handle a 1,300cc heavy bike of net weight 370kg, it requires a long-term training on the part of the rider. To achieve a unity between the bike and its rider, however, it depends not on neither one's own rank and, gender or physical strength, but an aggregation of experiences.

Private He Shi-wen was used to be a weightlifter, and her athletic background was helpful helps to control the bike. However, to ride in a formation demands much more coordination with her fellow riders. It is understandable that a perfect formation-riding drill on the national day celebration is done by countless practices and hard work of the riders.

Staff Sergeant Gao Jia-ping has taken part in for 4 times in the national day parades. He is not only a rider, but also a vehicle mechanic. In the course of his military career, SSG Gao continues to excel himself by acquiring taekwondo and judo black belts qualifications, and is working on getting vehicle maintenance certificates to develop himself as a cadre member with "multiple skills and diversified talents."

Major Yan Pei-wen noted, commander of the Quick Reaction Company, noted that his Company is an elite unit that takes honor and rigorous training seriously, and its members have developed a close bond in terms of its unique missions. Going through a progressive and rigid training regime, the Company will do its utmost to accomplish all combat and escort missions to demonstrate their esprit de corps and robust combat power.

2021 ROC (Taiwan) National Defense Report, Editorial Committee. Ministry of National Defense: ISBN: 978-986-5446-94-9. (7/7).

MILITARY BROCHURES FROM MEDIA TRIPS

564 Armored Brigade. Unknown Media Visit. Background: Based in Alian, Kaohsiung. Directly subordinate units a Tank Battalion x2 (CM11 Brave Tiger MBT/hybrid M60 chassis and M48 turret), Mechanized Infantry Battalion x2 (CM21/M113 variants and CM32 Clouded Leopard variants), Artillery Battalion x1 (155mm M109 Paladin Self-Propelled Howitzer). Brochure 1/4.

564 Armored Brigade. Unknown Media Visit. Brochure 2/4.

本旅戮力戰備整備工作，在全體官兵勤訓精練及軍團指導下，圓滿達成年度漢光演習、聯勇操演等各項演訓任務，為一戰力堅強之鋼鐵勁旅。平時於國南地區保持一個戰備部隊，隨時可遂行戰備任務，防止敵對我突襲，確保地區安全。

裝備整備

本旅以CM11戰車為主力裝備，平時均保持全妥善。CM11戰車為國軍目前主力戰車之一，具有高度機動力、強大火力、裝甲防護力、靈活通信及彈性編組等特性。

中型戰術輪車具適應多種道路及各種地形之越野能力，可有效遂行戰場人員及裝備運輸任務。

M48A5戰車推進橋，具機動性大及裝甲防護力，能於2～5分鐘內完成架橋與收橋動作，完全採用機械動力，無須人力架設。

標槍飛彈具射後不理、攻頂及直攻能力，不受日、夜間及天候影響，可精準摧毀敵之戰甲車，支援戰鬥部隊，開創勝利契機。

渦輪發煙車為美軍現役裝備，依作戰需要，施放抗紅外線石墨粉煙霧，可快速遮蔽與掩蔽我軍行動，且能有效阻抗精準武器打擊。

核生化偵檢車，具備偵測軍事戰劑與工業用化毒功能，可於災害現場，立即執行毒化物檢驗與判讀。

564 Armored Brigade. Unknown Media Visit. Brochure 3/4.

564 Armored Brigade. Unknown Media Visit. Brochure 4/4.

564 Armor Brigade; 8th Army Corps (第八軍團指揮部): Southern Taiwan.

Patch: Taiwan is currently upgrading their fleet of M60A3 main battle tank fire control system. National Chungshan Institute of Science and Technology (NCSIST) won the US$15.7 million contract in early 2022. FYI: the military decommissioned the ageing M41D Bulldog tank in early 2022.

Armor Training Command and Armor School. Hukou Township, Hsinchu. 2003 Media Visit. Brochure 1/4.

Armor Training Command and Armor School. Hukou Township, Hsinchu. 2003 Media Visit. Brochure 2/4.

Armor Training Command and Armor School. Hukou Township, Hsinchu. 2003 Media Visit. Brochure 3/4.

Armor Training Command and Armor School. Hukou Township, Hsinchu. 2003 Media Visit. Brochure 4/4.

Old Armor Patch.

22nd Han Kuang Exercise. July 20, 2006. Media Trip. Brochure 1/10.

演練裝備數量統計

區分	單位	射擊武器	數量 小計	數量 合計
聯合防空	空軍作戰指揮部	經國號戰機	4	15
		F-5戰機	4	
	海軍艦隊指揮部	濟陽級艦	1	
	空軍防空砲兵指揮部	愛國者飛彈	2	
		鷹式飛彈	4	
聯合截擊	海軍艦隊指揮部	艦射雄二飛彈	1	25
		車載雄二飛彈	1	
	空軍作戰指揮部	F-16戰機	9	
	海軍艦隊指揮部	濟陽級艦	1	
	陸軍第六軍團化兵群	渦輪發煙機	12	
	中山科學研究院	UAV無人載具	1	
聯合泊地攻擊	陸軍蘭陽地區指揮部	105榴砲	16	93
	陸軍二一砲指部、蘭指部	155榴砲	20	
	陸軍二六九旅	M109A2自走砲	8	
	陸軍五四二旅	M109A5自走砲	16	
	陸軍二一砲指部	M110A2自走砲	12	
	陸軍航空特戰指揮部	OH-58D戰搜直昇機	2	
		AH-1W攻擊直昇機	6	
	中山科學研究院	雷霆2000火箭	2	
	海軍艦隊指揮部	海鷗級艇	4	
	空軍作戰指揮部	F-16戰機	6	
	中山科學研究院	UAV無人載具	1	
灘岸戰鬥	空軍防空砲兵指揮部	[illegible]	2	127
		天劍一型飛彈車	2	
	海軍陸戰隊指揮部	樹欉飛彈車	2	
		AAVP7 AAVC7兩棲突擊車	19	
	陸軍航空特戰指揮部	OH-58D戰搜直昇機	2	
		AH-1W攻擊直昇機	8	
	陸軍二一砲指部	復仇者飛彈車	2	
		M110A2自走砲	12	
	陸軍二六九旅	M109A2自走砲	8	
	陸軍五四二旅	M109A5自走砲	16	
	陸軍二一砲、蘭指部	155榴砲	20	
	陸軍蘭陽地區指揮部	105榴砲	16	
	陸軍蘭陽地區指揮部 一七六旅、一七八旅	120迫砲	18	

22nd Han Kuang Exercise. July 20, 2006. Media Trip. Brochure 2/10.

演練裝備數量統計

區分	單位	射擊武器	數量 小計	數量 合計
灘岸戰鬥	陸軍蘭陽地區指揮部一七八旅	81迫砲	12	149
	陸軍六軍團反裝甲營	M113拖式飛彈車	8	
	陸軍蘭陽地區指揮部五四二旅	M88救濟車	2	
		CM11戰車	36	
		CM26甲車	1	
	陸軍蘭陽地區指揮部、陸軍六軍團化學兵群	連發煙火槍	24	
	陸軍蘭陽地區指揮部二六九旅	五〇機槍	25	
	後備九〇一旅	班用機槍	23	
		排用機槍	18	
城鄉守備	陸軍航空特戰指揮部	CH-47SD運輸直昇機	3	88
		UH-1H通用直昇機	12	
	空軍作戰指揮部	F-16戰機	4	
		C-130運輸機	8	
	陸軍蘭陽地區指揮部	CM21、CM22、CM23、M113甲車	16	
		悍馬車	21	
	蘭指部、宜蘭後備旅	班、排用機槍	24	
海上警戒兵力	海軍艦隊指揮部	濟陽級艦	2	20
		成功級艦	1	
		康定級艦	1	
		AP艦	3	
		錦江級艦	3	
		海鷗級艦	6	
	海巡署	PLC巡防艦	4	
靶勤作業兵力	海軍艦隊指揮部	AP艦	1	4
		LCU艦	3	
空中校閱編隊	陸軍航空特戰指揮部	AH-1W攻擊直昇機	6	33
		OH-58D戰搜直昇機	6	
		CH-47SD運輸直昇機	3	
	空軍作戰指揮部	經國號戰機	6	
		F-16戰機	6	
		幻象2000戰機	6	
總計十七個單位、三十七類五五四項武器裝備				

22nd Han Kuang Exercise. July 20, 2006. Media Trip. Brochure 3/10.

「國土防衛作戰演練」演練程序表

項次	項目	起迄時間	備考
一	恭迎	0850	
二	國土防衛作戰演練	0850 | 1010	
三	總統訓示	1010 | 1020	
四	行程	1020 | 1100	
五	城鄉守備戰鬥演練	1100 | 1140	
六	恭送	1140	
總計時間		170分鐘	

22nd Han Kuang Exercise. July 20, 2006. Media Trip. Brochure 4/10.

演練構想

基於「全民防衛」理念，結合年度漢光演習戰備任務訓練。針對敵軍「猝然突擊」，運用斬首威懾、癱瘓迫談、全面攻略等作戰模式；依國軍防衛作戰計畫，運用蘭陽地區現有之軍民總力及增援之兵、火力，按接戰程序，採實兵、實彈、全員、全裝方式，實施國土防衛作戰演練。

22nd Han Kuang Exercise. July 20, 2006. Brochure 5/10.

演練目的

一、驗證三軍聯合作戰指揮機制暨現有武器系統整合，檢討三軍聯合作戰效能。

二、驗證應急戰備能力及指、管、通、資、情、監、偵系統戰力整合成效、並蒐整相關參數。

三、驗證「後備守土、常備打擊」，作戰分區統一指揮三軍地面部隊，遂行國土防衛作戰效能。

四、藉聯合實兵、實彈作戰訓練，驗證作戰（火力）計畫之適切性與可行性，做為作戰區後續戰備整備之參據。

五、針對未來戰爭型態，驗證「精進案」執行成效，檢討三軍戰力整備之適切性。

六、藉國土防衛作戰演練等對外展示項目，彰顯國軍勤訓精練成效，凝聚全民國防共識。

七、驗證國軍新式武器及裝備系統戰力組建成效。

22nd Han Kuang Exercise. July 20, 2006. Media Trip. Brochure 6/10.

名稱 愛國者飛彈

簡介 自美國採購之防空武器，射程160公里，雷達導控及半主動歸向導航，為國軍主要之防空飛彈。

名稱 戰搜直昇機

簡介 戰搜直昇機座艙採併列式設計，右座為正駕駛、左側為副駕駛兼射手。巡航速度80哩，最大空速140哩，最大航程350哩，續航力2小時30分，可攜行地獄火飛彈4枚、海神火箭7枚、五〇機槍500發。

名稱 雷霆2000火箭

簡介 由中科院自行研發之新一代多管火箭系統，配置高精度之定位定向單元及全自動化之射控系統，具備機動性高、投射火力強等特點，可提供陸軍作為反登陸作戰之攻擊性武器。

名稱 攻擊直昇機

簡介 攻擊直昇機座艙採縱列式設計，後座為正駕駛、前座為副駕駛兼射手。巡航速度120哩，最大空速170哩，最大航程243哩，續航力3小時，可攜行3管20公厘機砲750發、海神火箭76枚、拖式飛彈8枚、地獄火飛彈8枚、響尾蛇飛彈2枚。

22nd Han Kuang Exercise. July 20, 2006. Media Trip. Brochure 7/10.

名稱 復仇者飛彈

簡介 復仇者飛彈為輕量化、機動性高以及轉移陣地迅速之地對空飛彈系統。車上配有2具發射器，共可裝8枚飛彈，有效射程為10公里，用以對付來襲的巡弋飛彈、無人飛行載具，以及打擊低空高速穿越的定翼機或直升機，堪稱全天候均可執行任務的短程野戰防空利器。

名稱 濟陽級艦

簡介 濟陽級飛彈巡防艦，最大速率為28節，巡航距離4500浬，排水量4200噸，1993年自美國陸續移交我國，並且於2004年由我國自行完成武三系統換裝，加裝了10枚標準一型飛彈，大幅的提升防空與反飛彈作戰能力，可同時遂行防空、反水面及反潛等三度空間作戰，有效的提升整體作戰效益。

名稱 檞樹飛彈

簡介 檞樹飛彈為一簡單可靠、反應靈敏、高效率，且具全方位、全天候的防空飛彈。採前視紅外線自動追蹤，可以在夜間、惡劣天候及煙霧彌漫狀態下實施追瞄射擊，具有射後不理的功能。最大速度為2.25馬赫，有效射程為8公里，能有效攔截低空、高速進襲的各型航空器。

名稱 錦江級艦

簡介 光華三號計畫錦江級巡邏艦係採國艦國造方式籌建，此型艦之任務為執行港口、海峽偵巡、外島運補護航及近岸防衛。

22nd Han Kuang Exercise. July 20, 2006. Media Trip. Brochure 8/10.

名稱 AAVP7兩棲突擊車

簡介 AAVP7兩棲突擊車為陸戰隊軍購獲得美軍現役裝備，性能優越，宜海宜陸，在兩次波灣戰爭中均有卓越表現，本車海上耐波力可承受激浪10呎，仍能執行任務，陸上爬坡最大坡度60%，側坡40%，輸出最大馬力525匹，全車可搭載21名全副武裝士兵，車首採船型設計，水上最大可巡航7小時，陸上時速可達72公里，最大巡航里程480公里，配備主要武器為12.7公釐重機槍及40公釐槍榴彈。

名稱 幻象２０００戰機

簡介 幻象戰機是輕型戰機，攔截１架高度２４３８０公尺、速度３馬赫的敵機只要５分鐘，其性能為同級戰機最優。

名稱 雄風2型飛彈

簡介 雄風2型飛彈是繼雄風1型飛彈後，發展較遠程艦對艦飛彈武器系統，其有可抗電子干擾之尋找目標器及自動接戰功能射控系統。

名稱 F-16戰機

簡介 F-16戰機主要性能為最大速度2.05馬赫，最大作戰半徑925公里，航程3870公里，其固定式武裝為MA-61二〇機砲1門（510發砲彈），滿載兩個油箱續航時間為兩小時，可攜帶響尾蛇飛彈、麻雀飛彈。

22nd Han Kuang Exercise. July 20, 2006. Brochure 9/10.

22nd Han Kuang Exercise. July 20, 2006. Media Trip. Brochure 10/10.

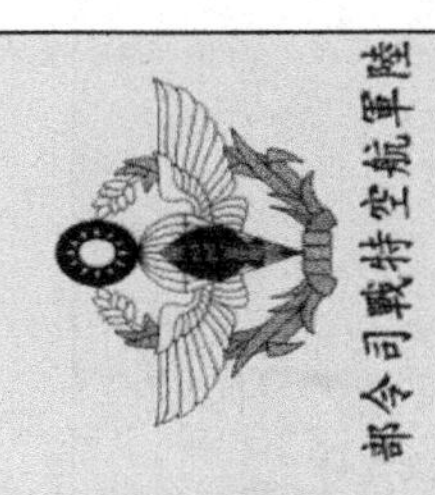

陸軍航空特戰司令部

新聞稿

中華民國九十三年一月六日

為了展示國軍春節期間加強戰備的成果，國防部於元月六日，在軍事發言人室黃穗生將軍率領下，邀請國內各家新聞媒體組成軍事記者參訪團，抵達陸軍航空特戰司令部「歸仁基地」參訪，藉媒體現地拍攝、報導，完整展現航特部在飛行軍官培養、擔任地區應急戰備及支援反恐任務等整備情形，讓國人瞭解陸軍訓練精實及戰備堅強的實況。

陸軍航空特戰司令部前身空降部隊係民國三十三年成軍於雲南昆明，歷經抗戰、中原戡亂各役、滇緬作戰及東山島戰役，戰績彪炳素有鐵帽子部隊之稱，並獲先總統 蔣公榮頒「忠義驃悍」四字相期許。轉進來台，經多次整編為空降司令部；民國六十三年併編特戰部隊成立空降特戰司令部；民國六十八年納編陸航部隊；基於地、空整體作戰及新一代兵力整建任務，於民國八十八年十月一日轉形為航空特戰司令部，編配有最先進的AH—1W的攻擊直昇機、OH—58D戰搜直昇機和CH—47SD中型運輸直昇機，可執行武裝偵搜、警戒、空中密接支援、反船團泊地攻擊、反裝甲作戰、空中運補及傷患後送等任務，結合特戰部隊更可遂行特攻、突擊、反恐支援與應變制變等任務，是一支具有空中、陸地、海上三度空間立體作戰能力的部隊，歷年來曾參與「漢光」、「聯興」、「聯勇」、「聯雲」等操演，完成「陸航出海攻擊」、「反恐任務」等重要任務規劃，可於台澎防衛作戰中，有效發揮地空整體作戰力量。

參訪活動在軍事記者採訪團報到領取參訪手冊及聆聽單位簡報、採訪須知後，由陸總部岳將軍、航空特戰司令部司令高中將全程陪同參觀部隊操演，操演活動區分上、下午二個階段實施：

第一階段首先由航空訓練指揮部飛行教官及學員，以十二架TH—67教練直昇機，採三機四批基本隊形編隊，低空慢速通過參觀台前揭開序幕，展示飛行初官培訓成果；隨後進行傘兵一般地形、特種地形、高空滲透空降作戰及花式跳傘等人員裝備、傘具介紹，並由高空突擊排及聞名中外的神龍小組，實施高空滲透突擊、花式定點著陸等高空跳傘技術；緊接著由空騎六○二旅特戰營突擊排搭乘UH—1H直昇機，在OH—58D戰搜直昇機、AH—1W攻擊直昇機掩護下，實施空中突擊作戰實兵演練，展現我戰備部隊反特攻、反突擊之快速反應能力。

第二階段於南停機坪實施靜態戰力展示開始，展示本軍各型作戰飛機，包括AH—1W攻擊直昇機、OH—58D戰搜直昇機、UH—1H通用直昇機、CH—47SD運輸機、TH—67教練直昇機及拖車、空管車、機動太康台、氣象車等地面支援裝具，在各型飛機及地面支援裝具組合運用下，構成陸軍地空整體作戰之強大戰力；隨後由高空突擊排實施「反劫持攻堅戰鬥」演練，模擬恐怖份子闖入市民活動中心，槍殺無辜市民，並挾持官員首長及人質，特戰部隊奉命執行「反恐任務」，以空中機動方式，實施立體平垂多點突入攻堅，完成反劫持任務，充分展示特戰部隊反恐制變能力。

縱觀此次參訪過程，航特部以精實的戰技、純熟的默契展現「地空整體」作戰及支援反恐任務的精實戰力，讓國人對航特部堅強戰力與平日勤訓精練、積極戰備整備的實況與訓練成效，留下深刻印象。

陸軍航空特戰司令部
聯絡電話：（○八）七五五○五九五
聯絡人：王良賢少校

Aviation and Special Forces Command. Exercise. Media Event. January 6, 2004. Brochure 1/1.

Army Aviation and Special Forces Command (陸軍航空特戰指揮部). 602nd Aviation Brigade. Hsinshe, Taichung. Attack Battalion x2 (AH-1W Super Cobra attack helicopter), Reconnaissance Battalion x1 (OH-58D Kiowa Warrior), Utility Battalion x1 (UH-1H Huey and UH-60M Black Hawk). 2013 Media Visit. Brochure 1/9.

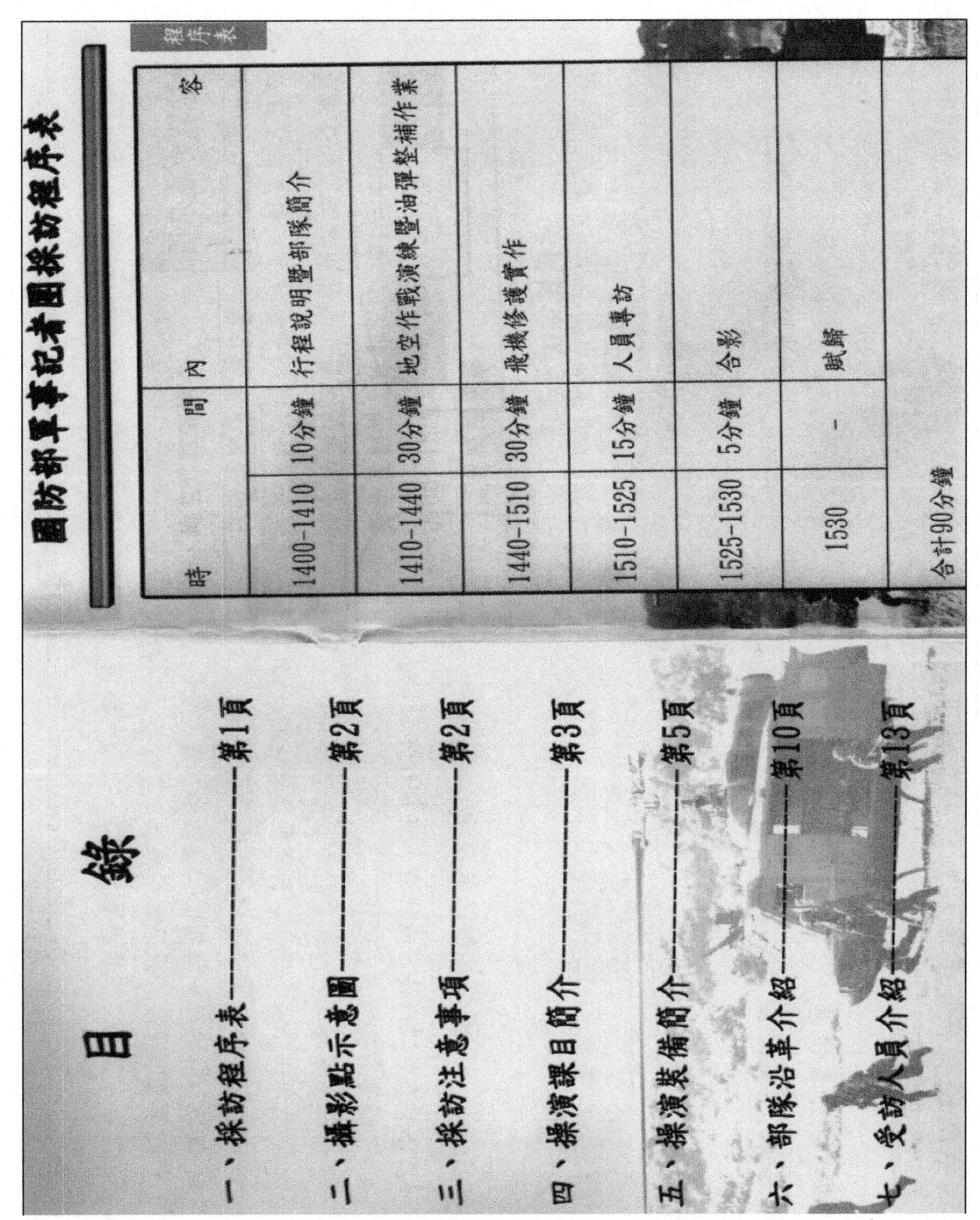

程序表

國防部軍事記者團採訪程序表

時	間	內容
1400-1410	10分鐘	行程說明暨部隊簡介
1410-1440	30分鐘	地空作戰演練暨油彈整補作業
1440-1510	30分鐘	飛機修護實作
1510-1525	15分鐘	人員專訪
1525-1530	5分鐘	合影
1530	-	賦歸

合計90分鐘

目錄

Army Aviation and Special Forces Command (陸軍航空特戰指揮部). 602nd Aviation Brigade. Hsinshe, Taichung. 2013 Media Visit. Brochure 2/9.

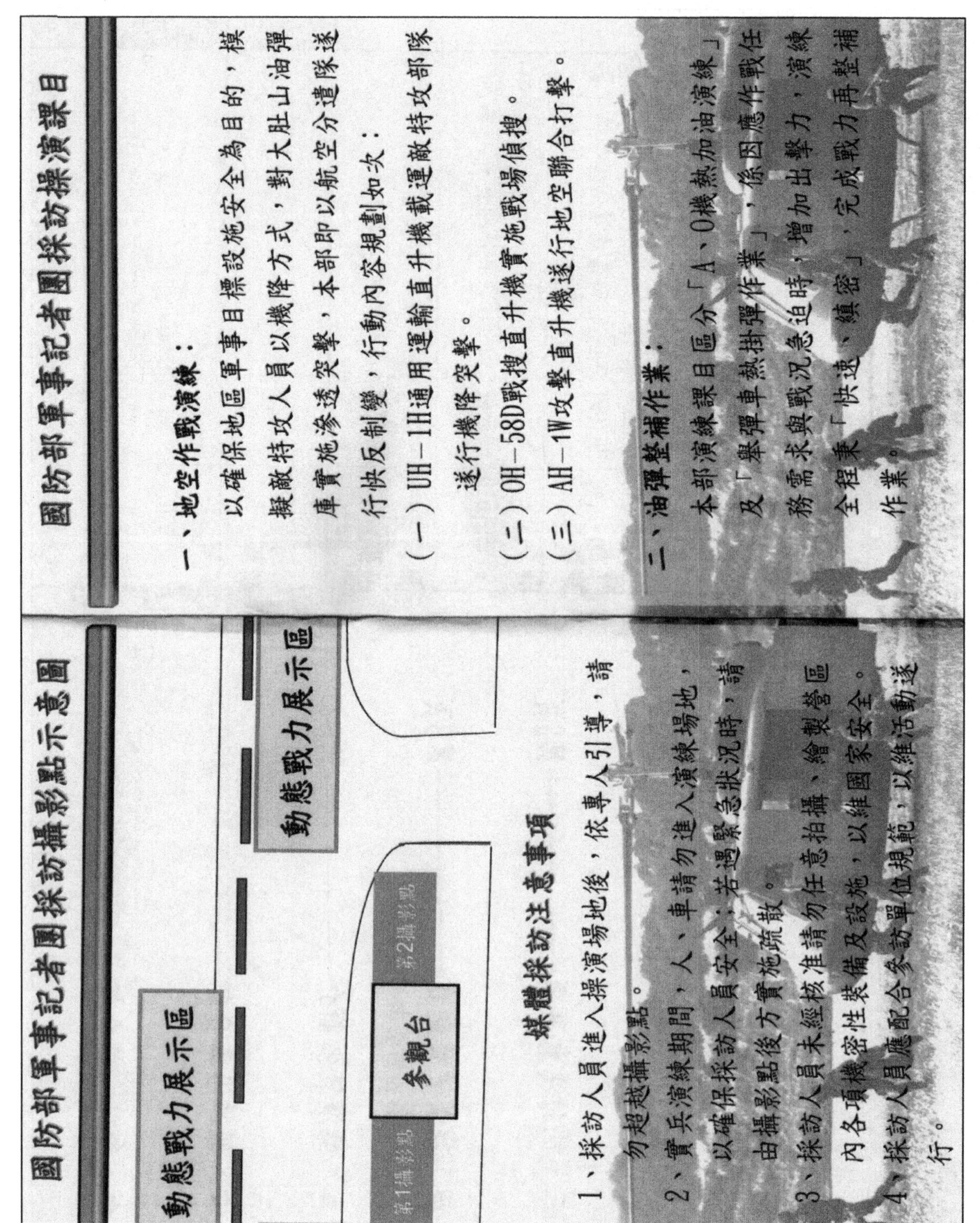

國防部軍事記者團採訪攝影點示意圖

動態戰力展示區

動態戰力展示區

第1攝影點

參觀台

第2攝影點

媒體採訪注意事項

1、採訪人員進入操演場地後，依專人引導，請勿超越攝影點。

2、實兵演練期間，人、車請勿進入演練場地，以確保採訪人員安全；若遇緊急狀況時，請由攝影點後方實施疏散。

3、採訪人員未經核准請勿任意拍攝、繪製營區內各項機密性裝備及設施，以維國家安全。

4、採訪人員應配合參訪單位規範，以維活動遂行。

國防部軍事記者團採訪操演課目

一、地空作戰演練：

以確保地區軍事目標設施安全為目的，模擬敵特攻人員以機降方式，對大肚山油彈庫實施滲透突擊，本部即以航空分遣隊遂行快反制變，行動內容規劃如次：

（一）UH－1H通用運輸直升機載運敵特攻部隊遂行機降突擊。

（二）OH－58D戰搜直升機實施戰場偵搜。

（三）AH－1W攻擊直升機遂行地空聯合打擊。

二、油彈整補作業：

本部演練課目區分「A、O機熱加油演練」及「舉彈車熱掛彈作業」，係因應作戰任務需求與戰況急迫時，增加出擊力，演練全程秉「快速、縝密」，完成戰力再整補作業。

Army Aviation and Special Forces Command (陸軍航空特戰指揮部). 602nd Aviation Brigade. Hsinshe, Taichung. 2013 Media Visit. Brochure 3/9.

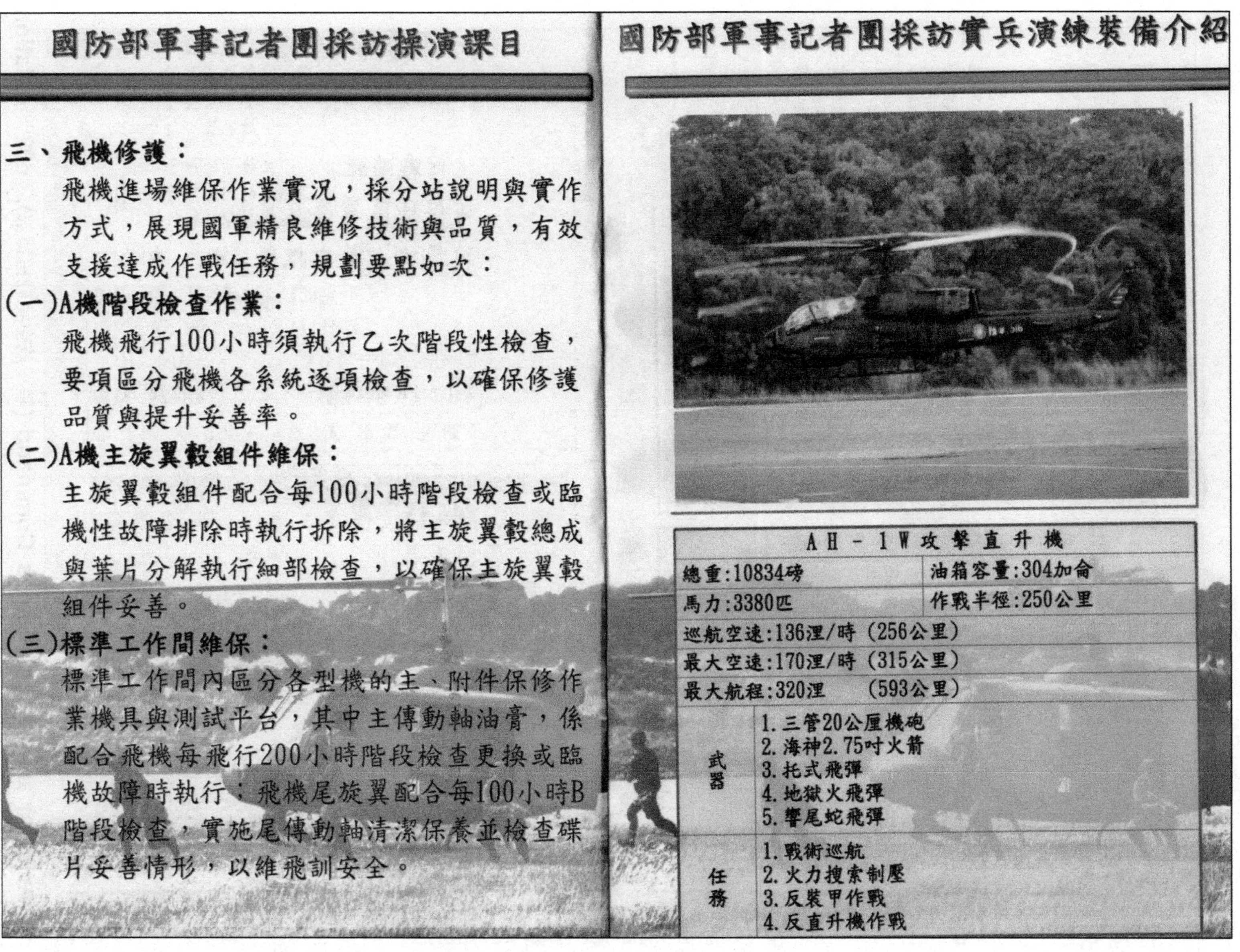

國防部軍事記者團採訪操演課目

三、飛機修護：

飛機進場維保作業實況，採分站說明與實作方式，展現國軍精良維修技術與品質，有效支援達成作戰任務，規劃要點如次：

(一)A機階段檢查作業：

飛機飛行100小時須執行乙次階段性檢查，要項區分飛機各系統逐項檢查，以確保修護品質與提升妥善率。

(二)A機主旋翼轂組件維保：

主旋翼轂組件配合每100小時階段檢查或臨機性故障排除時執行拆除，將主旋翼轂總成與葉片分解執行細部檢查，以確保主旋翼轂組件妥善。

(三)標準工作間維保：

標準工作間內區分各型機的主、附件保修作業機具與測試平台，其中主傳動軸油膏，係配合飛機每飛行200小時階段檢查更換或臨機故障時執行；飛機尾旋翼配合每100小時B階段檢查，實施尾傳動軸清潔保養並檢查碟片妥善情形，以維飛訓安全。

國防部軍事記者團採訪實兵演練裝備介紹

AH－1W攻擊直升機		
總重:10834磅		油箱容量:304加侖
馬力:3380匹		作戰半徑:250公里
巡航空速:136浬/時（256公里）		
最大空速:170浬/時（315公里）		
最大航程:320浬　（593公里）		
武器	1. 三管20公厘機砲 2. 海神2.75吋火箭 3. 托式飛彈 4. 地獄火飛彈 5. 響尾蛇飛彈	
任務	1. 戰術巡航 2. 火力搜索制壓 3. 反裝甲作戰 4. 反直升機作戰	

Army Aviation and Special Forces Command (陸軍航空特戰指揮部). 602nd Aviation Brigade. Hsinshe, Taichung. 2013 Media Visit. Brochure 4/9.

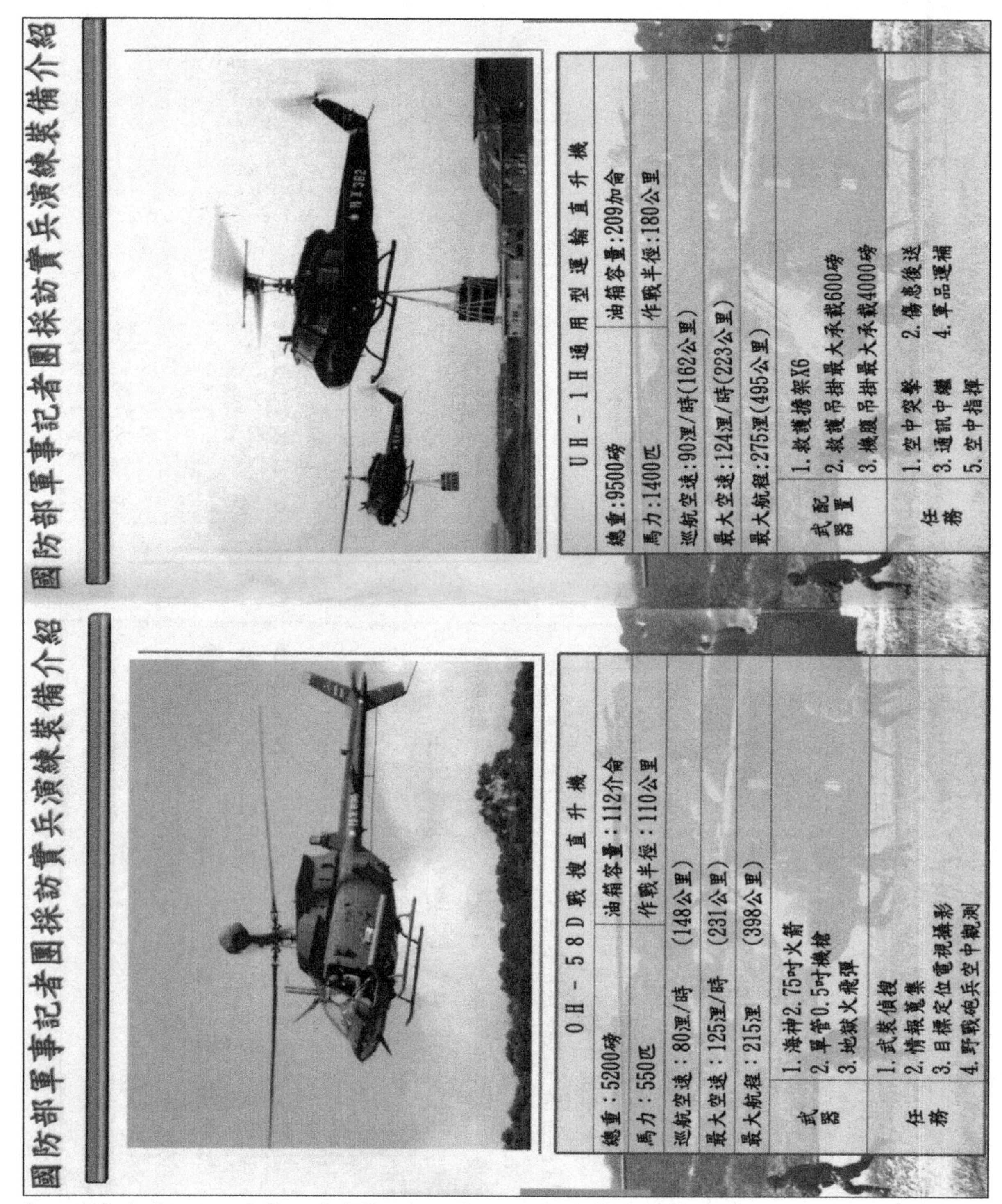

國防部軍事記者團採訪實兵演練裝備介紹

OH－58D戰搜直升機	
總重：5200磅	油箱容量：112介侖
馬力：550匹	作戰半徑：110公里
巡航空速：80浬/時 (148公里)	
最大空速：125浬/時 (231公里)	
最大航程：215浬 (398公里)	
武器	1. 海神2.75吋火箭 2. 單管0.5吋機槍 3. 地獄火飛彈
任務	1. 武裝偵搜 2. 情報蒐集 3. 目標定位電視攝影 4. 野戰砲兵空中觀測

國防部軍事記者團採訪實兵演練裝備介紹

UH－1H通用型運輸直升機	
總重:9500磅	油箱容量:209加侖
馬力:1400匹	作戰半徑:180公里
巡航空速:90浬/時(162公里)	
最大空速:124浬/時(223公里)	
最大航程:275浬(495公里)	
武器配置	1. 救護擔架X6 2. 救護吊掛最大承載600磅 3. 機腹吊掛最大承載4000磅
任務	1. 空中突擊 2. 傷患後送 3. 通訊中繼 4. 軍品運補 5. 空中指揮

Army Aviation and Special Forces Command (陸軍航空特戰指揮部). 602nd Aviation Brigade. Hsinshe, Taichung. 2013 Media Visit. Brochure 5/9.

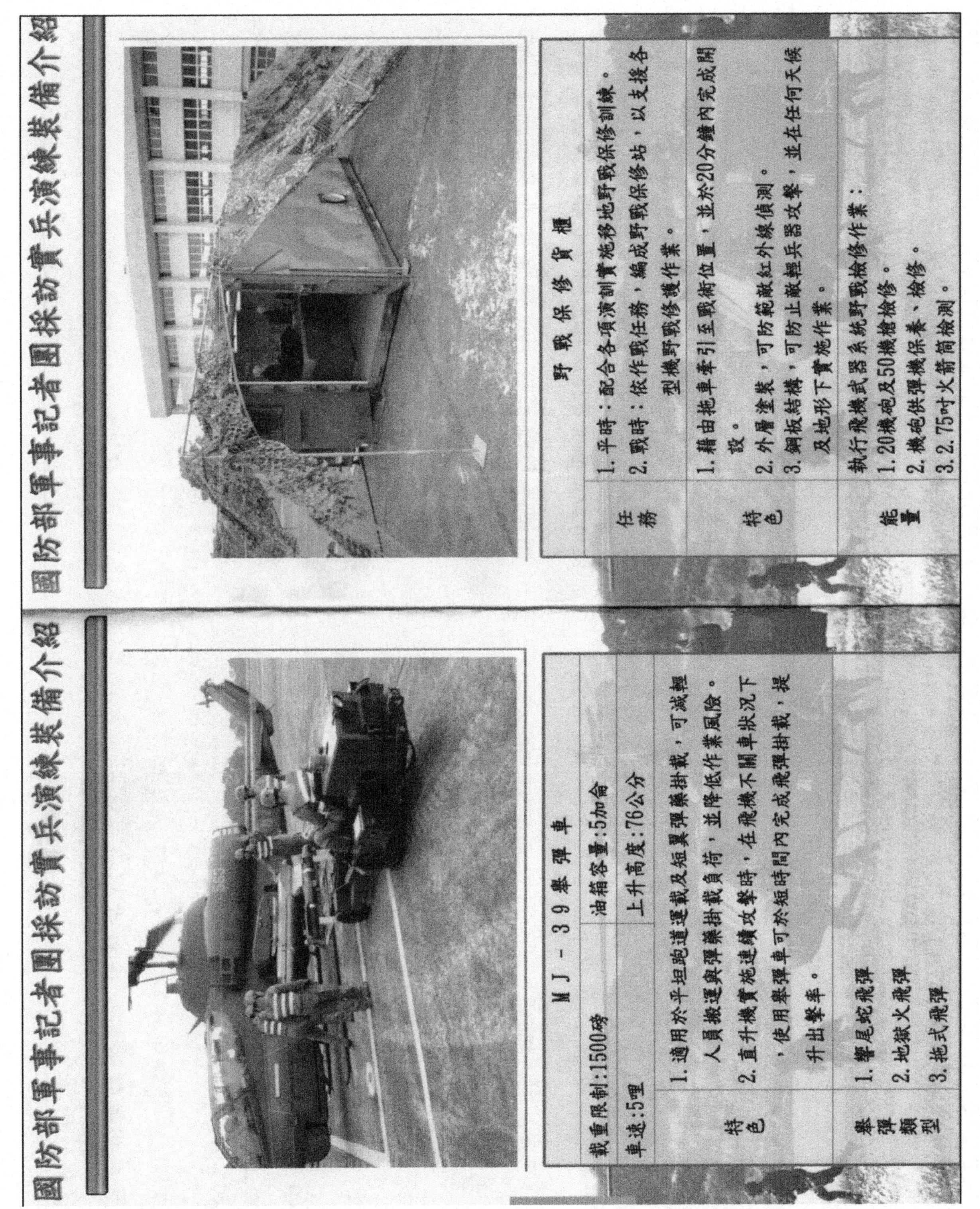

Army Aviation and Special Forces Command (陸軍航空特戰指揮部). 602nd Aviation Brigade. Hsinshe, Taichung. 2013 Media Visit. Brochure 6/9.

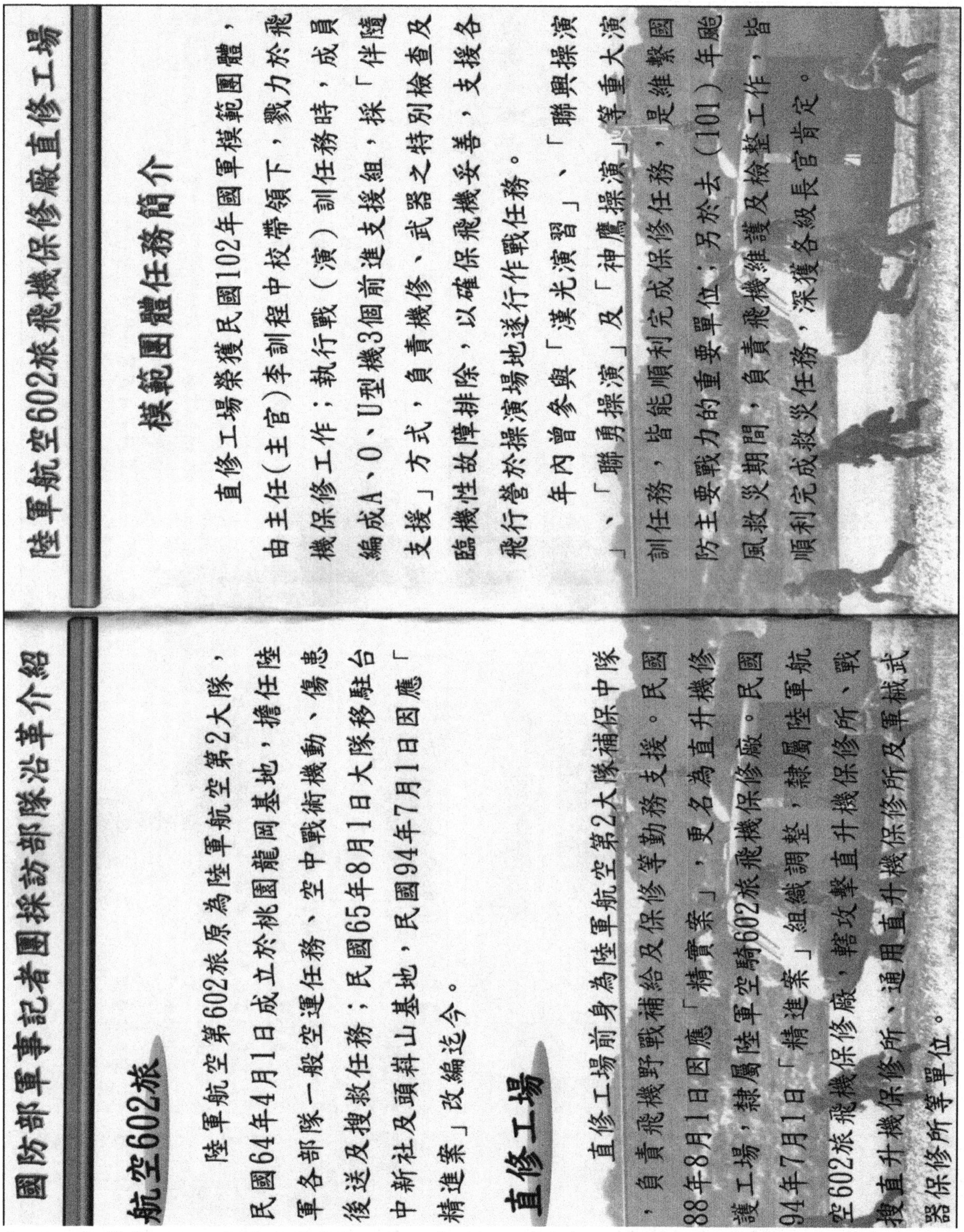

國防部軍事記者團採訪部隊沿革介紹

航空602旅

陸軍航空第602旅原為陸軍航空第2大隊，民國64年4月1日成立於桃園龍岡基地，擔任陸軍各部隊一般空運任務、空中戰術機動、傷患後送及搜救任務；民國65年8月1日大隊移駐台中新社及頭料山基地，民國94年7月1日因應「精進案」改編迄今。

直修工場

直修工場前身為陸軍航空第2大隊補保中隊，負責飛機野戰補給及保修等勤務支援。民國88年8月1日因應「精實案」，更名為直升機修護工場，隸屬陸軍空騎602旅飛機保修廠。民國94年7月1日「精進案」組織調整，隸屬陸軍航空602旅飛機保修廠，轄攻擊直升機保修所、戰搜直升機保修所、通用直升機保修所及軍械武器保修所等單位。

陸軍航空602旅飛機保修廠直修工場

模範團體任務簡介

直修工場榮獲民國102年國軍模範團體，由主任(主官)李訓程中校帶領下，戮力於飛機保修工作；執行戰（演）訓任務時，成員編成A、O、U型機3個前進支援組，採「伴隨支援」方式，負責機修、武器之特別檢查及臨機性故障排除，以確保飛機妥善，支援各飛行營於操演場地遂行作戰任務。

年內曾參與「漢光演習」、「聯興操演」、「聯勇操演」及「神鷹操演」等重大演訓任務，皆能順利完成保修任務，是維繫國防主要戰力的重要單位；另於去(101)年颱風救災期間，負責飛機維護及檢整工作，皆順利完成救災任務，深獲各級長官肯定。

Army Aviation and Special Forces Command (陸軍航空特戰指揮部). 602nd Aviation Brigade. Hsinshe, Taichung. 2013 Media Visit. Brochure 7/9.

單位隊徽介紹

陸軍602旅

外形為倒鑽石形，意涵為堅強、鞏固之意，象徵旅為最現代化航空部隊，編成攻無不克、戰無不勝之勁旅。

羅馬數字Ⅱ代表第2個編成之航空旅；H字母代表直升機部隊；展翅之圖形代表航空旅為機動力最高、打擊力最強、部隊訓練最精良之現代化部隊，亦代表航空旅已成為國軍部隊「量小、質精、戰力強」之新一代兵力。

Maintenance

飛機保修廠

翅膀底圖象徵我為航空部隊，羅馬數字「Ⅱ」為支援航空602旅飛機保修勤務，交叉的板手象徵「保修作為」相互支援及合作的特性；板手下方正方形及四端黑白交錯，象徵「誠樸、務實、精勤、創新」之核心價值。

軍事記者團蒞臨採訪受訪人員

單位：陸軍航空602旅飛保廠直修工場
級職：前中校主任
姓名：李訓程
事蹟：擔任飛保廠直修工場主官，參加漢光演習、聯興操演、聯勇操演及神鷹操演，均能戮力執行飛機保修作業，榮獲航特部年度精神戰力週評比第1名、軍歌競賽甲組第2名及國軍模範團隊。

單位：陸軍航空602旅攻擊第一營
級職：上尉飛行官
姓名：林冠良
事蹟：擔任攻擊直升機飛行官，曾參加漢光演習、聯興操演、聯勇操演及神鷹操演等重大任務，對飛行技能具有優異之表現，深受各級長官肯定。

單位：陸軍航空602旅飛保廠直修工場
級職：中尉組長
姓名：賴昱勳
特色：擔任攻擊直升機保修所保修組組長，曾參加漢光演習、聯興操演、聯勇操演及神鷹操演，圓滿執行各項飛機保修檢驗工作，並榮獲飛機修護丙級證照。

單位：陸軍航空602旅飛保廠直修工場
級職：一兵保修兵
姓名：謝凱俐(女)
特色：擔任戰搜直升機之保養修護兵，曾參加漢光演習、聯興操演、聯勇操演及神鷹操演，執行飛機保修工作，盡心盡力，為陸軍募兵制最佳典範。

Army Aviation and Special Forces Command (陸軍航空特戰指揮部). 602nd Aviation Brigade. Hsinshe, Taichung. 2013 Media Visit. Brochure 8/9.

Army Aviation and Special Forces Command (陸軍航空特戰指揮部). 602nd Aviation Brigade. Hsinshe, Taichung. 2013 Media Visit. Brochure 9/9.

Press Badge. *Defense News*. Wendell Minnick. Army Aviation and Special Forces Command (陸軍航空特戰指揮部). Airborne Training Center. Dawu, Pingtung County. December 10, 2010.

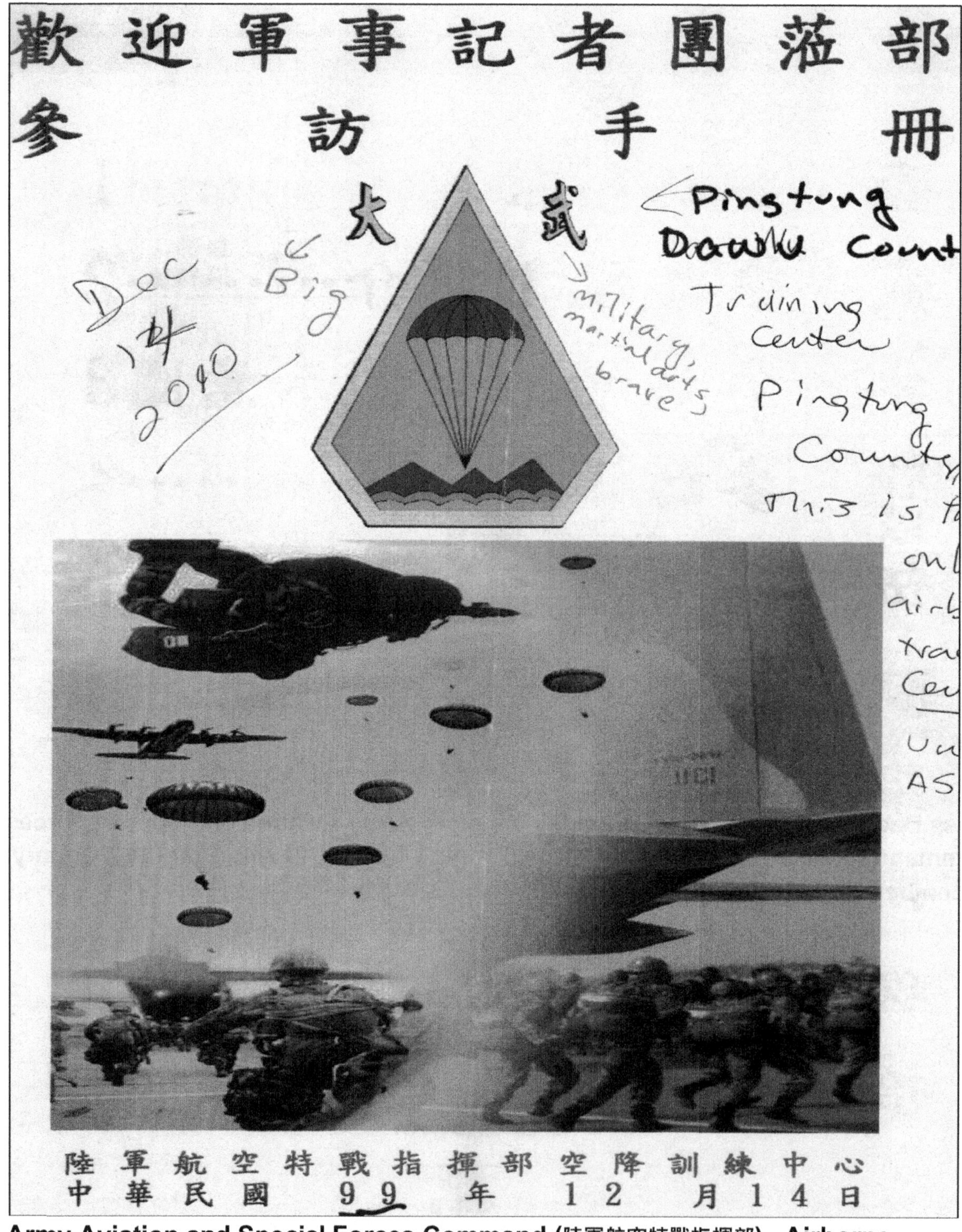

Army Aviation and Special Forces Command (陸軍航空特戰指揮部). Airborne Training Center. Dawu, Pingtung County. Media Event. December 10, 2010. Brochure 1/18.

目　錄

Army Aviation and Special Forces Command (陸軍航空特戰指揮部). Airborne Training Center. Dawu, Pingtung County. Media Event. December 10, 2010. Brochure 2/18.

壹、參訪行程表

歡迎軍事記者團蒞部參訪行程表

中華民國99年12月14日(星期二)

項次	時間	使用時間	行程概要	地點
一	0800-0820	20'	1. 恭迎 2. 參訪行程說明	439聯隊124停機坪
二	0820-0830	20'	拍攝學員穿、架傘及分組作業	439聯隊124停機坪
三	0830-0920	50'	車程(439聯隊-潮洲空降場)	
四	0920-1040	80'	參觀基本(高空)跳傘、動力飛行傘編隊飛行、人物專訪	潮洲空降場
五	1040-1050	10'	合影	潮洲空降場
六	1050-1140	50'	車程(空降場-大將餐廳)	
七	1140-1300	80'	用餐	大將日本料理餐廳
八	1300-1310	10'	車程(大將餐廳-大武營)	
九	1310-1400	50'	參觀傘訓五大科目(含高塔體驗)	傘訓場
十	1400-1410	10'	車程(大武營區-439聯隊)	
十一	1410-1420	10'	恭送(致贈紀念品、合影相片)	439聯隊124停機坪
合計	6小時30分鐘			

第1頁，共16頁

Army Aviation and Special Forces Command (陸軍航空特戰指揮部). Airborne Training Center. Dawu, Pingtung County. Media Event. December 10, 2010. Brochure 3/18.

參、參訪行程概要說明

本日參訪活動行程的第一站，將區分隨機拍攝組（登機拍攝）及潮洲拍攝組（地面跳傘實況）等2組，使各位記者小姐、先生能完整拍攝傘兵空跳實況。

到達潮洲空降場後，安排基本傘訓跳傘、神龍高空跳傘及動力飛行傘編隊飛行等動態操演，動態操演完後安排與本日操演官兵合影；另外為使各位能更加瞭解本部訓練成果，安排女神龍、神龍夫妻檔、世運會跳傘比賽選手、動力飛行傘、自摺自跳女性士兵等5位接受專訪。

中午用完餐後將前往傘兵的搖籃空訓中心大武營參觀基本傘訓五大科目訓練，屆時將開放高塔體驗，預計1400時結束參訪行程，1410時前往屏東機場，1420時登機返回台北松山機場，各位媒體先進在參訪過程中如遇有任何問題請洽服務人員，本部很高興能為各位服務，謝謝。

第2頁，共16頁

Army Aviation and Special Forces Command (陸軍航空特戰指揮部). Airborne Training Center. Dawu, Pingtung County. Media Event. December 10, 2010. Brochure 4/18.

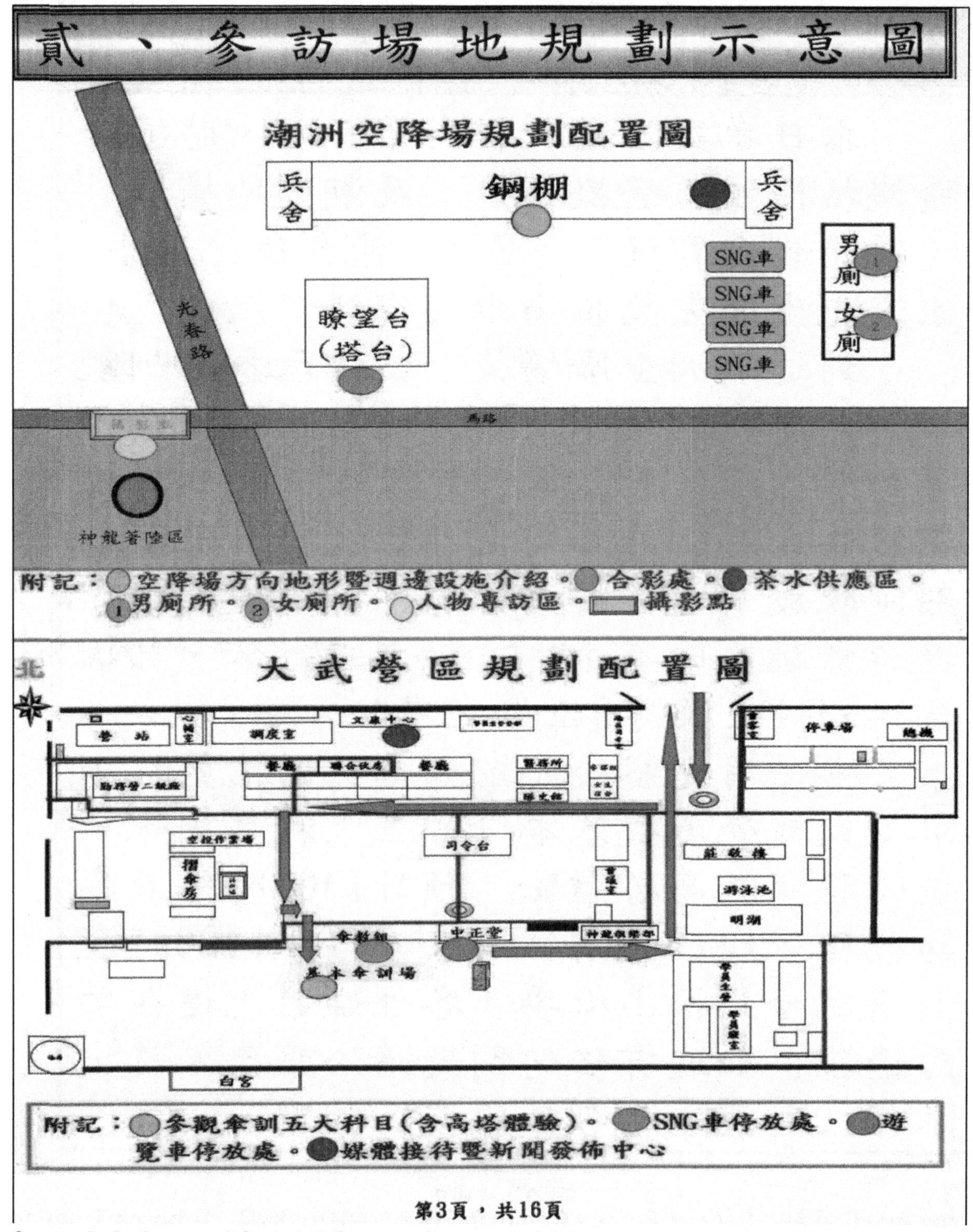

Army Aviation and Special Forces Command (陸軍航空特戰指揮部). Airborne Training Center. Dawu, Pingtung County. Media Event. December 10, 2010. Brochure 5/18.

參、參訪須知

一、採訪全程，請配合接待人員引導，以維行動安全及參訪活動順利遂行。

二、採訪期間，如有需要任何服務，請向隨行接待人員反映，如有受傷、體力不支、身體不適等狀況，請立即知會隨行接待人員，本部將立即安排醫療人員進行診治。

三、採訪期間如遇惡劣天候或突發狀況，即刻終止採訪行程，並請依工作人員引導離開。

四、任務操演時，請配合本部規劃攝影點進行拍攝，避免肇生危安。

五、體驗各項設施時，請配合專業教官引導進行著裝及操作，施作前，由各組教官進行安全檢查，操作期間請遵守各項安全規定。

第4頁，共16頁

Army Aviation and Special Forces Command (陸軍航空特戰指揮部). Airborne Training Center. Dawu, Pingtung County. Media Event. December 10, 2010. Brochure 6/18.

為確保你的安全，如有下列症狀請勿參加體驗：

一、高血壓(165-95MMHG)而有心臟或血管之合併症者。
二、先天性心臟病或心瓣膜疾患者。
三、有疝氣、精鎖靜脈曲張者。
四、骨折畸形以致運動功能障礙者。
五、肢體肌肉萎縮疾患者。
六、高度近視（800度以上）及視網膜疾患者。
七、航空性疾病者(耳聾、重聽)。
八、有慢性疾病者。
九、因病(傷)短期內無法復原。

第5頁，共16頁

Army Aviation and Special Forces Command (陸軍航空特戰指揮部). Airborne Training Center. Dawu, Pingtung County. Media Event. December 10, 2010. Brochure 7/18.

伍、受訪人員簡歷

相片	單位	級職	姓名	學歷	經歷	服役年資	跳傘次數
	保傘連	上兵	劉昱綺	麥寮高中 志願役士兵96-4梯	傘具整摺兵 （自摺自跳）	3年	9次
	傘教組	下士	楊雅婷	志願役儲備士官班96-3期	傘訓助教 （現役女神龍）	3年	106次
	傘教組	士官長	潘益龍	高空領士班14期 士官長正規班21期	傘訓助教 教官 （神龍夫妻、世運會選手）	14年	830次
	傘教組	士官長	陳昱廷	專業士官班6期 士官長正規班25期	傘訓助教 教官 （神龍夫妻）	13年	530次
	特勤中隊	士官長	徐仁輝	領士班84-9期 士官長正規班31期	特勤隊員 副分隊長 分隊長 （動力飛行傘學員）	15年	150次

第6頁，共16頁

Army Aviation and Special Forces Command (陸軍航空特戰指揮部). Airborne Training Center. Dawu, Pingtung County. Media Event. December 10, 2010. Brochure 8/18.

陸、各型人員傘具介紹

一般傘兵
攜帶傘具T-10B人員主傘、T-10R人員副傘，並攜帶T-91步槍一枝及個人裝備包一個。T-10B人員主傘為一般地形作戰時所使用之傘具，T-10R人員副傘的使用時機是當主傘發生故障時所使用。

機槍射手
攜帶傘具同一般傘兵並攜帶國造A6槍包，A6槍包現在所捆紮的軍品是T-74輕機槍一挺。

砲手
攜帶傘具同一般傘兵並攜帶國造GP通用包，可細紮通信器材及各類型輕兵器，現在所捆紮的軍品為六0迫擊砲一門。

特種地形跳傘人員
適用於山川、叢林、沙漠等特種地形。攜帶傘具MC-1B人員主傘、T-10R人員副傘及T91步槍一枝、個人裝備包一個，個人身上所穿著服裝為鐵漢裝，分為頭盔、面罩，皮手套及連身服裝，鐵漢裝在人體各關節部位均有加裝強化護墊可使傘兵在森林、叢林或不規則地形實施空跳任務時，不會遭到樹枝或突出石塊尖銳物所刮傷。

高空滲透人員
攜帶傘具MT-1X高空滲透傘，它可在25000呎以上高空實施空跳任務。頭上所帶的為氧氣面罩系統供應總成，裡面含有護目鏡、氧氣面罩，通信器及氧氣筒個人爆破袋一個。爆破袋內可裝TNT黃色炸藥4磅。

高空疊傘跳傘人員
攜帶傘具P-230高空運動傘，主要用於空中疊傘時使用，P-230高空運動傘在作戰時也可做為高空滲透的傘具，平時為高空跳傘人員訓練，以及參加國際比賽或國際表演時，所使用之傘具。

高空接力跳傘人員
攜帶傘具P-252高空運動傘，主要用於空中接力及定點著路時使用。

第7頁，共16頁

Army Aviation and Special Forces Command (陸軍航空特戰指揮部). Airborne Training Center. Dawu, Pingtung County. Media Event. December 10, 2010. Brochure 9/18.

柒、各型空投傘具介紹

裝備名稱：二號20呎低空拖曳
用途：可捆紮各類補給品及裝備
載重量：3780-16250磅(1718-7380公斤)
使用傘具：28呎拖曳傘X1
作業人數：5員
作業時間：4小時

裝備名稱：五號16呎組合墊板
用途：可捆紮悍馬車、155榴砲、救護車等各型裝備或各類補給品
載重量：5040-28000磅(2291-12727公斤)
使用傘具：G-11A投物傘X3、22呎拖曳傘X1
作業人數：7員
作業時間：8小時

裝備名稱： A-22空投包
用途：可空投各類軍品如：大米、麵粉、彈藥、油桶、通材、化材、81迫砲、機槍等。
載重量：700-2200磅(318-1000公斤)
使用傘具：G-12D投物傘X1
作業人數：4員
作業時間：1小時

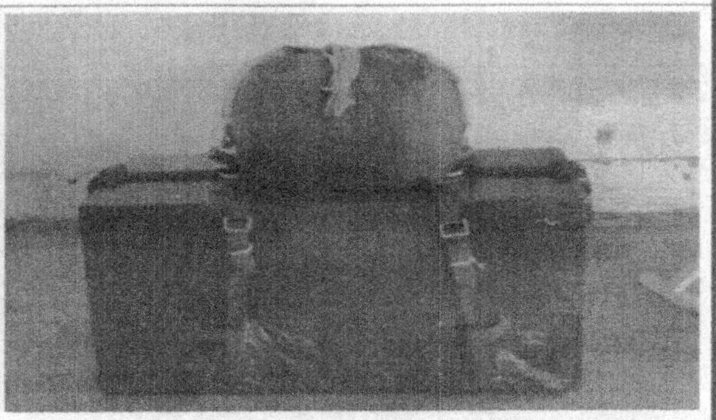

裝備名稱： A-21空投包
用途：可空投各類軍品如：大米、麵粉、彈藥、油桶、通材、化材、81迫砲、機槍等。
載重量：300-500磅(136-227公斤)
使用傘具：G-13D投物傘X1
作業人數：2員
作業時間：30小時

第8頁，共16頁

Army Aviation and Special Forces Command (陸軍航空特戰指揮部). Airborne Training Center. Dawu, Pingtung County. Media Event. December 10, 2010. Brochure 10/18.

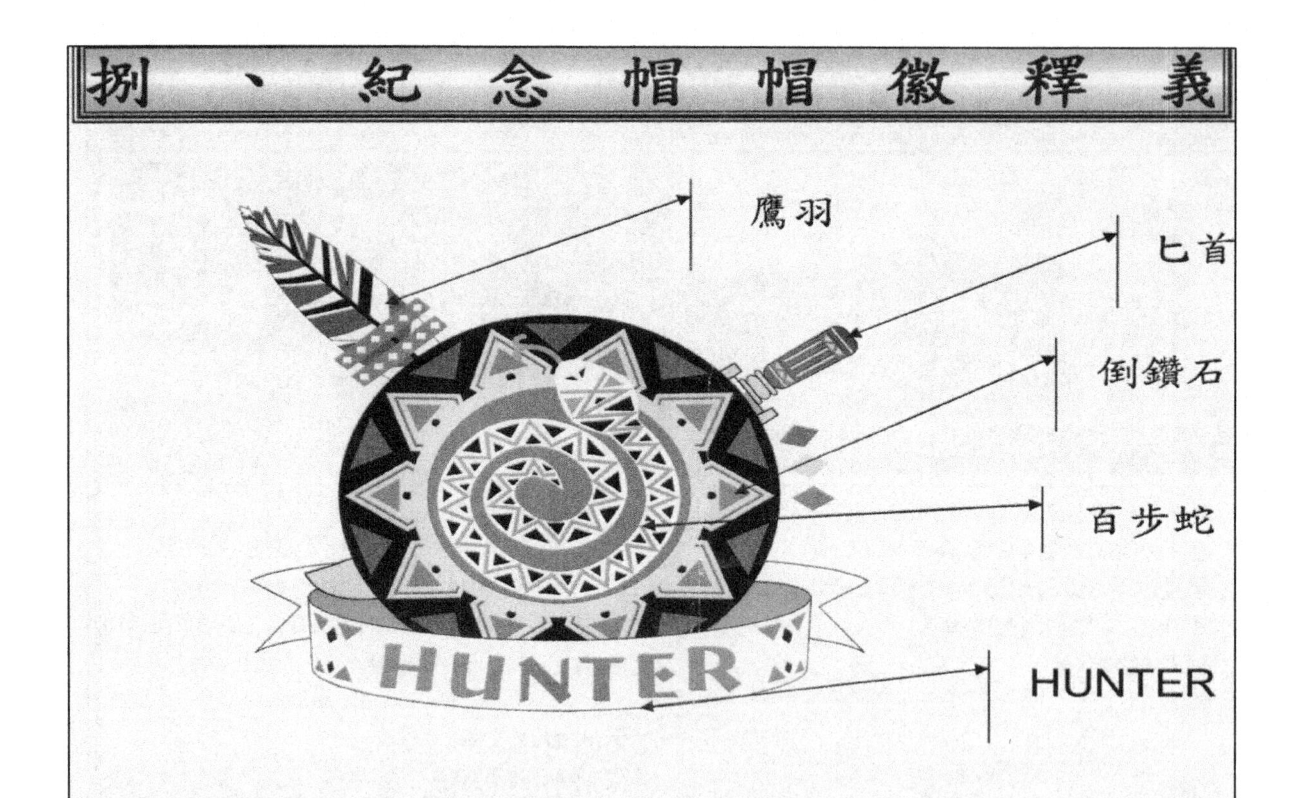

紀念帽帽徽釋義：

鷹羽：象徵航空部隊機動快速、反應快捷，並且備全天候打擊、運輸及偵搜能力。

匕首：象徵勇猛驃悍之特戰部隊，無論身處山地叢林、雪地、險峻高山之各種地形，均可披荊斬棘、快速突擊之戰力，讓敵人聞風喪膽之意志。

倒鑽石：象徵具有堅強抗敵意志，身處任何環境均無法摧破保國衛民之決心。

百步蛇：象徵具有反應靈敏、隱匿、偽裝、奇襲之能力，並隨時保持高度的警覺與絕佳的忍耐力

HUNTER(航特）：由陸航、特戰、空降、兩棲部隊結合一體，相輔相成、精誠團結之意。

第9頁，共16頁

Army Aviation and Special Forces Command (陸軍航空特戰指揮部). Airborne Training Center. Dawu, Pingtung County. Media Event. December 10, 2010. Brochure 11/18.

拾、空降場動態操演說明稿

項次	時間（時.分.秒）	說明稿
一	09.36.00	◎今日由本部陸軍航空特戰指揮部空降訓練中心執行本日跳傘任務，將為各位展示課目計有三項，第一為至本部服役之新進弟兄於地面基礎訓練完訓後所實施的基本跳傘訓練，跳出高度1250英呎、第二為享譽中外的「陸軍神龍小組」所展現的定點、接力及疊傘等課目，跳出高度分別為4000英呎及8000英呎，第三為本部動力飛行傘訓練班，將在高度300-600英呎中採編隊飛行方式展現精實訓練成果，請各位媒體先進及記者小姐、先生拭目以待。 ◎在執行跳傘任務有以下天候限制：地面風13海浬(含)以下、雲幕高1000呎(含)以上、水平能見度4800英呎(含)以上。 今日所使用的飛機為空軍C-130H力士型運輸機，可裝載64名武裝傘兵，將採兩機編隊方式執行任務，目前飛機約2分鐘後抵達上空。
二	09.40.00 / 09.42.00 試風教官	◎各位長官、各位記者小姐、先生： 現在1號機進入空降場，第1員人員跳出。 ◎每一次空中跳傘都會有一名教官先行跳出，其目的有二，第一：增加學員膽識，給予學員信心；第二：擔任試風人員，雖然於人員跳出前完成漂流計算，但在安全考量，尤其測試上空風是否有亂流、對流風等因素，試風人員必須經過長時間訓練，在空中不得實施任何操縱傘具動作，以提供地面導航人員正確漂流數據，所以必須要有過人毅力與膽識。然空降場指揮官必須依據上述一些數據果斷立即判斷下達決心，藉以修正下一波航向，本次擔任試風人員為本部保傘連同仁，是經過228小時摺傘訓練後於今日完成「自摺自跳」訓練課目。 本次擔任試風人員為保傘連上兵劉昱綺，是本部女性志願役士兵，至本部服務人員無論階級、性別都需完成5次空中跳傘訓練才可成為特戰健兒一份子。

第10頁，共16頁

Army Aviation and Special Forces Command (陸軍航空特戰指揮部). Airborne Training Center. Dawu, Pingtung County. Media Event. December 10, 2010. Brochure 12/18.

拾、空降場動態操演說明稿

三	09.42.00 / 09.50.00 基本跳傘階段	目前1號機已進入航道，導航人員正依據試風人員漂流重新計算航線，以使人員可以精準著陸於空降場中。 ◎現在第一名人員跳出，各位可以看到10名人員陸續由傘兵門跳出，本次操演共計有51名弟兄擔任。 ◎目前第二架機正採單機跟蹤隊形進場距離第1架機保持4000英呎距離，人員也將在導航人員引導下陸續跳出機門。 ◎現在第1架機的人員正準備著陸於空降場內，人員著陸時將採五點滾翻的動作，目的為降低著陸瞬間衝擊力，維護人員安全。 ◎要成為一位合格傘兵，除了必須有強健體魄更必須要有相當膽識，今天所執行基本空中跳傘學員，是經過52小時地面訓練。完成6大課目，經過層層考驗通過教官合格簽證始可登上飛機，執行5次空中跳傘，其中3次徒手、1次攜槍跳傘及1次夜間跳傘任務訓練，才可成為合格傘兵，現在目前因風向不同，地面指揮官正引導學員前往良好地形實施著陸。 ◎由於跳傘人員在空降著陸時會產生一股相當大的衝力，這股衝力並非人體任何一部位可單獨承受，需藉由人體肌肉較發達、彈性較佳的部位抵銷這股衝力，依序為前腳掌、小腿外側、大腿外側、臀部一側及背肌。 ◎目前第1架機已經進入五邊航線，同樣由導航人員引導進入，第2架機執行第三邊航向飛行，同時導航人員必須回報上一波次人員落點，藉以提醒機組員於第五邊時能準確進入航線。 ◎各位長官、各位記者小姐、先生： 目前第2架機已進入航向人員準備跳出，10名學員陸續人員跳出。 ◎現在空降場指揮官正引導學員檢查傘，注意四週動作，並告知目前地面風向、風速，跳出位置，提醒學員注意事項，使學員安全著陸於空降場內。（20"） ◎1號機任務結束脫離空降場目前爬升3000英呎返航，2號機於基本傘訓人員跳出後即將爬升4000英呎高度執行神龍小組高空跳傘操演。（20"）

第11頁，共16頁

Army Aviation and Special Forces Command (陸軍航空特戰指揮部). Airborne Training Center. Dawu, Pingtung County. Media Event. December 10, 2010. Brochure 13/18.

拾、空降場動態操演說明稿

四	09:50:00 | 09:51:00 高空跳傘階段定點	◎各位長官、各位記者小姐、先生： 緊接著操演的科目為神龍小組定點、接力及疊傘操演。 ◎利用飛機爬升的時間，由我為各位介紹神龍小組的由來，國軍有鑒於高空跳傘的技能，可以培養冒險犯難精神，且極富軍事價值，尤具振奮民心士氣的效果，所以在民國51年春著手編組神龍小組。 ◎民國52年正式核定「陸軍神龍小組」隊名，並積極推廣高空跳傘表演及研發滲透作戰技能，期間經歷數次組織更遞，目前是以空訓中心傘教組承襲「神龍小組」高空跳傘表演任務；另高空滲透作戰，則由陸軍特種部隊擔任。 ◎現在，請將目光轉向1點鐘方向，一架C-130H運輸機，正搭載12員享譽國內外，隸屬於陸軍航空特戰指揮部的神龍小組隊員。（20"） ◎本次的演練，將採取「縮短跳出距離方式」，以便各位貴賓能清楚目視人員滑降的情形。（10"） ◎今天的演練科目，為隊員在4000及8000英呎高空跳出，並配合施放國旗、彩帶及煙幕，依序滑降至目標區精準著陸，展現高超的操傘技巧。（17"） ◎神龍隊員必須要有超人般的膽識與毅力，接受耐力與意志力的層層考驗，才能不斷地突破，克服心理恐懼的障礙，才能成為一位忠義驃悍勇猛頑強的隊員。（15"）
五	09:51:00 | 09:55:00 高空跳傘階段定點	「人員跳出」 ◎各位長官、各位記者先生、小姐： 現在第一員已經跳出，緊接著所有隊員也依序跳出，我們隱約可以看到他們正以自由落體的方式滑降，並分別在5秒、8秒拉傘，現在第一波次所有隊員已完成張傘程序，由少校教官吳杰成擔任領隊。（20"） ◎當隊員在空中完成隊形排列以後，將配合施放國旗、彩帶，並依序滑降至目標區精準著陸，展現高超的操傘技巧。（15"） ◎神龍小組，他們平時除了擔任傘訓的教官，並且不斷研發高空跳傘技術，戰時亦可轉換為高空突擊的隊員，執行各種特定任務。

第12頁，共16頁

Army Aviation and Special Forces Command (陸軍航空特戰指揮部). Airborne Training Center. Dawu, Pingtung County. Media Event. December 10, 2010. Brochure 14/18.

拾、空降場動態操演說明稿

五	09:51:00 | 09:55:00 高空跳傘階段定點	◎現在我神龍小組隊員已完成張傘，第一名隊員正實施（盤旋），傘具每盤旋一圈下降100-150尺，其主要目的加大隊員之間間隔，最後對正風向採取L行方式進航著陸。（18"） ◎今天我神龍小組隊員著陸目標區，為空降場中央沙堆，以四面鋪設直徑5公尺白色布板，中央放置海綿墊著陸於上方。各位可以看到目標區旁放置一紅白相間風筒，是用來給空中隊員判定風向、風速所使用，藉敏銳觀察力精準著陸，現在隊員已將青天白日滿地紅國旗拉出，後續隊員也將彩帶拋出。（25"） ◎各位長官、各位記者小姐、先生： 第一員帶著我國青天白日滿地紅國旗的隊員已開始進航，這位正是少校教官吳杰成，曾多次參加國內各項操演任務，表現極為優異。個人跳傘次數為560次， 現在他藉著純熟的操傘技巧，以及豐富的跳傘經驗，使傘具很平穩的緩緩下降，逐漸接近目標區，「精準著陸」。（25"） ◎緊接在後的為張正雄上士，跳傘次數為420次，目前擔任高空跳傘種子教官也是本部動力飛行傘師資，去年曾經完成接裝訓練亦取得民航局所頒發動力飛行傘執照。表現十分優異，並以其優異的跳傘技術與經驗，曾參與過漢光演習等多項大型演訓任務。（25"） ◎第三名為上尉教官陳豐智，跳傘次數為289次，為高空跳傘資深教官。除多次參加國內各項演訓任務外，也培育出無數傘兵健兒。（10"） ◎而現在我們可以看到，在場中唯一的女性隊員，正是「女神龍」楊雅婷下士，楊下士是本次操演中唯一的女隊員，跳傘次數106次，堪稱女中豪傑，現在她也即將同樣緩緩下降準備著陸，「精準著陸」。（15"）
六	09:55:00 | 09:55:30	◎各位長官、各位記者小姐、先生： 目前飛機已進入航向現在高度8000英呎，本波次要為各位操演科目為四人接力「炸彈開花」，由四位神龍小組隊員於跳出前拉開繫於腳上煙霧，採自由落體方式至6000英呎分向四個方向實施俯衝如同花朵綻放光芒，第二組為四人疊傘，隊員將於五秒鐘拋傘，利用手上控制繩靈活操控傘具使傘具於空中相結合。（40"）

第13頁，共16頁

Army Aviation and Special Forces Command (陸軍航空特戰指揮部). Airborne Training Center. Dawu, Pingtung County. Media Event. December 10, 2010. Brochure 15/18.

拾、空降場動態操演說明稿

七	09:55.30 | 10:01:30 高空跳傘階段 接力疊傘	◎現在隊員跳出，各位在上方看到有一道彩色煙霧，這是操演「四人接力」課目，四名隊員已完成接力隊形，現正以時速180公里落速下降。（15"） ◎現在高度6000英呎，隊員利用空中姿態滑行分向四個方向轉身俯衝，如同花朵綻放光芒。本次為少校黃志成、士官長彭勇智、曾英志、潘益龍四人所組成的隊形。（20"） ◎四名隊員已完成本次操演科目陸續張傘，如同定點組一樣，在空中他們會分別依照重量排出先後著陸順序。（10"） ◎各位貴賓，四名疊傘隊員現在利用靈活操傘技術將傘具組合，第一個要帶來科目為四人「採肩式」疊傘，這是一種非常高超及困難動作，不僅有靈活操傘技術更必須達到人傘合一境界。（18"） ◎高空跳傘必須先期掌握著陸場的海拔高度、依據當時不同空層的風向、風速、氣流、觀察雲朵的高度與飄動情形，研判上空氣流及感應氣流的變化，作出正確判斷與傘具操控，才能確保精準的著陸與安全。 ◎各位現在疊傘組已完成四人疊傘組合，目前正朝向上風待命。分別由士官長洪威棣、吳明雄、上士黃克強、吳宗祐四人所組成隊形，他們將以精湛技術至4000英呎分開為兩組，表演「扇型」花式變換至著陸。（25"） ◎疊傘由下至上為1.2.3.4項，每一位隊員並需先行靠近，一項一項往上疊，最下面一員控制方向，所以在疊傘組合上必須要有良好默契。（15"） ◎現在接力隊員已準備進航，因使用傘具不同，他們將以跟蹤隊形進入目標區。第一員準備著陸的是潘益龍士官長，跳傘次數為637次，為高空跳傘資深教官也是2009年高雄市運會跳傘比賽國手。除多次參加國內各項演訓任務外，也培育出無數傘兵健兒。（20"） ◎緊接著預定著陸的為彭勇智士官長，跳傘次數為620次，除擔任高空跳傘種子教官外，更是各項演訓任務的指導員。（20"）

第14頁，共16頁

Army Aviation and Special Forces Command (陸軍航空特戰指揮部). Airborne Training Center. Dawu, Pingtung County. Media Event. December 10, 2010. Brochure 16/18.

拾、空降場動態操演說明稿

七	09:55.30 | 10:01:30 高空跳傘階段接力疊傘	◎第三名準備著陸的是少校教官黃志成，個人跳傘次數為800次。他的雙手各握有一條操縱繩，憑著靈活的操傘技巧，使傘具平穩的緩緩下降。(13") ◎最後準備著陸的是原住民資深士官長曾英志，個人跳傘次數為1152次，是本次跳傘次數最多的一員。擔任神龍小組特業教官已達19年，現在他也即將緩緩下降準備著陸，「精準著陸」。 「疊傘分開時稿」 ◎現在高度已到達4000英呎，4項傘隊員分兩組，第一組由士官長洪威棣、上士黃克強保持「採肩式」隊形至著陸，第二組士官長吳明雄及上士吳宗祐將變換「扇型」隊形，所謂「扇型」是隊員採平行方式互相抓住對方套袋，如同一把扇子形狀。(20") ◎目前隊員們已準備進航，第一組人員將採小角度方式進入，讓各位記者先生、小姐感受傘具震撼力，第二組隊員目前也準備進航，他們將在300呎高度分開，請各位記者小姐、先生儘速捕捉畫面。 ◎目前高空跳傘人員已全數著陸，接下來實施動力飛行傘操演。
八	10:02.00 | 10:10.00 動力飛行傘階段	◎接下來要為各位展示的課目為動力飛行傘師資班所帶來的動力飛行傘編隊飛行，各位記者小姐先生請將您的目光轉向一點鐘的方向，五具動力飛行傘已起飛，目前高度300呎，持續爬升至500英呎，本次操演的人員由領隊士官長徐仁輝帶領士官長李紹瑜、周建順、賴琮信、潘天亮等五員為各位展現「右跟蹤」、「大雁型」及「菱形」編隊飛行。(30") ◎動力飛行傘可區分為基礎、越野、地貌、海上飛行，亦可行小部隊執行敵後空中突擊、偵察、搜索、警戒、連絡、搜救、運補等多種任務，其編組極具彈性，並可與特種作戰部隊實施編組，執行地空聯合作戰、特種地區偵搜及突擊作戰等任務。 ◎現在五具動力飛行傘機具已到達500英呎高空，目前正準備以「右跟蹤」的隊形以衝場的方式向各位致敬，請各位嘉賓給予傘兵勇士熱烈的掌聲。現在已通過司令台上空。(15")

第15頁，共16頁

Army Aviation and Special Forces Command (陸軍航空特戰指揮部). Airborne Training Center. Dawu, Pingtung County. Media Event. December 10, 2010. Brochure 17/18.

拾、空降場動態操演說明稿

八	10:02.00 | 10:10.00 動力飛行傘階段	◎跟蹤隊形為基本隊形之一，其主要目的是第一員領隊必須有敏銳的觀察力，確實的掌握情資帶領隊員降落至待機位置，準備下一個戰術行動。目前已編隊完成正準備通過參觀台上空，每一位隊員將採節節高的方式通過，避免受到前一員的尾流造成傘具的不穩定狀態。 ◎現在動力飛行傘已飛越各位上空，即將前往潮東地區重新編隊，將以大雁型的方式通過。（30"） ◎今日所使用的動力機具為二行程氣冷式25匹單缸馬力引擎，啟動方式為電子式啟動，人員於高空中可熄火滯空靜音滑行，(如遇突發狀況可立即發動引擎實施脫離)；油箱容量為12公升，推力85公斤，可續行60公里，時速22-45公里，最大滯空時間為120分鐘。傘衣材質為達克龍多元脂布料製成，每條傘繩可達120斤拉力。 ◎現在飛行傘已通過參觀台，將前往大腳森林重新整裝編隊，執行下一個課目，就是「菱形」隊形。（8"） ◎菱形隊形也是基本隊形之一，除了要有過人的膽識，更要有高超的操控技術，不論是左右操縱繩及油門的控制都必須掌握相當的精確。（10"） ◎動力飛行傘可依任務不同實施不同架次編隊飛行可區分由2~4機所組成的分遣隊， 5~8機所組成的戰鬥隊及兩個分遣隊所組成特遣隊。 ◎目前已重新編隊完成，將由士官長李星慶帶領下將以菱形隊形通過參觀台。（8"） ◎動力飛行傘不受地形、地障限制，滯空時間長，不需運輸機即可實施地貌飛行，隱密性高、機動性強，並可擔任臨時性任務或獨立遂行小部隊特攻作戰，以提供指揮官戰術運用。 ◎動力飛行傘師資班為本部年度新開設班隊，目前屬於師資培訓階段，去年已通過民航局的認證考試，將推廣至本部各單位使用。 ◎動力飛行傘的編隊飛行已操作完畢，所有的動力飛行傘隊員將採五邊進航頂風著陸的方式降落在各位記者小姐先生的面前，請各位記者嘉賓掌聲鼓勵。 ◎本日動態操演已全部結束，請各位記者先生、小姐隨服務人員引導實施人物專訪，再次謝謝各位平日給予國軍支持與愛護，最後敬祝各位身體健康、萬事如意，謝謝大家！（15"）

第16頁，共16頁

Army Aviation and Special Forces Command (陸軍航空特戰指揮部). Airborne Training Center. Dawu, Pingtung County. Media Event. December 10, 2010. Brochure 18/18.

Press Badge. Defense News. Wendell Minnick. Chiayi Air Base; Tsoying Naval Base; Kinmen Island (ARB 101). 27 January 2016. During the Kinmen visit, the media witnessed a demonstration by the 101st Amphibious Reconnaissance Battalion[1] (Army Frogmen), under the Army Aviation and Special Forces Command (陸軍航空特戰指揮部).

[1] **Comparable to the Marine Corps Amphibious Reconnaissance Patrol (ARP).**

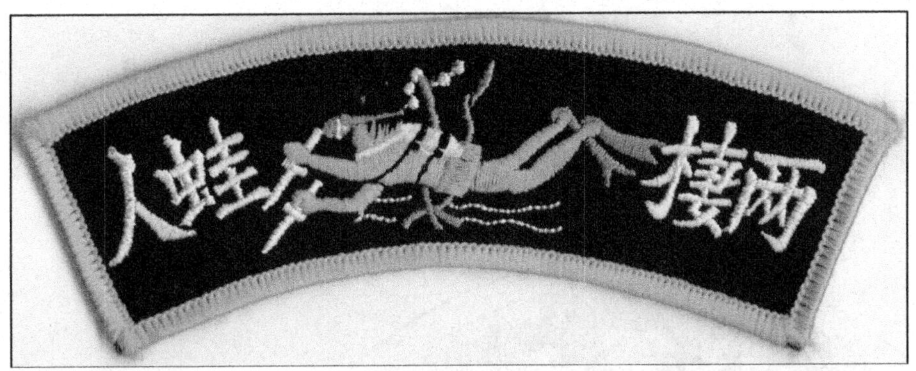

Army Aviation and Special Forces Command (陸軍航空特戰指揮部). 101st Amphibious Reconnaissance Battalion; nicknamed "Army Frogmen" or "Sea Dragon Frogmen" (中華民國陸軍 101 兩棲偵察營). Detachments spread across outer islands: 1st Reconnaissance Company - Kinmen; 2nd Reconnaissance Company – Penghu; 3rd Reconnaissance Company – Matsu; 4th Reconnaissance Company – Dongyin. Considered similar to US Navy SEAL Teams, but without helicopter support. The 101st was notorious during the Cold War for commando raids along the coast of mainland China. It is the most famous unit in Taiwan's military history and often confused with the Taiwan Marine Corps Amphibious Reconnaissance Patrol unit.

Army Aviation and Special Forces Command (陸軍航空特戰指揮部). Official Emblem: 101st Amphibious Reconnaissance Battalion (中華民國陸軍 101 兩棲偵察營).

Army Aviation and Special Forces Command (陸軍航空特戰指揮部). Official Emblem: 101st Amphibious Reconnaissance Battalion (中華民國陸軍 101 兩棲偵察營).

Today, the ARB has become famous for giving demonstrations to visiting foreign media and government delegations. However, they have largely become thing of nostalgia and have not seen active operations against mainland China in decades.

ARB 101 demonstration of beach infiltration. Kinmen Island (ARB 101). 27 January 2016. Photo by Wendell Minnick.

Veterans of the Army 101 Amphibious Reconnaissance Battalion Association (VAARBA). Non-Governmental Organization.

VAARBA

Veterans of the Army 101 Amphibious Reconnaissance Battalion Association (VAARBA). Non-Governmental Organization.

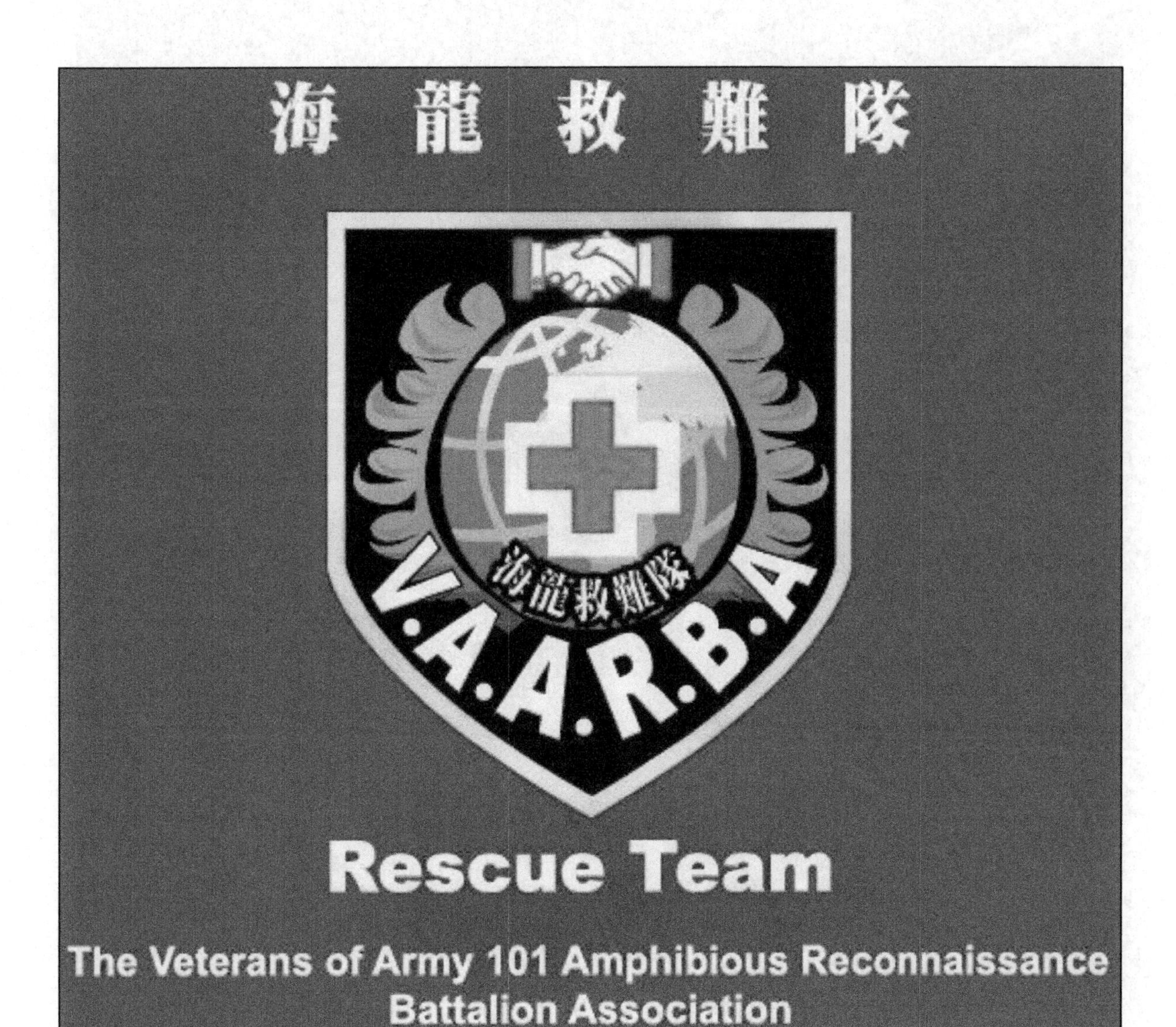

Rescue Team: Veterans of the Army 101 Amphibious Reconnaissance Battalion Association (VAARBA). Non-Governmental Organization.

Being a member of the ARB is a significant achievement amongst Taiwanese and has sparked a variety of fan art available in military surplus stores around Taiwan. They have been compared to the US Navy SEAL Teams, but exaggerations of this type reflect the ARBs legendary performance during the Cold War.

Marine Corps Training Center. Tsoying, Kaohsiung. Media Event. August 27, 2013. Brochure 1/10.

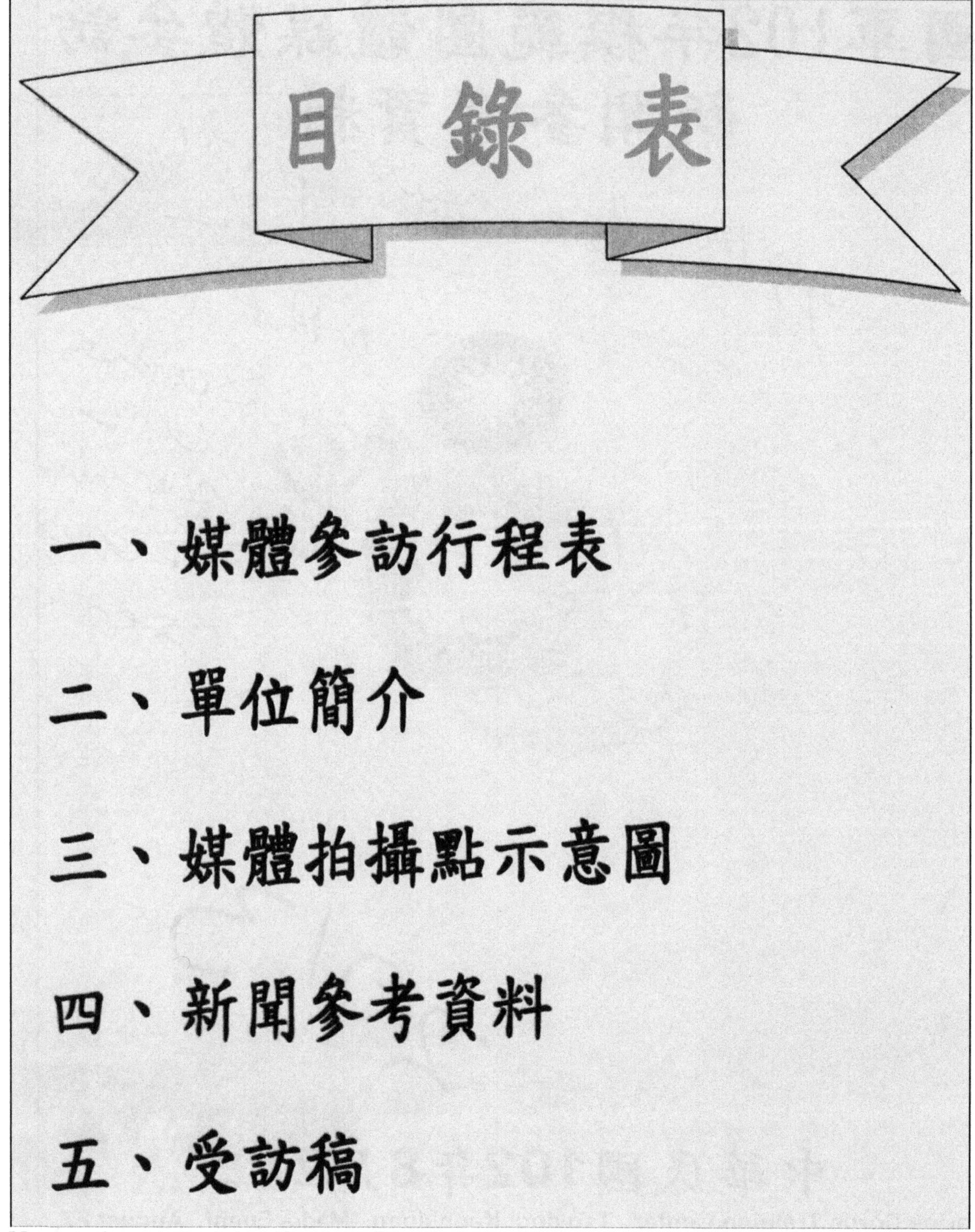
目錄表

一、媒體參訪行程表

二、單位簡介

三、媒體拍攝點示意圖

四、新聞參考資料

五、受訪稿

Marine Corps Training Center. Tsoying, Kaohsiung. Media Event. August 27, 2013. Brochure 2/10.

海軍陸戰隊辦理 102 年國軍模範團體媒體參訪行程表					
日期	起訖時間	行程	使用時間	地點	備考
民國一〇二年八月二十七日（星期二）	0900 | 0910	參訪行程說明 暨 模範單位簡介	10 分鐘	火牛營區 參觀台	合計 90 分鐘
	0910 | 1010	特戰突擊 實兵演練	60 分鐘	火牛營區 教練場	
	1010 | 1025	人員受訪	15 分鐘	火牛營區 教練場	
	1025 | 1030	合影	5 分鐘	火牛營區 教練場	
	1030 | 1200	發稿及休息	90 分鐘	四海一家	

Marine Corps Training Center. Tsoying, Kaohsiung. Media Event. August 27, 2013. Brochure 3/10.

國軍模範團體-陸戰九九旅步1營步1連簡介			
主官	單　　　　位	級　　　　職	姓　　　　名
	陸戰九九旅步1營步1連	上　尉　連　長	顧　　家　　安
沿革	連前身為陸戰99師657團409營第1連，於民國86年「精實案」改編為守備旅步1營步1連；94年4月1日配合「清泉案」改編為「海軍陸戰隊陸戰99旅步1營步1連」，99年由小港少康營區移防至林園句踐營區迄今。		
任務	平時負責責任區域「災害防救」及「重要目標防護」應(制)變任務，戰時負責第四作戰區高雄作戰分區，依令執行作戰任務。		
兵力編組	連部轄3個步兵排、1個火力排，全連編制約100員。		
主要裝備	T91步槍、機槍、機砲、、榴彈發射器等各式武器，另有中型戰術輪車、悍馬車等戰術車輛。		
能力	一、為基本戰鬥單位，具摩托化作戰能力，可於任何時間、天候、地形下遂行特種戰鬥。 二、具兩棲作戰能力，可裝載海軍登陸艦船遂行兩棲突擊、外島增援、規復作戰。 三、陸上作戰時可編組AAV7兩棲突擊車遂行機械化、濱海、渡河作戰，另可藉運兵直升機執行空中機動任務。		
重要事蹟	一、海軍司令部評比101年迄今督導第四作戰區跨區增援及戰備任務、進訓恆春兵科基地及「天秤」颱風地區災後復原等任務，執行成效為陸戰隊連級戰鬥部隊最佳單位。 二、完成102年聯興102-1兩棲登陸操演。 三、102年執行本隊「全志願」編裝實驗，執行成效良好。 四、101年基本體能、射擊、游泳鑑測，合格率均達80%以上。 五、101年7月迄今均無肇生訓練危安與軍紀安全事件。 六、102年榮獲國軍模範團體殊榮。		

Marine Corps Training Center. Tsoying, Kaohsiung. Media Event. August 27, 2013. Brochure 4/10.

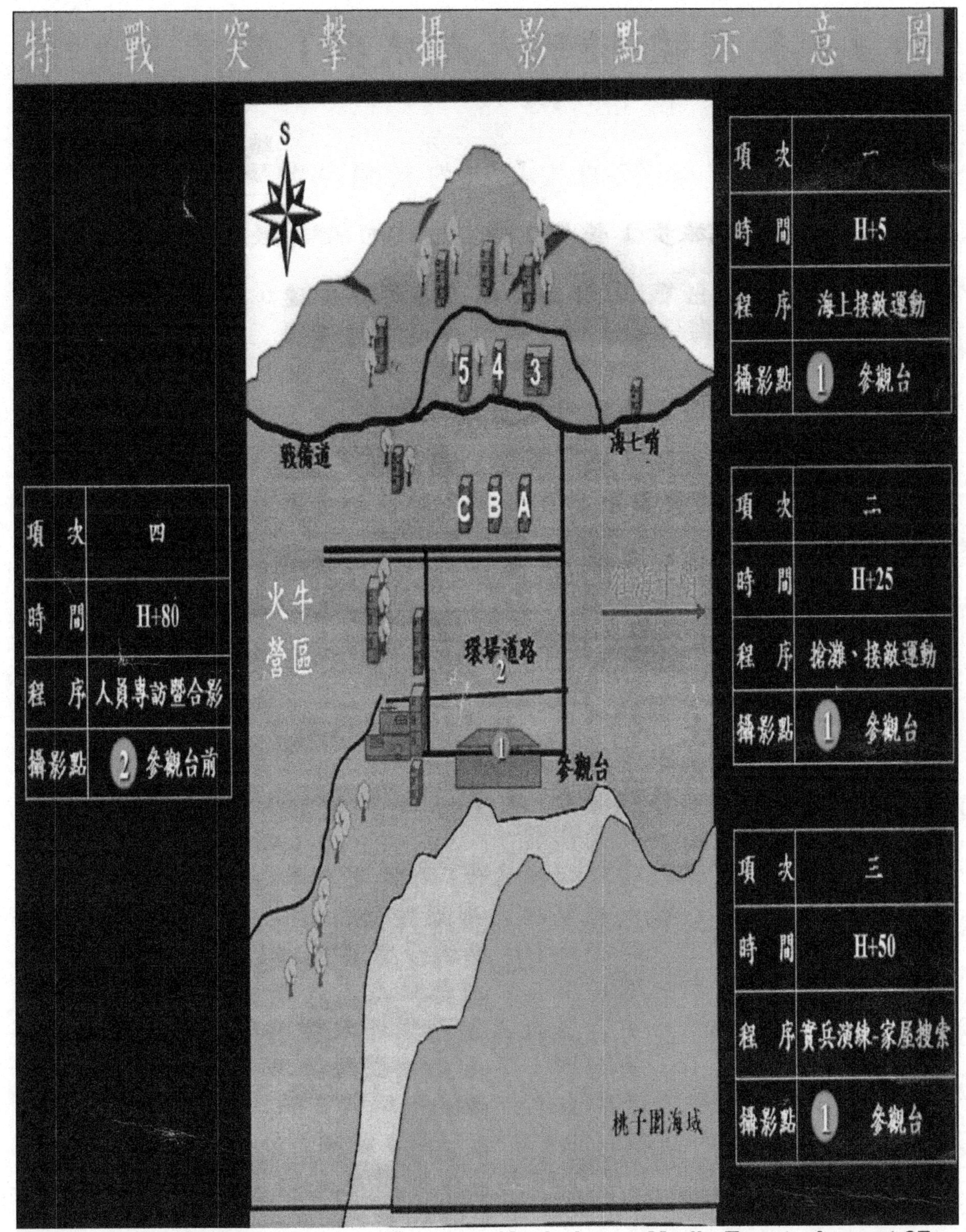

Marine Corps Training Center. Tsoying, Kaohsiung. Media Event. August 27, 2013. Brochure 5/10.

海軍陸戰隊新聞參考資料

102年8月27日

國防部於8月27日安排國內新聞媒體軍事記者，至高雄左營基地採訪當選國軍模範團體的海軍陸戰隊陸戰九九旅步一營步一連，該連展現高昂的士氣與堅實的訓練成果，獲得媒體高度的讚許。

當天國內新聞媒體軍事記者訪問團一行人，於上午9時抵達火牛營區，受到陸戰隊官兵的熱烈歡迎。在參訪過程中，軍事記者們親身體驗及目睹各階段戰鬥演練，對於陸戰隊隊員精誠團結、勇猛剽悍的表現，深表讚佩，更確信陸戰健兒們必能帶給全國同胞最安全的保障。

海軍陸戰隊九九旅步一營步一連，平時輪值地區戰備部隊，擔任「重要目標」防護及應變制變任務，並隨時支援地方「災害防救」，確保民眾生命財產安全；戰時任作戰區打擊部隊，並依令增援外、離島作戰。步兵連為基本戰鬥單位，其屬性可於任何時間、天候、地形下遂行戰鬥，摩托化編裝與兩棲專長訓練，強化多元作戰能力，編配AAV7兩棲突擊車，可編組機械化部隊，遂行兩棲突擊、濱海作戰，並可藉運兵直升機實施空中機動，是一支具三棲投射能力的精銳部隊。

Marine Corps Training Center. Tsoying, Kaohsiung. Media Event. August 27, 2013. Brochure 6/10.

陸戰隊指揮部指出，步一連是本隊「全志願役」編裝實驗單位，去年因執行作戰區跨區增援及高雄作戰分區戰備任務、進訓恆春兵科基地及「天秤」颱風地區災後復原等任務，執行成效經海軍司令部評比為陸戰隊連級戰鬥部隊最佳單位。在今年度執行聯興兩棲登陸操演、基本體能、射擊、游泳鑑測，合格率均達80%以上；去年7月迄今均無肇生訓練危安與軍紀安全事件，執行各項任務均有卓越成效，本次當選102年國軍模範團體是實至名歸，堪為國軍各部隊之表率。

陸戰隊指揮部進一步表示，陸戰隊全體官兵將賡續秉持不怕苦、不怕難、不怕死的三不怕精神，賡續戮力戰備訓練工作，建立堅實的可恃戰力，作為國家安全之屏障。

Marine Corps Training Center. Tsoying, Kaohsiung. Media Event. August 27, 2013. Brochure 7/10.

海軍陸戰隊陸戰九九旅步一營步一連
102 年國軍模範團體人員受訪稿

級職	上尉連長	學歷	海軍官校 97 年班 陸校正規 100 年班	主要事蹟	完成恆春兵科基地、雨棲基地、漢光 28、29 號演習、長青演習、聯勇操演、聯興演習、天秤颱風災害防救等任務。
姓名	顧家安				

本人於民國 100 年 7 月 8 日接任九九旅「步一連」連長迄今，對於本連獲選為國軍模範團體之殊榮，心中感到無比榮耀，同時也深深覺得自己我肩膀上的責任更加重了。

在接任連長後，看到本連，不論是幹部還是弟兄們，在各項演訓任務上，大家都全心全力的投入與付出，為的就是能使九九旅「步一連」這個招牌，成為全連共同的榮耀。

時至今年 2 月，為因應募兵制的推動，本連有幸擔任陸戰隊全志願役實驗連隊的主官，目前全連均由志願役人員所組成，連上目前不管在生活管理、部隊訓練等方面，均在建立未來陸戰隊全志願役部隊的雛形，期藉由完善的規劃，能使社會有志青年瞭解，成為一個陸戰隊員，會是自己一生中最正確的抉擇。

最後我也期盼在歷任連長所打造的堅實基礎，進一步發揮出九九旅「步一連」堅強的戰力，努力維護國軍模範團體這份得來不易的榮耀。

Marine Corps Training Center. Tsoying, Kaohsiung. Media Event. August 27, 2013. Brochure 8/10.

海軍陸戰隊陸戰九九旅步一營步一連
102 年國軍模範團體人員受訪稿

級職	士官督導長	學歷	屏東高工畢業 指職士官6期	主要事蹟	完成恆春兵科基地、兩棲基地、漢光28、29號演習、長青演習、聯勇操演、聯興演習、天秤颱風災害防救等任務。
姓名	黃國雄				

身為九九旅「步一連」資深士官幹部，除了士官督導長之職責外，還必須兼顧到連隊各層面之事務。尤其以如何加強戰備整備及持續維繫部隊戰力為重點，期能達成長官所交付的任務。

我們單位是全志願役人員所編成的連隊，在日常的生活管理或是戰備訓練方面，因為有一致的目標，所以人人精誠團結、上下一心，才能獲得今天這個屬於大家的共同榮耀，我以九九旅「步一連」為榮。

我們所有士官幹部都有一個共同努力的目標，就是讓每位弟兄都能嚴格的自我要求，完成職責內所應盡的本份，成為一名優秀的陸戰隊隊員，主動積極的完成任務，展現陸戰隊勇猛剽悍之精神。

Marine Corps Training Center. Tsoying, Kaohsiung. Media Event. August 27, 2013. Brochure 9/10.

海軍陸戰隊陸戰九九旅步一營步一連
102 年國軍模範團體人員受訪稿

級職	上兵	學歷	高英工商 畢業 社青 100-2 梯	主要事蹟	完成恆春兵科基地、兩棲基地、漢光 29 號演習、聯勇操演、聯興演習、天秤颱風災害防救等任務。
姓名	陳義東				

個人參加志願役士兵的行列，加入九九旅「步一連」服務迄今已近 3 年，深深體會榮譽是本連優良的傳統，每位官兵都為了連上榮譽，竭盡全力的付出，不論是演訓或救災任務，大家都能全力以赴，不分彼此秉持海軍陸戰隊「不怕苦、不怕難、不怕死」的三不怕精神，持續於工作崗位上努力，完成個人所肩負的任務。

本人身為九九旅「步一連」的一份子，非常高興本連能夠獲得年度國軍模範團體，這份榮譽，不屬於個人，而是我們大家共同努力所得來的，能夠在我們連上服務是本人的榮幸。身為海軍陸戰隊隊員，必須謹記「一日陸戰隊、終身陸戰隊」之精神，終生捍衛陸戰隊的榮譽。

Marine Corps Training Center. Tsoying, Kaohsiung. Media Event. August 27, 2013. Brochure 10/10.

MARINE CORPS

Marine Corps Training Base. Hengchun Base, Pingtung. There have been political calls to remove the image of mainland China from military patches. Expect some changes in the near future.

Taiwan Marine Corps. 66th Regiment. Motto: "Vanguard". Responsible for protecting Taipei. There have been political calls to remove the image of mainland China from military patches. Expect some changes in the near future.

Taiwan Marine Corps. 99th Regiment. Motto: "Iron Fist". Tsoying, Kaohsiung. There have been political calls to remove the image of mainland China from military patches. Expect some changes in the near future.

Marine Corps. Special Service Company (中華民國海軍陸戰隊特勤隊). Shoushan Camp, Gushan District, Kaohsiung City; 高雄市鼓山區壽山壽山營區. The editor has only interacted with this unit two times during the past twenty years and the unit is perhaps the most secretive of Taiwan's special operations units. Half of the team members are Aboriginal people. It is a battalion-level formation, but the actual strength is closer to an infantry company and is divided into six district teams. Each district team has its own expertise, e.g. boating, explosives, etc. The teams cross train the other teams.

SSC Uniform Pin

Emblem seen at SSC Hq.

Unit 1: 特勤隊第一區隊

Unit 2: 特勤隊第二區隊

Unit 3: 特勤隊第三區隊

Unit 4: 特勤隊第四區隊

Unit 5: 特勤隊第五區隊

Unit 6: 特勤隊第六區隊 (Explosives Experts/Door Breaching)

Marine Corps
Amphibious Reconnaissance Patrol (ARP)

Marine Corps. 1st, 2nd and 3rd Amphibious Reconnaissance Patrol (ARP).[2] 兩棲偵搜大隊徽，低識度版.

[2] **Comparable to the Army's Amphibious Reconnaissance Battalion (Army Frogmen).**

Taiwan Marine Corps. 1st Amphibious Reconnaissance Patrol (ARP). Taiwan's military never wore berets, but this ARP patch remains the only patch that continues to use a beret in the Taiwan military. Older patches for other units, including the Military Police, also included berets during the Cold War for unexplained reasons (probably to symbolize Western units like the US Army Green Berets and French Foreign Legion).

Taiwan Marine Corps. 2nd Amphibious Reconnaissance Patrol (ARP).

Taiwan Marine Corps. 3rd Amphibious Reconnaissance Patrol (ARP).

Taiwan Marine Corps. Training, Maintenance and Headquarters. Amphibious Reconnaissance Patrol (ARP).

Modern US Military/Defense Contractor Interaction

This challenge coin was minted celebrating the transition from contract civilians serving as US Defense Attachés to the return of active-duty Defense Attachés at the American Institute in Taiwan (AIT), the de facto US Embassy in Taipei. Coin designed by former US Navy Attaché (Deputy Chief - 2001-2005) Michael McCallus. Coin presented to the editor by US Army Attaché (Deputy Chief and Senior Advisor & Army Affairs Officer) Brad Gerdes around 2005. Contractors proved unreliable after the 1979 US government policy to end US-Taiwan military relations. This side of the challenge coin is an exact duplicate of the China Service Medal awarded to US Navy and US Marine Corps during World War II.

Flip side of the coin shows the original Flying Tigers (American Volunteer Group) along with the original shield from WW II.

The shield on the AIT coin reflects the US Military Assistance Advisory Group in Taiwan (1951-1978).

China Service Medal – World War Two.

US Defense Contractors

Note to reader: The US has a vast network of defense contractors working in Taiwan to fulfil training/technical service contracts on military equipment. These are not US military personnel, but retired US military personnel working for companies such as Lockheed, Northrop, Boeing, etc. These types of contractors can be found anywhere in the world where US military equipment is sold. However, due to their cautious nature, they tend to congregate in the same locations for recreation. In Taipei, they are well known to frequent two pubs in the seedy Combat Zone (Shuangcheng Street): My Place Pub and Malibu Pub. The Malibu also serves as the *de facto* Veterans of Foreign Wars (VFW) meeting place. In Kaohsiung, the most popular pub is the Bottoms Up Saloon. Those interested in the history of the US military pub scene need go no further than the 50th anniversary edition of the 1969 *Taipei After Dark* by Andrew Harris (Amazon books).

A fan patch celebrating the history of the US Army's 1st Special Forces Group (Green Berets) Operational Detachment Alpha (ODA) or Detachment 26 stationed at Longtan District, Taoyuan, from 1960-1973. Taiwan's military also participated in special operations with Green Berets in the Vietnam War via Political Warfare Special Task Force (PWSTF). They were formed during the early part of the Vietnam War after requests were made by Saigon for special units who could counter the intervention of Chinese troops in the war, particularly translating Chinese prisoners captured by the South Vietnamese military.[3] The Political Warfare Regiment created a variety of task forces that included elements from

[3] Tourison, Sedgwick. *Secret Army, Secret War.* Naval Institute Press, 1995. Later republished in 1997 by St. Martin's: *Project Alpha: Washington's Secret Military Operations in North Vietnam.*

different special operations units, some of which operated inside enemy territory.[4] They also taught South Vietnamese personnel on the political war system; comparable to the US Special Forces (Green Beret): Unconventional Warfare (UW), Direct Action (DA), Counter-Insurgency (COIN), Special Reconnaissance (SR), Information Operations (IO), Psychological Warfare (PW), and Security Force Assistance (SFA). China sent around 300,000 troops into Vietnam and Laos from 1960 to 1970; numbers conflict, but China lost an estimated 1,000 troops during the war. According to a former CIA analyst at the US Embassy in Saigon, the Taiwan embassy failed to evacuate 1,000 personnel, many intelligence/military assets.[5]

During the 1980s, the US Congress shut down US Special Forces training at the School of the Americas after claims torture methods were taught. The School taught soldiers and intelligence assets involved in anti-Communist insurgencies in Central America. When shuttered, the training was moved to Taiwan's Fuhsing Kang College (Political Warfare Academy). As Taiwan moved toward democracy in the 1990s, these units and tactics were phased out.

---

[4] Leary, William M. *Perilous Missions: Civil Air Transport and CIA Covert Operations In Asia.* Smithsonian Institute Press, 2002. See also: Pocock, Chris. *The Black Bats: CIA Spy Flights Over China From Taiwan, 1951-1969.* Schiffer Military History Book, 2010; Conboy, Kenneth and James Morrison. *Feet to the Fire: CIA Covert Operations in Indonesia, 1957-1958.* Naval Institute Press, 1999.

[5] Snepp, Frank. *Decent Interval.* University Press of Kansas, 1977. Page 421.

中華民國陸軍第三特遣隊
STF
鉄 血
STF-001

P.W S.T.F
2
STF
遣特
戰政

PWSTF Club/Veterans Group.

A member of former President Ma Ying-jeou's personal protection unit under the National Security Bureau. At the time of this photograph, they were trained by US-based Blackwater via a contract secured by the late USMC Col. (Ret.) Paul D. Behrends (1958-2020). The contract for outside security training was prompted by the 2004 assassination attempt on former President Chen Shui-bian. The security officer's face was intentionally obscured by the editor to protect his identity. Photograph by Wendell Minnick, Editor. *See*: Weinberger, Sharon. "The Life and Death of the Most Notorious Man in Washington." *New York* Magazine (1 January 2021).

Front and back of Blackwater challenge coin presented to the editor by Behrends at the 2006 US-Taiwan Defense Industry Conference in Denver, Colorado. The annual conference is sponsored by the Arlington-based US-Taiwan Business Council in different cities around the United States.

Commemorating the 25th anniversary of the 1995-1996 Taiwan Strait Crisis of March 8th 1996. Often dubbed the Third Taiwan Strait Crisis (July 21, 1995 to March 23, 1996). The US sent two aircraft carrier groups to the area in response to Chinese short-range ballistic missile tests (DF-15/M-9) to intimidate Taiwan. The patch only mentions ships listed for the March 8th missile launch: CVA-62 USS *Independence*, CG-52 USS *Bunker Hill*, DD-975 USS *O'Brien*, DD-966 USS *Hewitt*, and the FFG-41 USS *McClusky*.

"March Madness 96" commemorative patch for the 1995-1996 Taiwan Strait Crisis (Third Taiwan Strait Crisis) from July 21, 1995 to March 23, 1996. The US sent two aircraft carrier groups to the area in response to Chinese short-range ballistic missile tests (DF-15/M-9) to intimidate Taiwan. The patch lists only two USS Navy assets: Carrier Air Wing Five (CVW-5) and USS *Independence* (CV-62). On the top of the patch is the triangle insignia for the CVW-5. At the bottom is the First Navy Jack flag with a rattlesnake with gold scales and the "Don't Tread On Me" slogan against red and white stripes. The First Navy Jack flag is flown by the oldest US Navy ship in service (CVA-62 USS *Independence* at the time of the crisis). "It's The Only Game In Town" with four dice is unclear (possibly a reference to the Asian game of Liar's Dice, but that game requires five dice).

Taiwan Domestic Civilian Militia Training

Taiwan Military and Law Enforcement Tactical Research and Development Association (TTRDA)

台灣軍警戰術研究發展協

Angst amongst some civilians over Taiwan's reduced military readiness resulted in the creation of the TTRDA for military and first responder training for civilians. The facility has become so popular that Taiwan special operation forces visit the facility for close quarter combat (CQC) exercises. Consisting of a mix of former, reserve and active-duty special operations soldiers, current members of Taiwan's SWAT police units, and those with special skills. The organization was created in 2015 as a non-governmental organization (NGO). Though these are airsoft guns, live fire is difficult in Taiwan due to strict gun laws, but some are members of the Taiwan Shooting Sports Association-Ding Fwu and the Chinese Taipei Shooting Association. The Taiwan government has expressed concern over the creation of a para-military organization, but has not moved to shut TTRDA down so long as it remains an NGO and uses non-lethal methods to train. To accuse the organization as just another LARPing (Live Action Role Play) club, the members of this organization believe that China will invade in the near term and are frustrated by the Taiwan government's continued reduction in military training. Consult: https://www.facebook.com/TTRDA/

"LE" = Law Enforcement. Taiwan Military and Law Enforcement Tactical Research and Development Association (TTRDA). 1/6.

Taiwan Military and Law Enforcement Tactical Research and Development Association (TTRDA). 2/6.

Taiwan Military and Law Enforcement Tactical Research and Development Association (TTRDA). 3/6.

Taiwan Military and Law Enforcement Tactical Research and Development Association (TTRDA). 4/6.

TTRDA 台灣軍警戰術研究發展協會
戰術射手推廣賽-南部分區賽(賽程#03)

賽程#3
內容說明：你在餐廳喝咖啡時遭遇匪徒搶劫，你必需要在被匪徒殺害前發起攻擊，擊退入侵者，並保護路人。

是否隱藏用槍：必須。
最少發數：16 發
接戰要求：每個紙靶 2 發胸
槍口安全指向：採用 180 度規則
計分方式：Vickers Count

開始：
手槍上膛、入套、於 P1 位置坐下，左手拿取報紙，右手拿咖啡杯。

聲響：
於 P1 接戰 T1-T2，戰術優先。
移動至 P2 前接戰鋼靶(至少四發)及 T3，於 P2 接戰 T4 不可由掩體上方出槍，至 P3 於掩體後方任選姿勢及左右側，接戰 T5-T6，戰術優先。

請注意：至 P2 不得重複接戰 T1-T4：不可於沒有掩體保護下更換彈匣。

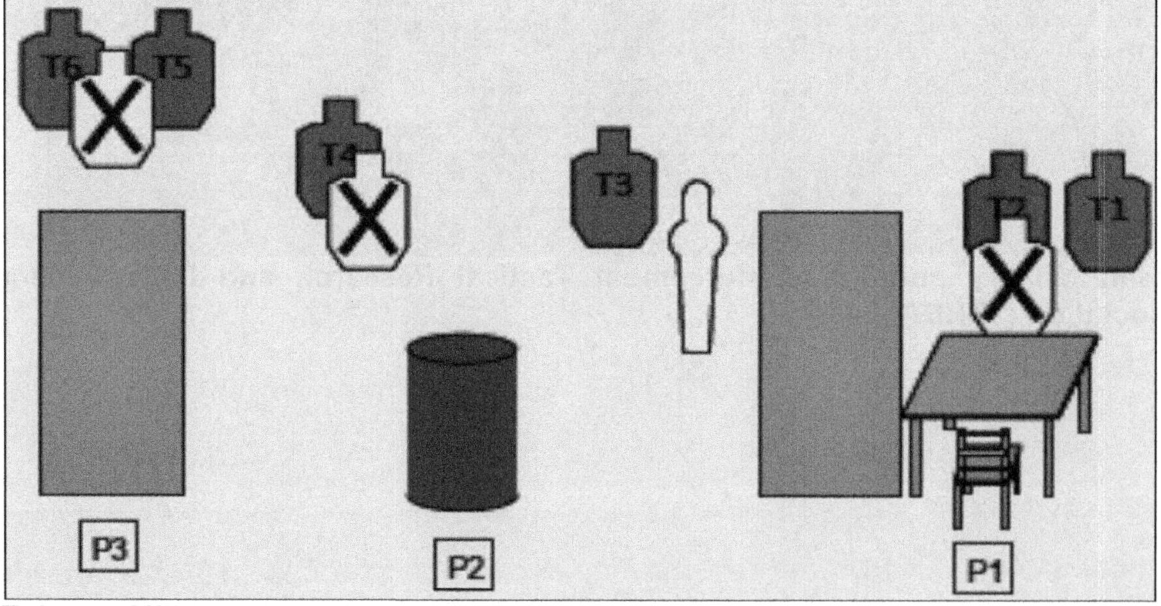

T6 T5 T4 T3 T2 T1

P3 P2 P1

Taiwan Military and Law Enforcement Tactical Research and Development Association (TTRDA). 5/6.

Faces distorted to protect identities of active-duty military/police personnel. Taiwan Military and Law Enforcement Tactical Research and Development Association (TTRDA). 6/6.

Camp 66 Military Club.

Camp 66 (2019-2021) was shuttered after only two years when the land was reclaimed for new condominiums. Camp 66 owner, "Bear", is attempting to restart the camp at another location. This is one example of fandom interest in military affairs in Taiwan. However, despite this example of live action role play (LARP), most of the population of draft age males have negative opinions of serving in Taiwan's military. This particular camp was unusual as it was a museum, pub, restaurant, close quarter combat (CQC) airsoft shooting course, and obstacle course for Polaris Off-Road MRZR-4, Ranger 570 and RZR (pronounced "razor) 570 EP5 rovers for rent. Guns are illegal in Taiwan. Overall, Camp 66 was family friendly and will be missed. The US Marine Corps procured x144 MRZR vehicles in 2016; and the US Special Operations Command procured x1500 RZR in 2013. Taiwan has showed no in the Polaris system; in 2012 an unarmored, 1,225kg four-wheel-drive, three-seat Special Combat and Assault Vehicle (SC-09A) was locally manufactured. Airborne was the first unit to test the vehicle, which came with puncture-proof wheels, an anti-blast fuel tank, night-vision equipment and a searchlight. The vehicle had right passenger and rear gun mounts that can be fitted with a MK-19 40mm grenade launcher and T74 machine gun. A third gun rack, which can accommodate three T91 assault rifles, was located in the rear compartment. None of the vehicles were ordered by the military despite successful tests. Note to reader: Camp 66 was named in tribute to U.S. Route 66.

Camp 66 Military Club. 2019 Taipei Aerospace and Defense Technology Exhibition (TADTE). Brochure 1/16.

Camp 66 Military Club. 2019 Taipei Aerospace and Defense Technology Exhibition (TADTE). Brochure 2/16.

Camp 66 Military Club. 2019 Taipei Aerospace and Defense Technology Exhibition (TADTE). Brochure 3/16.

Camp 66 Military Club. 2019 Taipei Aerospace and Defense Technology Exhibition (TADTE). Brochure 4/16.

Camp 66 Military Club. 2019 Taipei Aerospace and Defense Technology Exhibition (TADTE). Brochure 5/16.

Camp 66 Military Club. 2019 Taipei Aerospace and Defense Technology Exhibition (TADTE). Brochure 6/16.

Camp 66 Military Club. 2019 Taipei Aerospace and Defense Technology Exhibition (TADTE). Brochure 7/16.

Camp 66 Military Club. 2019 Taipei Aerospace and Defense Technology Exhibition (TADTE). Brochure 8/16.

Camp 66 Military Club. 2019 Taipei Aerospace and Defense Technology Exhibition (TADTE). Off-Road Vehicle (ORV) Obstacle Course. Brochure 9/16.

Camp 66 Military Club. 2019 Taipei Aerospace and Defense Technology Exhibition (TADTE). Ranger 570, MRZR-4, RZR-570 EPS. Brochure 10/16.

Camp 66 Military Club. 2019 Taipei Aerospace and Defense Technology Exhibition (TADTE). Brochure 11/16.

Camp 66 Military Club. 2019 Taipei Aerospace and Defense Technology Exhibition (TADTE). Naval Section. Brochure 12/16.

Camp 66 Military Club. 2019 Taipei Aerospace and Defense Technology Exhibition (TADTE). Combat Section. Brochure 13/16.

Camp 66 Military Club. 2019 Taipei Aerospace and Defense Technology Exhibition (TADTE). Brochure 14/16.

Camp 66 Military Club. 2019 Taipei Aerospace and Defense Technology Exhibition (TADTE). Vietnam War Section. Brochure 15/16.

Camp 66 Military Club. 2019 Taipei Aerospace and Defense Technology Exhibition (TADTE). Brochure 16/16.

RECOMMENDED PERIODICALS (TRADITIONAL CHINESE ONLY)

Note to reader: The legendary *Defense Technology Monthly* (*尖端科技軍事雜誌*) magazine went out of business around 2019.

***Illustrated Guide for Weapons and Tactics*. Issue: 112, August 2020. Note to Reader: This is in the editor's opinion the best magazine in Taiwan on army/land warfare issues both historical and current for Taiwan. This particular issue (#112) has the best order of battle (OB) available via open source. 1/5.**

Illustrated Guide for Weapons and Tactics. 2/5.

兵器戰術圖解

ILLUSTRATED GUIDE FOR WEAPONS & TACTICS

出　　版 2020年08月10日
出 版 者 兵器戰術圖書有限公司
發 行 人 王玉瑛
總 編 輯 劉文孝
特約撰述 陳慶鎧 王清正 劉雨青
資料編輯 林宏一
美術編輯 林書豪 張詠翔 陳建中 吳宗恒
法律顧問 江燕偉律師
發 行 址 台北市林森北路409號9樓之16
Telephone 886-2-2100-1220
Facsimile 886-2-2541-9080
網　　址 http://www.shop2000.com.tw/wingweb/
電子信箱 wing.web@msa.hinet.net
登 記 證 新聞局局版臺業字第5087號
印 刷 所 通威印刷有限公司
總 經 銷 創新書報份有限公司
國內訂價 新台幣229元

***Illustrated Guide for Weapons and Tactics.* 3/5.**

如何獲得中國之翼的出版訊息？

中國之翼除了出版中文軍事圖書之外，也是目前台灣地區最大的外文軍警圖書集散中心，已經進口數以千計的相關資料及商品。若您希望爾後均能獲得中國之翼更多更新的訊息，以下有3種方式可供選擇：

1. 請附60元回郵寄至本公司，即可訂閱未來半年中國之翼郵寄的《新品目錄》。
2. 請瀏覽本公司的網站 www.shop2000.com.tw/wingweb，可供您選購商品，也可查閱資料。
   您也可以直接發一通 E-MAIL 到中國之翼的電子信箱 wing.web@msa.hinet.net，本公司就會將您的資料登錄，爾後每週都會將最新的訊息發送至閣下的電子信箱。
3. 直接光臨中國之翼門市部，地址：台北市林森北路409號9樓之16，開放時間：週一～週五中午12：00～晚間6：30，週六及例假日上午10：00～下午6：00 （週日公休），其他時間請來電預約。
   服務電話：(02) 2100-1220 傳真：(02) 2541-9080

◆ 海外地區讀者預定《兵器戰術圖解》◆

收件地區	海運		空運	
	一年6期	二年12期	一年6期	二年12期
港澳地區	1700	3200	2280	4200
亞太地區	2200	4300	2580	4900
歐美地區			2700	5200
大陸地區	1700	3200	2280	4200

◆ 國內地區讀者預定《兵器戰術圖解》◆

	掛號		平郵	
	一年6期	二年12期	一年6期	二年12期
會員	1220	2440	1100	2200
非會員	1370	2740	1250	2500

（二年訂戶加贈10片本社採訪光碟）

郵購注意事項：

對於代理之進口商品，為確保郵寄安全，煩請訂購者按郵局公訂標準支付掛號郵資。

⊙ 書籍類— 249g 以下 36元
250g—999g 55元
1kg 以上 80元 （郵資合併計算）

⊙ 模型類－基本包裝掛號費80元，2盒以上每增一盒請附加10元包裝材料費。

⊙ 光碟類－無論數量，一律80元。

⊙ 海報類－（包含裝紙筒）無論數量，一律100元。

大量採購，請先洽本社報價。

會員參加辦法：

如果閣下對本社大多數商品均具有興趣，而且採購預算也較高，則可考慮每次購物均以「基數金額」（1500元為一基數、3000元為二基數，以此類推，會費如同禮卷可抵購物金額）辦理預約，可享下列之優惠：

1. 「郵購目錄」優待期限內之商品，均可享8折優待。
2. 超過優待期限之商品仍可享9折優待。
3. 可以電話搶先預訂商品。

注意事項：

1. 存款餘額不足以購買任何商品時，請繼續繳付下一基數之預約款，以繼續享有會員之優惠。
2. 所有餘額用盡之後，即喪失價格優惠資格，而須比照一般顧客計價。
3. 軍警學生可憑證（郵寄影本至本社存檔）比照存款會員優待。
4. 凡個人或團體一次存款達20,000元，即可比照經銷商享有之折扣。

年（拾大）

郵政劃撥存款收據
注意事項

一、本收據請妥為保管，以便日後查考。
二、如欲查詢存款入帳情形，請檢附本收據及已填妥之查詢函向任一郵局辦理。
三、本收據各項金額、數字係機器印製，如非機器列印或經塗改或無收款郵局收訖章者無效。

請寄款人注意

一、帳號、戶名及寄款人姓名、通訊處請詳細填明，以免誤寄。
抵付票據之存款，務請於交換前一天存入。
二、本存款單金額之幣別為新台幣，每筆存款至少須在新台幣十五
元以上，且限填至元位為止。
三、倘金額塗改時請更換存款單重新填寫。
四、本存款單不得黏貼或附寄任何文件。
五、本存款金額業經電腦登帳後，不得申請撤回
六、本存款單備供電腦影像處理，請以正楷工整書寫並請勿折疊。
帳戶如需自印存款單，各欄文字及規格必須與本單完全相符。
如有不符，各局應婉請寄款人更換郵局印製之存款單填寫，以利處理。
七、本存款單帳號與金額欄請以阿拉伯數字書寫。
八、帳戶本人在「付款局」所在直轄市或縣（市）以外之行政區域存款，需由帳戶內扣收手續費。

交易代號：
0501、0502現金存款、0503票據存款、2212劃撥票據託收
本聯由儲匯處存查284,000束（100張）96.12.210×110mm（80g/m2模）保管五

Illustrated Guide for Weapons and Tactics. 4/5.

Illustrated Guide for Weapons and Tactics. **Issue: 112, August 2020. Note to Reader: This is in the editor's opinion the best magazine in Taiwan on army/land warfare issues both historical and current for Taiwan. This particular issue (#112) has the best order of battle (OB) available via open source. 5/5.**

***Asia-Pacific Defense Magazine*. Volume 164; February 2022. Note to Reader: This is a Taiwan published magazine and covers both Taiwan and the region. The magazine sends journalists to air shows and defense exhibitions around Asia, including the Zhuhai Airshow. 1/4.**

ASIA-PACIFIC DEFENSE MAGAZINE
亞太防務雜誌

發行人 Publisher
鄭宗漢 John Cheng

總編輯 Editor in Chief
鄭繼文 Kevin Cheng

法律顧問 Legal Adviser
詮泰法律事務所 Cyuantai Law Offices
林傳源律師 Arron C.Y. Lin

編輯顧問 Edit Adviser
楊仕樂 Shih-Yueh Yang
莊中毅 Chuang Chung-I

行銷顧問 Marketing Adviser
楊忠翰 Arthur Yang

採訪編輯 Project Editor
沈振宇 Chenyu Shen
傅明德 Melvin Fu
C.H King

視覺設計 Art Director
雅圖創意設計有限公司 Art2studio

特約主筆 Writers by Arrangement
林煒舒、黃東、滕昕雲、詹皓名、
謝仲平、曹瀛生

攝影 Photographer
黃瑞志

特約攝影 Photographer by Arrangement
顏際隆、吳根賢、徐小丹、周民孝、方明道
管延境

特派員 Correspondent
Johnson Yahg (USA)
Vincent Martens (Europe)
杜礎圻(香港)

出版：嶧儒文化創意國際企業有限公司
地址：10049台北市中正區紹興北街31巷15弄8號2F
電話：886-2-2322-3176
傳真：886-2-2322-3759
email: apd.onlineservice@gmail.com

YUE-RU CULTURE CREATIVE INTERNATIONAL,INC.
No.8-2, Alley 15, Lane 31, Shaoxing N. St.,
Zhongzheng District, Taipei 10049, Taiwan
TEL: 886-2-2322-3176
FAX: 886-2-2322-3759

製版印刷 龍岡印刷有限公司

國內總經銷 創新書報股份有限公司
電話：886-2-2917-8022
傳真：886-2-2915-6275
同德書報有限公司
電話：852-3551-3388
傳真：852-3551-3300

中華郵政台北雜字第1586號執照登記為雜誌交寄
雜誌如有缺頁、破損、裝訂錯誤，請寄回更換

封面故事 Cover Story 國軍春節戰備巡弋

14 2022年國軍春節加強戰備媒體參訪活動
ROC Forces in Chinese New Year Festivities

焦點議題 Top Issues

4 空軍的迫切危機
The ROCAF's Urgent Crisis

特蒐報告 Special Reports

6 層出不窮的共諜案
The Endlessly Emerging Espionage Cases in Taiwan

發燒話題 Hot Topics：難產的震海專案

18 海軍新一代飛彈巡防艦案
Taiwan's New Generation Frigate Project

專題策劃 Monograph：盤點2021年解放軍海軍

26 2021年度解放軍海軍行動回顧（上）
Forward to 2022: Review PLAN Footprints Part1

臺海軍情 Taiwan Strait Issues

35 國軍新版軍事訓練役評析
"Late" is better than "Never"
40 F-16戰機讓臺美關係更緊密
The F-16 Fighter Stories in ROCAF
48 鯤龍量產 戰翼待展
AG-600 Mass Production Starts, Operational Role Expected
52 恐將失控的海巡造艦計劃（下）
ROCCG's Fleet Expansion Plan without Sufficient Manpower Part2

Asia-Pacific Defense Magazine. 2/4.

第164期 目錄

亞太防務・封面特報

軍情動態 Intelligence

58 亞太小北約現身
AUKUS - the Indo-Pacific Pact of the Future

64 普京舞劍 意在攻烏？
2021-2022 Russo-Ukrainian Crisis

武器大觀 Armaments

68 美國空軍「下一代制空」計畫
Next Generation Air Dominance Program

74 「香草」無人機
Vanilla Unmanned

防務快門 Military Shutter

77 歷史性的全女性編隊飛行
Historic All-Female formation flight

80 裝訓部戰車排專精訓練
ROC Army's Tank Platoons Conduct Combat Firing Exercise

84 第40批亞丁灣護航編隊
40th Chinese Naval Escort Taskforce Departs for Gulf of Aden

封面攝影／APD

1月5日至7日，國防部舉辦「國軍春節加強戰備媒體參訪活動」，參訪活動的首站是空軍嘉義基地，駐防該基地的空軍四聯隊於2021年11月18日舉行「F-16V戰機接裝成軍典禮」，目前是配備最先進戰機的空軍戰鬥機部隊。在這次的媒體參訪活動中，基地安排了12架F-16V戰機大象走路、緊急起飛、飛行員頭盔瞄準系統介紹等課目，展現部隊的戰備訓練現況。

編輯室報告 Editor Desk

烏克蘭問題持續僵局，依舊是國際關注焦點。在過去的一個多月來，俄羅斯與美國、北約分別舉行多場雙邊對話，但各方都堅持立場，使得緊張局勢未能獲得緩解。

自二〇一四年的烏克蘭危機以來，北約與俄羅斯之間就處於對峙狀態，雙方每年都舉行相互針對性軍演，而且雙方機艦經常近距離接觸，國際間都擔心會擦槍走火爆發衝突。期間，烏克蘭東部經常爆發零星的衝突，基本上仍舊是處於內戰狀態。隨著西方增加對烏克蘭援助，並派出軍事顧問協訓和供應新武器裝備，讓烏克蘭政府逐漸有底氣，對東部地區盧甘斯克和頓內次克人民共和國的態度也日趨強硬，增加武力鎮壓，而且日益急迫加入北約組織。

烏克蘭政府的作為刺激俄羅斯的敏感神經，在二〇二一年兩度增兵俄烏邊境，造成烏克蘭情勢急轉直下，國際間都非常擔心俄烏會爆發戰爭，進而導致北約與俄羅斯的衝突。烏克蘭情勢令人憂心，就連一向不太關心國際情勢的臺灣社會也關注該情勢的發展，臺灣媒體甚至將烏克蘭問題類比於兩岸目前的僵局。雖然就大國競爭角度來看，烏克蘭和臺灣都是美國用以壓制俄羅斯和中國大陸的好牌，但烏克蘭問題與兩岸問題其實有著很大的差異，筆者在此限於篇幅不予論述，日後會在雜誌內做完整的論述。

本期雜誌有多篇關於兩岸軍事、安全議題的精采文章，F-16V戰機尤其是關注焦點。一月五日，國防部「國軍春節加強戰備媒體參訪活動」邀請媒體採訪空軍嘉義基地，見證F-16V戰機的戰備狀況；十一日，編號六六五〇的F-16V戰機在執行訓練任務中失事墜毀，飛行員不幸殉職，該機在媒體參訪活動也參加操演，令人震驚與不捨。本期規劃多篇與F-16V戰機有關的文章，希望讀者們滿意編輯部的安排。

Asia-Pacific Defense Magazine. 3/4.

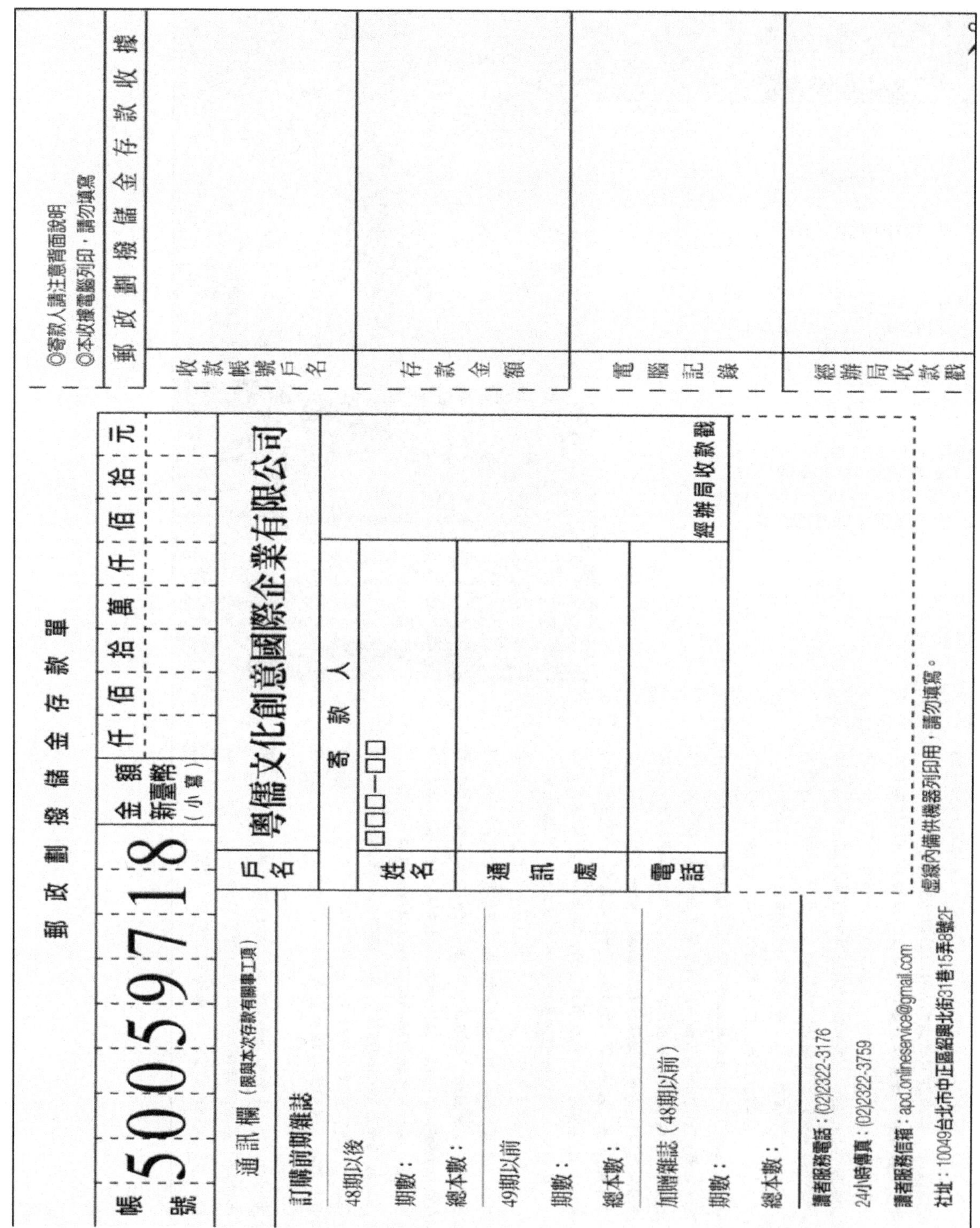

郵政劃撥儲金存款單

帳號 50059718

金額 新臺幣（小寫） 仟 佰 拾 萬 仟 佰 拾 元

戶名 粵儒文化創意國際企業有限公司

寄款人

姓名

通訊處 □□□—□□

電話

經辦局收款戳

通訊欄（限與本次存款有關事項）

訂購前期雜誌

48期以後

期數：

總本數：

49期以前

期數：

總本數：

加贈雜誌（48期以前）

期數：

總本數：

讀者服務電話：(02)2322-3176

24小時傳真：(02)2322-3759

讀者服務信箱：apd.onlineservice@gmail.com

社址：10049台北市中正區紹興北街31巷15弄8號2F

虛線內備供機器列印用，請勿填寫。

◎寄款人請注意背面說明

◎本收據電腦列印，請勿填寫

郵政劃撥儲金存款收據

收款帳號戶名

存款金額

電腦記錄

經辦局收款戳

***Asia-Pacific Defense Magazine*. 4/4.**

全球防衛雜誌
450
軍事家
FEB 2022
NT$200
JHMCSII
聯合頭盔瞄準系統二代
次世代顯示裝置
春節加強戰備 陸軍機步排城鎮戰對抗演練
空軍首度展示聯合頭盔瞄準系統 海軍塔江號巡邏艦首度海上操演
ISSN 1010-3228

Defence International. **Volume 450; February 2022. Note to Reader: This is a Taiwan published magazine and covers both Taiwan and the region. The magazine sends journalists to air shows and defense exhibitions around Asia, including the Zhuhai Airshow. http://www.wingweb.com.tw/ 1/6.**

想得到第一手最新武器動態介紹嗎
請至FB官方網站按讚.不定時會有免費好康贈獎活動
軍事家 全球防衛雜誌 Defence International

第七十五卷六期 2022 FEB

軍事家

全球防衛雜誌社有限公司
台北中文版 局版北市誌字第 2199 號
地 址／北市忠孝東路 1 段 35 號 3 樓
電話：（886-2）-2391-5105（代表號）
傳真：（886-2）-2391-8049

軍事家 軍事雜誌
香港國際中文版
Defence International
3F.35, Sec. 1, Chung Hsiao E. Rd.
Taipei, Taiwan.
TEL：（886-2）2391-5105（代表號）

發行人暨執行長 General Managing
黃慧津 Hui-Chin Huang

媒體總監 Vice General Managing
陳永盛 Vincent Chen 分機 14
diic.vincent@gmail.com

資深編輯 Senior Editor
陳國銘 Smith Chen 分機 15
mig21@mail2000.com.tw
資深編輯 Senior Editor
宋玉寧 Jerry Song 分機 13
jerrys6439@gmail.com
採訪編輯 Project Editor
Tsungfang Tsai
採訪編輯 Project Editor
羅界山 Chieh-Shan LO
專案編輯 Project Editor
吳凱文 Kevin Wu

美術編輯 Art Chief Editor
陳子毅 morries Chen 分機 16

特約攝影 Special Correspondent
強振國 Johnson Chiang
特約專業軍事作者 Special Correspondent
Mark Munzel （加拿大） 尼斯（德國）
Raymond Sun（港澳）托瑞曼斯（比利時）

客服部 Supervisor
王雅韻 Emilie 分機 10
hhuichin@ms6.hinet.net

行政郵務 Reader Service Specialist
林恩弘 ELLIS 分機 12

台灣總經銷／創新書報股份有限公司
地址：新北市新店區寶橋路 235 巷 6 弄 6 號 2 樓
電話：（886-2）2917-8022

香港總經銷：同德書報有限公司
地址：九龍觀塘大業街 34 號楊耀松第五工業大廈一樓
電話：852-35513388

編輯室報告

農曆新年將至，在此先祝各位讀者在新的一年都虎虎生風。

國軍按往例，在過年前夕安排陸、海、空三軍加強戰備的參訪。在 F-16AM/BM 接裝典禮上首度曝光聯合頭盔瞄準系統後，空軍此次正式對外介紹該型頭盔。其實在東亞地區，日本航空自衛隊、南韓空軍早已配備聯合頭盔瞄準系統，即便是解放軍也有自製的類似頭盔顯示／瞄準系統，空軍終於開始配發，總算是追上世界的腳步。

不過，在採購數量上由於成本高昂，空軍未來還需取得適當的平衡點。此外，高性能新裝備代表需要更多的訓練，相對於經國號、幻象 2000-5 戰機，也需擔負更多樣的任務，顯示未來 F-16 型機飛行員的壓力將更加沉重。

陸軍步訓部展示與北測、南測中心同時配備的 Cubic 公司第四代 MILES 系統，的確使地面作戰的訓練比較接近實戰情景，但用於演練的城鎮戰訓練場地中，建物、門窗、街道設備等，均已與現代的台灣都市實況差距不小，未來應該編列預算更新。

在沱江級巡邏艦塔江號的成軍後，此次是海軍第一次公開艦上艙間、艦橋及武器裝備等，而中科院研發已久的海劍二飛彈發射器，也終於近距離曝光。海劍二飛彈比海軍現有海欉樹防空飛彈的接戰性能優異，將大幅提升艦艇的短程防空戰力。不過塔江號艦上的海劍二飛彈仍舊配置斜射彈箱，而不是採用飛彈垂直發射系統，顯然還有很大的進步空間。

陸軍於 2021 年國慶日展演中首度公開了陸射劍二飛彈系統， Flak Talk 專欄也借鏡美軍野戰防空系統的發展經驗，探討該套防空系統的未來性。此外，本期還有以大利海軍第 1 直升機大隊、JHMCS II 頭盔等多篇精采文章，歡迎讀者在年節假期閱讀。

***Defence International*. 2/6.**

***Defence International.* 3/6.**

封面故事

國軍 2022 年春節加強戰備，分別安排空軍第 4 聯隊的 F-16AM/BM 展示，陸軍步訓部的機步排城鎮戰對抗，以及海軍第 131 艦隊的塔江、沱江號巡邏艦海上操演。DIIC/MNA

64 金戈鐵馬
戰車排專精訓練
ROC Army Tank Field Training

66 武器新知
戰場傷患後送王者
歐洲軍用全地形人員運送車
T0RSUS Praetorian/Terrastorm

70 中國速報
圍堵 VS. 突穿 美國與中國爭奪
「巴士海峽至菲律賓海」關鍵航道！
Contain VS Break Through Bashi Channel

76 七海艦船
海軍沱江級發展借鏡
美國海軍獨立級濱海戰鬥艦
Independence Class Littoral Combat Ship

83 中國速報
解放軍東部戰區 將領異動的意義與觀察
Meaning and Observation of PLA Promotion General in Eastern Theater

86 中國速報
美國陸軍《中國戰術》手冊揭密
美軍眼中的解放軍戰術戰法
The Introduction to US Army's Chinese Tactics Manual

92 中國速報
中國晉級潛艦 對美攻擊首選
PLAN Type 094 Submarine

96 采風錄
- 探索頻道：飛車逃生大挑戰
- 國家地理頻道：海難調查簿
- 國軍人才招募：南亞 ROTC 大學

100 新武器新系統
- 現今中小型無人機防護

Defence International. **4/6.**

郵政劃撥存款收據注意事項

一、本收據請詳加核對並妥為保管，以便日後查考。

二、如欲查詢存款入帳詳情時，請檢附本收據及已填妥之查詢函向各連線郵局辦理。

三、本收據各項金額、數字係機器印製，如非機器列印或經塗改或無收款郵局收訖章者無效。

請寄款人注意

一、帳號、戶名及寄款人姓名通訊處各欄請詳細填明，以免誤寄；抵付票據之存款，務請於交換前一天存入。

二、每筆存款至少須在新台幣十五元以上，且限填至元位為止。

三、倘金額塗改時請更換存款單重新填寫。

四、本存款單不得黏貼或附寄任何文件。

五、本存款金額業經電腦登帳後，不得申請撤回。

六、本存款單備供電腦影像處理，請以正楷工整書寫並請勿摺疊。帳戶如需自印存款單，各欄文字及規格必須與本單完全相符；如有不符，各局應婉請寄款人更換郵局印製之存款單填寫，以利處理。

七、本存款單帳號與金額欄請以阿拉伯數字書寫。

八、帳戶本人在「付款局」所在直轄市或縣(市)以外之行政區域存款，需由帳戶內扣收手續費。

交易代號：0501、0502 現金存款 0503 票據存款 2212 劃撥票據託收

本聯由儲匯處存查 210×110mm（80g/m²）保管五年

請沿虛線剪下

全球防衛雜誌訂閱方案 ※ 僅限直接訂戶享續訂優惠

單本定價 $ 200 元 / 期

1・訂閱 一年期

原價 $ 2,400 元；訂閱價 $ 2,000 元 / 12 期

2・訂閱 兩年期

原價 $ 4,800 元；訂閱價 $ 3,800 元 / 24 期

3・前期補購活動：

201912（D424 期）以前，任選 15 本，優惠價 $ 900 元（含運）

新舊訂戶若有 特殊需求 或 機關單位團體 訂購，歡迎致電

訂戶服務部 ： 02-2391-5105- 分機：10 或 12 洽詢

***Defence International.* 5/6.**

●郵政劃撥訂購方式

請沿虛線剪下

郵政劃撥儲金存款單

帳號：1 0 8 9 9 3 0 1

金額 新台幣（小寫）：仟 佰 拾 萬 仟 佰 拾 元

通訊欄（限與本次存款有關事項）

1. 訂閱「全球防衛雜誌」
請自 ______ 年 ______ 月寄出
贈品 ______ 掛號：□否 □是
訂戶：□否 □是會員編號 ______
生日：______ 年 ______ 月 ______ 日
身份證字號：______
職業 ______

2. 前期「全球防衛雜誌」買5送1
第 ______ 期
共 ______ 本。

3. 買特刊請書寫代號即可：

4. 第2、3項郵 總金額未滿500元，須另付60元郵資。

戶名：全球防衛雜誌社有限公司

寄款人

姓名

通訊處：□□□-□□

電話：公 宅

經辦局收款戳

虛線內備供機器印錄用請勿填寫

◎寄款人請注意背面說明。
◎本收據由電腦印錄，寄款人請勿填寫。

郵政劃撥儲金存款收據

收款帳號戶名

存款金額

電腦記錄

經辦局收款戳

請沿虛線剪下

●信用卡訂購方式

（請放大傳真）傳真 (02) 2391-8049 訂戶部 電話 (02) 2391-5105 10、12

特約商店名稱：全球防衛雜誌社有限公司 特約商店代號：0101603772

信用卡號：
□□□□-□□□□-□□□□-□□□□-□□□

有效期限：______（月）/ ______（西元年）

商品項目：

消費日期：______ 年 ______ 月 ______ 日

□雜誌、書籍

金 額：

□其它：______

授 權 碼：□□□□□□

持卡人同意依照信用卡使用約定，一經使用或訂購物品，均應按所示之全部金額付款予發卡銀行。

訂購人基本資料：□ 新訂戶 □ 續訂戶，會員編號：NO.______

訂閱日期：自 ______ 年 ______ 月 寄出

郵寄方式：國內：□ 1.平信 □ 2.掛號 （掛號費：$240元／年）

國外：□ 1.航空 □ 2.掛號 （掛號費：$780元／年）

收件人：______ E-mail：______

出生日期：______ 年 ______ 月 ______ 日 □男 □女

電話：（公）______（宅）______ 職業：______

寄書地址：______

發票：□ 二聯式

□ 三聯式 統一編號□□□□□□□□ 抬頭：

煩請填妥完整資料後，傳真至（02）2391-8049，並來電告知（02）2391-5105 轉分機 10 或分機 12

（幣別：新台幣）

	新 訂	續 訂
國內地區	一年 2000	一年 1900
	兩年 3800	兩年 3610

PS：若需掛號一年加 240 元 兩年加 480 元

郵寄地區	航	空
	一年	二年
港澳地區	3700	7000
亞太地區	4100	7800
歐美地區	4900	9300

PS：若需掛號一年加 780 元 兩年加 1560 元
國外贈品須自付郵資

ATM 轉帳（匯款）帳號：610102006804
台北富邦銀行師大分行 代號 (012)
匯款前請先來電，匯款後請來電確認寄書地址

※ 中華郵政已全面停止國際海運郵寄業務

***Defence International.* 6/6.**

***Defense Shutter*. Volume 64; February 2022. Note to Reader: This is a Taiwan published magazine and covers both Taiwan and the region. The magazine sends journalists to air shows and defense exhibitions around Asia, including the Zhuhai Airshow. This publication is also available via e-book: pubu.com.tw/store/119935 ; ebook.hyread.com.tw/magazine.jsp?jid=195). 1/5.**

發　行　人	林真真	
社長兼總編輯	陳東龍	
顧　　問	莊中毅	張正義
美術設計	羅其泓	
專案記者	黃睿堅	Magic Wu
	林佩儀	范碩芬
專案攝影記者	謝江如	游勝忠
	傅楓宸	Steven Weng
	周民孝	
美國特派員	Tom H Lee	
加拿大特派員	Tina Lin	
出　　版	東愷圖書有限公司	
服務電郵	service@ewdefense.com	
服務傳真	02-2517-8437	
印　　刷	白紗科技印刷股份有限公司	
國內總經銷	創新書報股份有限公司	

***Defense Shutter*. 2/5.**

DEFENSE SHUTTER

www.ewmib.com

專題企畫

台灣真是寶島
強國們就愛在台灣周邊玩
戰爭遊戲 .6

國防大聲公

外行領導內行且作繭自縛的
軍事情報作為 .16

軍事剪影

國軍第四季戰備訓練
實兵演練 .14

Defense Shutter. 3/5.

東北亞防務

盤點中國大陸無人武器平台
及現役飛彈家族 .20

有模有樣

Eurofighter Typhoon 7L-WB,
Austrian Air Force, 2019 .26

P-51D
中華民國空軍第四飛行大隊
第二十一中隊 .29

戰記

四號戰車奮戰東戰場
二十三歲的上校！
十小時激戰擊毀敵軍三百輛戰車
和一百輛裝甲車 .34

空戰記

單日三殺米格的空戰英雄
Ran Ronen .51

冷戰故事輯

RB-47E 北海歷險記 .56

***Defense Shutter*. 4/5.**

《防務快門》雙月刊訂閱辦法

壹、單本訂價：新台幣240元
每單數月25日上架

貳、長期訂閱客戶優待辦法
訂閱1年6期1320元新台幣，每期優惠價為新台幣220元（含郵資）。
訂閱3年18期（含）以內，每期優惠價仍為新台幣200元（含郵資）。
訂閱3年以上另有優惠辦法，請以服務電郵聯繫。
郵資部分以本島郵寄標準為主，外（離）島及國外郵資另議。

參、訂戶匯款資料
受款行：中國信託城中分行
帳　號：107540348617
戶　名：陳東龍
完成匯款後請傳真個人姓名、聯絡電話地址及匯款帳號
末五碼至 (02)25178437。

東愷圖書有限公司
公司統編：24472752
公司負責人：陳東龍
服務傳真：（02）25178437
電子信箱：rex65100@gmail.com

***Defense Shutter*. 5/5.**

Fun Patch: The Republic of China Military Emblem Book **(English/Traditional Chinese)**. January 2022. ISBN: 978-957-43-9589-7. This book is the only military patch book in Taiwan. It only covers Air Force/Navy old and new patches. There are no Army aviation patches. 1/2.

FUN PATCH

The Republic of China Military Emblem Book

編　　著　劉聲宇 王庭萱
編　　輯　王庭萱
攝　　影　劉聲宇 王庭萱
出 版 者　王庭萱
電子郵件　allzdog01@gmail.com
地　　址　新竹市東區經國路一段379巷42號1樓
電　　話　03-5341458
初版首刷　2022年1月
定　　價　新台幣700元

平裝　ISBN 978-957-43-9589-7

Fun Patch: The Republic of China Military Emblem Book **(English/Traditional Chinese). January 2022. ISBN: 978-957-43-9589-7. This book is the only military patch book in Taiwan. It only covers Air Force/Navy old and new patches. There are no Army aviation patches. 2/2.**

ABOUT THE EDITOR

Wendell Minnick, 2022.

WENDELL MINNICK, BS, MA
Journalism Profile

Show Coverage:

- Aero India – 2007, 2009, 2011
- Association of United States Army (AUSA) Annual Meeting (2006)
- China International Aviation and Aerospace Exhibition (Zhuhai Airshow) - 2006, 2010, 2012, 2014, 2016
- Defence and Security 2017 (Bangkok) - 2017
- Defence Services Asia (DSA-Kuala Lumpur) - 2008, 2014, 2016, 2018
- Defence Technology Asia International Conference and Exhibition (Singapore) - 2007
- Dubai Airshow - 2007, 2009
- Global Security Asia - 2007
- International Defence Exhibition (IDEX – Abu Dhabi) – 2007
- International Maritime Defense Exhibition (IMDEX – Singapore) – 2007, 2009, 2011, 2013, 2015, 2017
- Langkawi International Maritime and Aerospace Exhibition (Malaysia) - 2017
- Singapore Air Show – 2006, 2008, 2010, 2012, 2014, 2016, 2018
- Seoul Air Show – 2009, 2017
- Shangri-la Dialogue (Singapore) – 2008, 2009, 2010, 2011, 2012, 2013, 2015, 2016, 2017, 2018
- Taipei Aerospace and Defense Technology Exhibition – 2001, 2003, 2005, 2007, 2009, 2011, 2013, 2015, 2017, 2019
- U.S.-Taiwan Defense Industry Conference – 2006 (Speaker)

Position: North/East Asia Correspondent – European Defense/Security
Company: Mittler (Germany)
Business: Military Outlet
Duration: October 2018 to October 2020

Position: Senior Asia Correspondent – Shephard
Company: Shephard Military Media (London)
Business: Defense Industry Media Outlet (www.shephardmedia.com)
Duration: October 2016 to October 2018

Position: Asia Bureau Chief – *Defense News*
Company: *Defense News* (Army Times Publishing)
Business: Defense Industry Media Outlet (www.defensenews.com)
Duration: June 2006 to October 2016

Job Scope – *Defense News*

- Over 1,000 articles on the defence industry and military affairs in Asia and Middle East.
- Recruited and managed local reporters in Japan, Pakistan, Philippines, Singapore and South Korea.
- Interviewed top tier defense industry officials and government officials in the region.

Have been widely quoted on defense and military issues by the international media, including *AFP*, *Aljazeera*, *AP*, *BBC*, *Deutsche Welle*, *New York Times*, *Reuters*, *Sky News*, *Time,* and *Wall Street Journal*.

Position: Taiwan Correspondent – *Jane's Defence Weekly*
Company: Jane's Information Group
Business: Information solutions provider in defense, risk and security
Duration: June 2000 to June 2006

Position: Freelance Journalist
Duration: 1989 to 2006

Job Scope

- I have penned articles for a variety of magazines and newspapers while living and travelling around Asia. I have resided in Hong Kong, South Korea and Taiwan for extended periods. Publications:

  - *Afghanistan Forum*
  - *Army Magazine*
  - *Asian Profile*
  - *Asian Thought and Society*
  - *Asia Times*
  - *BBC*
  - *C4ISR Journal* (*aka* - Command, Control, Communications, Computers (C4) Intelligence, Surveillance and Reconnaissance (ISR)
  - *Chicago South Asia Newsletter*
  - *Defense News* (Staff – Asia Bureau Chief)
  - *Far Eastern Economic Review*

- *International Peacekeeping*
- *Jane's Airport Review*
- *Jane's Defence Upgrades*
- *Jane's Defence Weekly*
- *Jane's Intelligence Review*
- *Jane's Missiles and Rockets*
- *Jane's Navy International*
- *Jane's Sentinel Country Risk Assessments*
- *Japanese Journal of Religious Studies*
- *Journal of Asian History*
- *Journal of Chinese Religions*
- *Journal of Oriental Studies*
- *Journal of Political and Military Sociology*
- *Journal of Security Administration*
- *Journal of the American Academy of Religion*
- *Kentucky Farmer Magazine*
- *Military Intelligence Professional Bulletin*
- *Military Review*
- *National Interest*
- *New Canadian Review*
- *New World Outlook*
- *Pacific Affairs*
- *Powerlifting USA*
- *South Asia in Review*
- *Taipei Times*
- *Topics* (Taiwan – AMCHAM)
- *Towson State Journal of International Affairs*
- *The Writer*

Book

Spies and Provocateurs: A Worldwide Encyclopedia of Persons Conducting Espionage and Covert Action, 1946-1991. North Carolina: McFarland, 1992. The book was well received in the U.S. intelligence community, including positive book reviews in *Cryptolog*, *Cryptologia*, *Military Intelligence Professional Bulletin*, *Periscope* (AFIO) and *The Surveillant*. The book was also profiled in the 1995 release of the *Whole Spy Catalog: A Resource Encyclopedia for Researchers*.

Editorial projects include the following published under my name and available on Amazon:

China Market Outlook for Civil Aircraft, 2014-2033 (2016)
Chinese Aircraft Engines (2016)

Chinese Air-Launched Weapons & Surveillance, Reconnaissance and Targeting Pods (2019)
Chinese Anti-Ship Cruise Missiles (2019)
Chinese C4I/EW (Vol. 1/2) (2022)
Chinese Fighter Aircraft (2016)
Chinese Fixed-Wing Unmanned Aerial Vehicles (2016)
Chinese Helicopters (2016)
Chinese Radars (2017)
Chinese Rocket Systems (2016)
Chinese Rotary/VTOL Unmanned Aerial Vehicles (2016)
Chinese Seaplanes, Amphibious Aircraft and Aerostats/Airships (2016)
Chinese Space Vehicles and Programs (2016)
Chinese Submarines and Underwater Systems (2019
Chinese Tanks and Mobile Artillery (2018)
Directory of Foreign Aviation Companies in China (2014)
I Was a CIA Agent in India: Analysis (2015)
List of Foreign Companies and Identities of Taiwan Local Agents (2019)
More Chinese Fixed-Wing UAVs (2019)
More Chinese Rotary & VTOL UAVs (2019)
Taiwan Cyber Warfare (2018)
Taiwan Space Vehicles (2018)
The Chinese People's Liberation Army: Analysis of a Cold War Classic (2015)
Unicorn: Anatomy of a North Korean Front, Casinos, Immigration, Trade Sanctions and Violations (2019)

NOTE ON FIELD JOURNALISM

Having served as a military correspondent for *Jane's Defence Weekly* (2000-2006) and *Defense News* (2006-2016) there are a number of things that should be noted. First, it is normal to wear utility vests to avoid carrying cumbersome backpacks. Vests normally bear identification via Velcro-backed patches on utility vests[6]. These vests, with a variety of pockets, have come to serve as a sort of "uniform" for international field journalists. As a former journalist, a profile of these identification labels might be of interest to future generations.

---

[6] There has been debate since Vietnam on whether utility vests get you killed when mistaken as an enemy soldier. However, not wearing a utility vest will result in carrying a rugged backpack that resembles a military pack. It is best to be educated; recommended reading: Dunnigan, James F. *How to Make War: A Comprehensive Guide to Modern Warfare in the 21st Century* (2003); Hedges, Chris. *What Every Person Should Know About War* (2003); Pedelty, Mark. *War Stories: The Culture of Foreign Correspondents* (1995); and Pelton, Robert Young. *The World's Most Dangerous Places* (1998). For investigative research there is no better guide: Bazzell, Michael. *Open Source Intelligence Techniques: Resources for Searching and Analyzing Online Information* (semi-annual editions published).

Original Press Identification provided by employer.

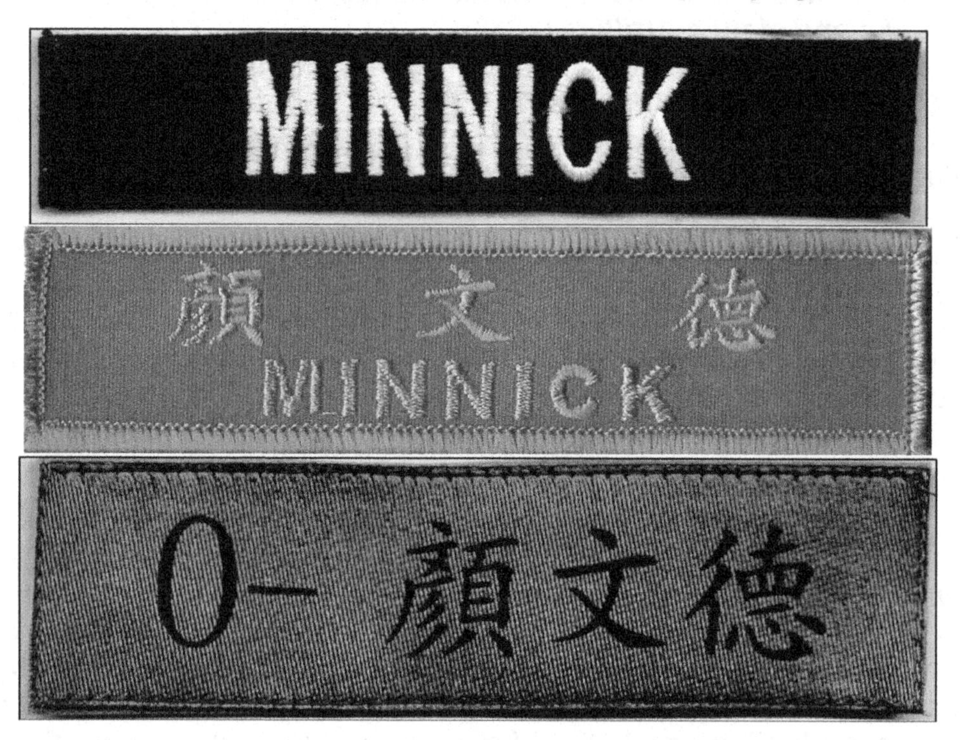

DEFENSE NEWS

Defense News, Asia Bureau Chief, 2006-2016.

JANE'S DEFENCE WKY

Jane's Defence Weekly, Taiwan Correspondent, 2000-2006.

OTHER BOOKS IN THIS SERIES

TAIWAN UNMANNED AERIAL VEHICLES

***Taiwan Unmanned Aerial Vehicles*. US $19.99. Amazon Books.**

Taiwan Space Vehicles. Available on Amazon. US $19.95. Amazon Books.

List of Foreign Companies and Identities of Taiwan Local Agents

MINISTRY OF NATIONAL DEFENSE

Wendell Minnick, Editor

List of Foreign Companies and Identities of Taiwan Local Agents. **US $19.95. Amazon Books. The book is a complete list of Taiwan sales agents for foreign defense companies.**

Taiwan Cyberwarfare: Information Communication and Electronic Force Command Cyber Warfare Wing. **US $19.95. Amazon Books.**

***Chinese Radars*. US $19.95. Amazon Books.**

Chinese Rocket Systems: Multiple Launch Rocket Systems Brochure. **US $19.99. Amazon Books.**

***Chinese Land-based Air Defense Systems: Anti-Aircraft Guns, Surface-to-Air Missiles, Electronic Warfare, and Radar Systems*. US $19.99. Amazon Books.**

CHINESE SEAPLANES, AMPHIBIOUS AIRCRAFT AND AEROSTATS/AIRSHIPS

- PRODUCT BROCHURES -

WENDELL MINNICK EDITOR

Chinese Seaplanes, Amphibious Aircraft and Aerostats/Airships. **US $19.95. Amazon Books.**

CHINESE ROTARY/VTOL UNMANNED AERIAL VEHICLES

PRODUCT BROCHURES

WENDELL MINNICK

***Chinese Rotary/VTOL Unmanned Aerial Vehicles*. US $19.95. Amazon Books.**

***More Chinese Rotary and VTOL UAVs.* US $19.95. Amazon Books.**

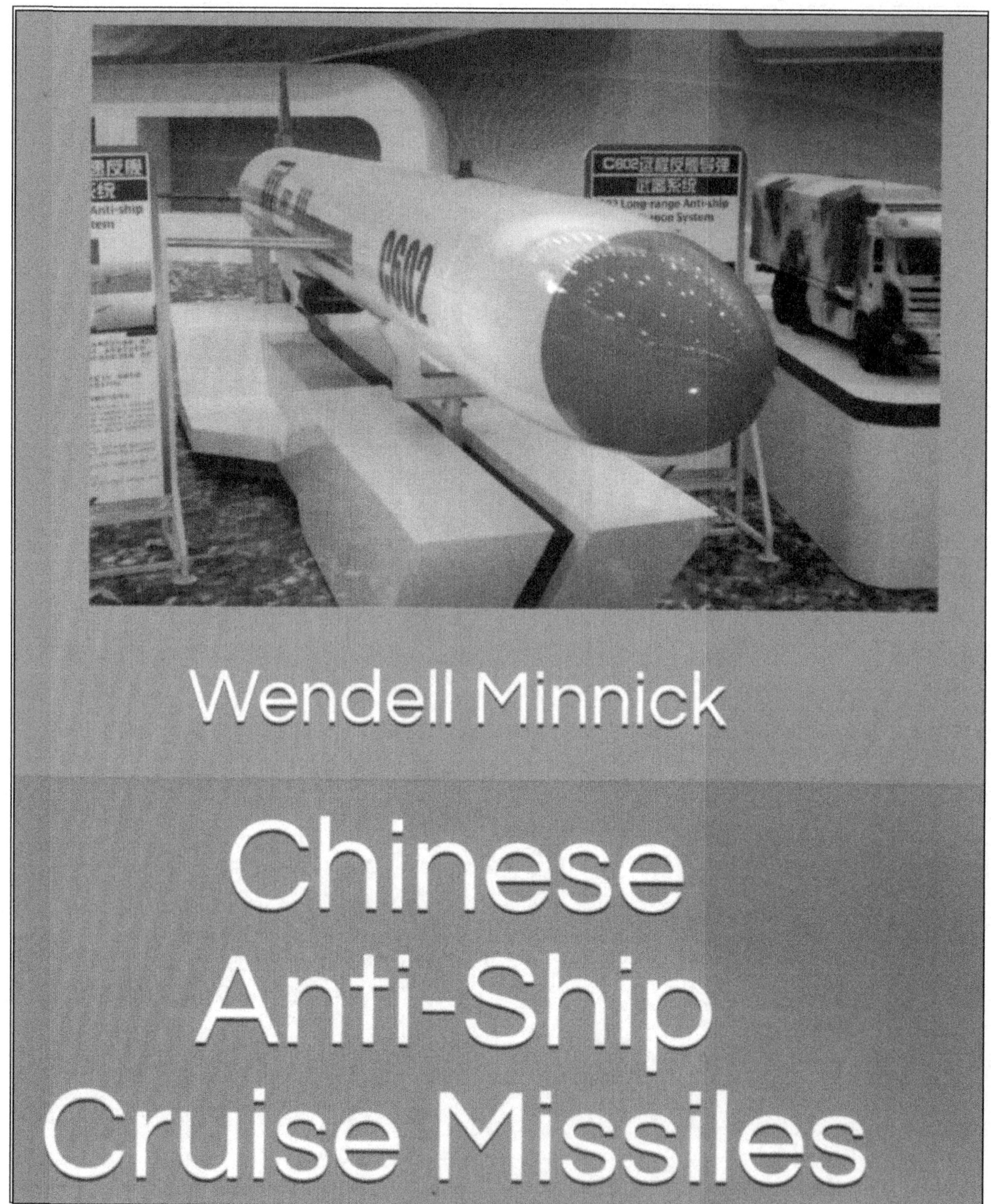

***Chinese Anti-Ship Cruise Missiles*. US $19.95. Amazon Books.**

CHINESE FIXED-WING UNMANNED AERIAL VEHICLES

RODUCT BROCHURES

WENDELL MINNICK

EDITOR

Chinese Fixed-Wing Unmanned Aerial Vehicles. **US $19.95. Amazon Books.**

Chinese Air-Launched Weapons & Surveillance, Reconnaissance and Targeting Pods

Wendell Minnick, Editor

***Chinese Air-Launched Weapons and Surveillance, Reconnaissance and Targeting Pods*. US $19.95. Amazon Books.**

***Chinese Space Vehicles and Programs.* US $19.95. Amazon Books.**

Chinese Tanks and Mobile Artillery. **US $19.95. Amazon Books.**

Chinese Submarines and Underwater Warfare Systems

Wendell Minnick

Chinese Submarines and Underwater Warfare Systems. **US $19.95. Amazon Books.**

***Chinese Anti-Ship Cruise Missiles*. US $19.95. Amazon Books.**

***Chinese C4I/EW* (Vol. 1/2). US $19.99 a piece. Amazon Books.**

CHINESE FIGHTER AIRCRAFT

PRODUCT BROCHURES

EDITED BY WENDELL MINNICK

***Chinese Fighter Aircraft*. US $19.95. Amazon Books.**

***Chinese Radars*. US $19.95. Amazon Books.**

CHINA MARKET OUTLOOK FOR CIVIL AIRCRAFT, 2014-2033

INCLUDING COMMERICAL AIRCRAFT PRODUCT BROCHURES

WENDELL MINNICK
EDITOR

***China Market Outlook for Civil Aircraft, 2014-2033*. US $19.95. Amazon Books.**

The Chinese People's Liberation Army: Analysis of a Cold War Classic – 1950.
US $9.99. Amazon Books.

INDEX

Numerals

.30 Carbine 339-340
.44 magnum 123-124, 148-149
.45 caliber 29, 31
.50 calibre 77, 79, 182
1 inch multi-option fuze 246
105mm 75, 77-79, 98, 182, 246, 364-368, 375-384
105mm-1 inch Multi-Option Fuze 246
10th Army (Central) 325-328
119A2/A5 77
12.7mm 29, 32, 77, 79
120mm 94, 96, 105-106, 182, 242-245, 375-384
155mm 74-79, 375-384
185180 Flare 155-154
198275 Flare 159-160
198464 Flare 155-156
1st Special Forces Group (Green Berets) 453-454
2.4 GHz GaAs (Gallium Arsenide) Monolithic Microwave Integrated Circuit (MMIC) Chip Set 312-317
2.75-inch rocket 258-259, 276
202nd Area Military Police Company 363
202nd Arsenal 69-81, 82-83, 85-87, 94, 97-98, 102, 108-113, 116, 118-122, 148-149
203mm (8 inch) 76-77, 79, 375-384
203rd Arsenal 29, 32, 69, 118-122
204th Arsenal 153-154, 157-162, 165-172
205th 28-47, 52-53, 58-59, 62, 114-115, 127-229, 133-135, 147, 150, 175-178, 182, 184-188
209th Arsenal 23-26, 361
20mm 29, 31, 77, 79, 128, 182, 225-226, 240
223 Remington Rifles 54-57
224418 167-170
22nd Han Kuang Exercise 375-384
2D 301-304
333rd Mechanized Infantry 359-360
35mm 75, 188, 295-300
35th Han Kuang Exercises 270-271
36th Han Kuang Exercise 359
37C-GPS/GIS Kinematic Control System 10
38 MK12 97-98
3D 10, 25, 221, 232, 234, 238, 283-288

4.2 inch 96
40/70 L62 74, 75, 289-294
401st Factory 190-191, 204-209, 213-216
402nd Factory 19-21, 192-200, 210-212, 217-219
40mm 31-32, 74-75, 77, 79, 81-89, 102, 108-113, 116, 182, 235, 237, 241
5.56mm 29, 32, 43-44, 46-47, 52-57, 76-77, 114-115, 182, 187
542nd Armor Brigade 357-358
6-Axis 351-352
601st Aviation Brigade 260-264, 271
602nd Aviation Brigade 265-270, 386-394
603rd Aviation 305
60mm 96, 244-245
60th Armament Factory 28-31
62A Mask Filter 169-170
65K2 77, 195-197
6th Army (North) 325-328, 357-358
7.62mm 8, 29, 31-32, 46-47, 58-59, 62, 74-75, 77, 102, 114-116, 125-128, 339-340
76 M32 (M41A1) 75
76mm L62 75
8 inch (203 mm) 76-77, 79, 375-384
81mm 90-93, 96, 105-106, 242-245, 375-284
8th Army (South) 325-328, 359-360
99th (Vanguard) Marine Corp Regiment 414-446
9mm 29, 31-37, 41-42, 123-124, 127, 131-132, 148-149, 182-186, 339-340

A

A-21 Cargo Parachute 395-413
A-22 Cargo Parachute 395-413
AAA Armor Force 341-346
AAVP7/C7 (*see* Amphibious Assault Vehicle)
Aboriginal Tribes 406, 433-446
Activated Carbon 15
Adaptative Motion Cuing Control System 51-52
Adjoining Multi-Functional Tent 182
Adlink 301-304
Advanced 81/120mm Mobile Mortar Weapon System 105-106
Advanced Ballistic Products 339-340
Advanced Semi-Auto Sniper Rifle 62
Aeronautical Structure and Materials 329-335
Aeronautical Systems Research Division (ASRD) 247-250, 279-280, 325-328
Aerospace Industrial Development Corp. (AIDC) 353
Agricultural Survey 325-328

AH-1W Cobra Helicopter 258-259, 271-272, 375-384, 386-394
AIDC (*see* Aerospace Industrial Development Corp.)
air defense 108-113, 227-231, 271, 283-294
Air Force 247-248, 283-288
air-to-ground rocket 258-259
Airborne Training Center 395-413
Aircraft Carrier 460-461
Airsoft 468-484
Albatross (*see* Chung Shyang)
Alian 364-368
American Institute in Taiwan (AIT) 447-451
American Volunteer Group 447-451
Amphibious Assault Vehicle (*AAV*) *375-384*
Amphibious Reconnaissance Battalion (ARB/101st) 414-422, 442-447
Amphibious Reconnaissance Patrol (ARP) – Marines 347, 414-432, 442-446
amplifier 312-317
analog data traffic 312-317
Antelope Air Defense System 227-231
Antenna 279-280, 283-288
Anti-Airborne 363
Anti-blast fuel tank 468-484
Anti-Communist Insurgencies 453-454
Anti-cut clothing 341-346
Anti-Infrared 253
Anti-jamming 312-317
Anti-landing mission 359
Anti-Material (AM) 92-93
Anti-Near-Infrared Paint 253
Anti-Personnel (AP) 92-93
Anti-Tank 17, 100, 182, 224
Anti-Visible 253
AP/AM Mortar Round 92-93
APFSDS (*see* Armor Piercing Fin Stabilized Discarding Sabot)
AR-26 (*see* T65)
AR15 Rifle 54-57
ARB (*see* Amphibious Reconnaissance Battalion)
ARM Mode 230
Armaments Bureau 5-219, 361
Armor Ceramic Tiles for Vehicles 340
Armor piercing 77, 79, 127, 128, 224
Armor Piercing Fin Stabilized Discarding Sabot (APFSDS) 77, 79
Armor School 370-373
Armor Training Command and Armor School 370-373
armored vehicle (*see also* Clouded Leopard) 10, 23-24, 102, 224, 340-346, 375-384

Army Aviation Training Command (AATC) 305
Army Field Air Defense Command and Control System 283-288
Army Infantry Training Command 65
Army Surveying Instruments Department 20
ARP (*see* Amphibious Reconnaissance Patrol)
ART FlightLab 301-304
artillery 246, 283-288, 357, 364-368, 375-384
AS-96/98 shoe 16
Asia Sun (Taiwan) Inc. 341-346
Asia-Pacific Defense Monthly 491-494
Asian body shape 123-124, 213-214, 318-324
ASMZ-TEDA-impregnated activated carbon 15, 163-164
ASRD (*see* Aeronautical System Research Laboratory)
assassination 459-460
ASTi (Advanced Simulation Technology inc.) Audio System 301-304
Atmospheric Radiation Measurement (ARM) 230
Attenuate Radar Signature 253
Audio and video life detector 179-180
Audio System 301-304
Augmented Reality (AR) Military Exercise Simulation 338
Autonomous Flight UAV 325-328
Aviation and Special Forces Command (ASFC) 63-65, 144-146, 260-270, 273-276, 305, 325-328, 375-384, 395-422
AW-1W Cobra 265-270
Az Track Accuracy 289-294

B

ballistic computer 102
ballistic helmet 15, 129, 131-131, 182, 339-340
ballistic inserts/plates 16, 123-124, 339-340
ballistic tiles 339-340
ballistic vest 16, 121-124, 128, 182, 339-346
barrel 60-61, 88-89, 100
bars/pubs 452
Battery Charger Unit (CBU) 17
Battle Command and Control System 283-288
Battlefield Damage Assessment 325-328
Battlejacket 15
Bear (nickname) 468-484
Behrends, Paul D. 459-460
beret (hat) 442-447

Beretta 33-40, 92, 182 (*see* also T75 pistol)
Big x Reality 337
BIIR Bromobutyl Rubber 169-170
binoculars 201-203
Biological Warfare (*see* Nuclear, Biological, Chemical)
Blackwater 458-459
Boeing 452
boots 16, 133-135, 182
boresight 103-104
Bottoms Up Pub 452
Brave Tiger 364-368
Bravo Flight 275
Breaching Doors 433-446
Broadband detecting radar 254
Bromobutyl Rubber 169-170
Browning M2E2 (*see* T90)
Built-In Test (BIT) 230, 312-317
Bullet Proof 118-120, 183-186, 341-346 (*see also* ballistic)

C

C-130 Cargo Transport 375-384, 395-413
C-27 87
C-4 explosive 71, 72
C208 Fiberboard Plate 241
C27 84
C31/C31A1 Point-Initiating/Self-Destruct (PIBD/SD) Grenade Fuze for High Velocity – High Explosive Dual Purpose Cartridge (HV-HEDP) 236-237
C4 91, 93-94, 242-245
C4 Point Detonating Fuze for Mortars 242-243
C49 Electronic Time Fuze for Mortars 244-245
C4ISR (Command, Control, Communications, Computers, Intelligence, Surveillance and Reconnaissance) 325-328
C9 97-98
camera 102, 114-115, 179-180, 277-278, 318-324
Camouflage 12, 148-149, 182, 253-254
Camouflage Net 12, 182, 253
Camp 66 Military Camp 468-484
cannon 74, 75
canteen 169-170
carbon 15
carbon fiber composite 279-280

carbon fiber telescoping pole 179-180
carbon impregnated with Copper, Silver, Zinc, Molybdenum (ASZM) 15
Carrier Air Wing Five (CVW-5) 461
catapult 329-335
Cayago Luxury Sea Toys 347
CB-103 Combat Boots 134, 182
CB-99 Camouflage Waterproof Canvas Boots 16, 33, 182
CCD Camera 102, 179-180, 318-324
Cell Phone (*see also* Smart Phone) 312-317
Central Processing Unit (CPU) 318-324
ceramics 340-346
CG-52 USS Bunker Hill 460-461
CH-47SD Chinook Helicopter 271, 375-384
challenge coin 447-451, 459-460
CHANG Chung-shing 362
Changhua County 270-271
Chemical Systems Research Division 253-254
Chemical Warfare (*see* Nuclear, Biological, Chemical)
CHEN Kai, Col. 357
CHEN Shui-bian, President 459-460
Chiayi Air Base 414-422
Chiayi County 271, 414-422
China Service Medal 447-451
China Shipbuilding Corp. (CSBC) 353
China Steel Corp. (CSC) 361
Chinese Taipei Shooting Association 462-467
chip signal processing 318-324
Christine projectors 301-304
Chung Shan Volunteer Team 362
Chung Shyang II (Albatross) 325-328
CIA (*see* US Central Intelligence Agency)
Civilian-Military Industrial Development Center 362
Close Quarter Combat (CQC) 9, 43-44, 51, 190-191, 462-484
Cloud Deployment Scalable Reserved 281-282
Clouded Leopard (armored fighting vehicle) 10, 12, 23-25, 102, 105-106, 116, 255-257, 340, 359, 361, 364-368
Clouded Leopard Mortar System 23-24
Clutter Suppression 234
CM-11 Brave Tiger Main Battle Tank (M60A3 hull/M48A3 turret) 364-368, 375-384
CM-21 Mechanized Infantry Combat Vehicle 375-384
CM-22/CM23 120mm/81mm Mortar Carrier 375-384
CM-26 Command Post Carrier (M577) 375-384
CM-32 (*see* Clouded Leopard)
CM151 Warhead 258-259

CMK66 motor 258-259
CNS 11981 Z2064 159-160, 161-162
Coast Guard 318-32*8*
Coaxial machine gun 102
Coaxial Turret Weapon 116
COBRA Trammel Launcher 178
Code Division Multiple Access (CDMA) 289-294
Cold War 414-422, 442-447
Collimator Sight 191
Collimator System 306-311
Collision Avoidance (CA) Protocol 312-317
Collision Sense Multiple Access (CSMA) Protocol) 312-317
Color HD Camera 277-278
Combat Boots 16, 133-135, 182
Combat Zone (Shuangcheng Street) 452
Combined Arms Battalion 357, 359
Command and Decision System (CDS) 289-300
Command Post Carrier (M577) 375-384
Communication broadcasting Vehicle 279-280
Communications Kaohsiung and Pingtung Operations Team 356
Communications Relay 325-328
compass 347
Composition A/B/C/D 91, 182
composition of glass fiber and honeycomb 254
computer 301-304
Conboy, Kenneth 453-454
Conductive Metal Fiber 253
copper 15
Counter-Insurgency (COIN) 453-454
Cr-Mo-V Steel 88-89
Cruise Missiles 232, 234
CS-MPQ-78 289-300
CS/TVS-32 318-324
CSBC Corp (*see* China Shipbuilding Corp.)
CSMA (*see* Collision Sense Multiple Access)
CTV/TVS-32 318-324
Curvature Screen 301-304
CVA-62 USS Independence 460-461
Cybernation 279-280

D

D-1-1 98
D-1-D 94
D/A Conversion 301-304
Dafar International, Inc. 347
Data Message Security Channel 312-317
Dawu 395-413
DD-966 USS Hewitt 460-461
DD-975 USS O'Brien 460-461
DDC (*see* Detector, Dewar, Cooler)
Decontamination (*see* Nuclear, Biological, Chemical Warfare)
DECT (*see* Digital Enhanced Cordless Telecommunications)
Defense Autonomy 341-346
Defense International 495-500
Defense News 414-422
Defense Shutter 501-505
Defense Sketch 354-363
Defense Technology Monthly 485
Illustrated Guide for Weapons and Tactics 486-490
delay fuze 242-243
delayed action of 0.027 to 0.063 s (SQ Fuze) 242-243
Delayed Fuze 242-243
Denver, Colorado 459-460
Detachment 26 (Green Berets) 453-454
Detector, Dewar, Cooler (DDC) 318-324
DF-15/M-9 Missile 460-461
DF602 Seabob Dive Jet 347
Digital Camouflage Broadband Radome 254
Digital data traffic 312-317
Digital Enhanced Cordless Telecommunications (DECT) 312-317
Digital Pattern Tactical Backpack 15
Ding Fwu 462-467
Direct Action (DA) 453-454
Direct Sequence Spread Spectrum (DSSS) 312-317
DM111 Mechanical PD (point detonating) fuze 242-243
DM93/M776 Mechanical Time Fuze 244-245
DMA (Dynamic Mechanical Analysis) Fiber Interface Transmission 306-311
Don't Tread On Me 461
Dongyin 414-422
Drone (*see* Unmanned Aerial Vehicle/UAV)
DSSS (*see* Direct Sequence Spread Spectrum)
Dual transmitters 232
Dual Transmitters for Higher MTBF (Mean Time Between Failure) 234
Dynamic Mechanical Analysis (*see* DMA)

E

E-Jet 347
Early Warning 227-231, 283-300
Early Warning Command and Control 289-294
Early Warning Command and Decision System (CDS) 289-300
Early Warning Radar 283-288
Early Warning System (EWS) 227-231
earthquakes (*see* natural disasters)
ECCM (*see* Electronic Counter-Countermeasures)
ECue 660-8000 301-304
Egress Procedure Trainers (EPT) 247-248
Ejection Seat Trainer (EST) 248
electric antenna 279-280
electric server 102
Electric Shock Unit 178
Electro Optical (EO) 18, 21, 225-226, 277-278, 289-300, 325-328
Electro-optical military maps 18
Electro-Optical Target Tracking Fire Control 225-226
Electro-Servo Control System 106
Electro-servo motion base 351-352
Electromagnetic Compatibility (EMC) 312-317
Electromagnetic interference (EMI) 312-317
Electronic Compass 347
Electronic Counter Measures (ECM) 289-294
Electronic Counter-Countermeasures (ECCM) 232, 234, 283-288, 295-300, 312-317
Electronic Scan 283-288
Electronic Time (ET) 244-246
elevator mechanism 279-280
EN402 161-162
EN403 159-160
Enhanced Modular Ballistic Helmet (EMBH) 131-132
EPT (*see* Egress Procedure Trainers)
Excellent Optical Company Award 21
Explosive Ordnance Disposal (EOD) 128
Explosives 91, 128, 182, 243, 246, 258-259, 433-446
External Battle Command and Control System 283-288
Eye-Safe Laser Range Finder 102

F

F-16 Fighter 375-384
F-5 Tiger Fighter Aircraft 247-248, 375-384
F-5E Tiger Fighter Unit Training Device 247-248
Fabrino 306-311
Factory Reset Protection (FRP) 224
Fan Patch Book: The Republic of China Military Emblems Book 506-507
Fault Diagnosis 312-317
FCS ECue 660-8000 301-304
FFAR (*see* Folding-Fin Aerial Vehicle)
FFG-41 USS McClusky 460-461
FFS (*see* Function Flight Simulator)
fiber composite 279-280
Fiber Interface Transmission 306-311
Fiber Reinforced Plastics (FRP) 222-224
Fiber Transmission 318-324
Fiberboard Plate 241
Field Movable Shower 151-152
Filter Canister, Type 87 15
Final Operational Capability (FOC) 264
Fire Control Radar System 235
Fire Control Shelter 289-294
Fire control system (FCS) 102
Fire Protection Equipment 159-160
Fire Retardant Polyester 253
Fire-fight spot system 279-280
Firepower allocation 283-288
First Navy Jack Flag 461
First Responder Training 462-467
Fishery Resources Survey 325-328
Flame-Resistant Working Uniform 182
Flare 11, 153-156
Flashless Powder (SPDF) 97-98
Flight Trajectory 301-304
Flightlab 301-304
FLIR (*see* Forward Looking Infra Red)
Flying Tigers 447-451
Flying Vehicle Recovering Parachute 136-137
FM radio 312-317
FMJ-RN (*see* full metal jacket flat nose)
FN Minimi (*see* T74/T75 Squad Machine Gun)
FOC (*see* Final Operational Capability)

focal plane array (*FPA*) 318-324
Fokker Kcur 306-311
Folding Machine Gun (FMG) 148-149
Force-to-force training drill 184-186
Forward Looking InfraRed (FLIR) 230
Frame Transfer CCD 318-324
Free Range Practice 348-350
French Foreign Legion 442-447
Frequency Hopping Radio 312-317
Frogmen 414-422
FRP (see Factory Reset Protection) 224
Fuhsing Kang College 453-454
Full color camera 114-115
full metal jacket flat nose (FMJ-FN) 148-149
Function Flight Simulator (FFS) 301-304
fuze 91, 93, 236-237, 242-246

G

G-11A Cargo Parachute Assembly 395-413
G-12D Cargo *Parachute* Assembly 395-413
GaAs MMIC (Gallium Arsenide Monolithic Microwave Integrated Circuit) 312-317
Gallium Arsenide (GaAs) 312-317
GAO Jia-pine, Staff Sgt. 363
Gas Mask 163-170
Geographic Information System (GIS) 10, 306-311, 338
Gerdes, Brad 447-451
GIS (*see* Geographic Information System)
Glass fiber 254
Global Positioning System (GPS) 10, 283-288, 312-317, 325-328
Global System for Mobile (GSM) Communications 312-317
GPS (see Global Positioning System)
Green Berets 442-447, 453-454
Green Laser 103-104
Grenade 12, 29-32, 77, 81-89, 102, 182, 236-237, 241, 244-245, 468-484
Grenade Fuze for High Velocity – High Explosive Dual Purpose Cartridge (HV-HEDP) 236-237
Grenade Launcher 77, 87-89, 102, 468-484
Grenade Machine Gun Barrel 88-89
Ground Control Stations (*GCS*) 325-328
Group Quality Award 21
GSM (*see* Global System for Mobile Communications)

Gushan District 433-446
Gyrostabilizer 102

H

Halo Effect 301-304
Han Kuang Exercise 270-271, 359, 375-384
Hand Signal Flare 153-156
Hand-held Infrared Camera 318-324
HC-47 helmet 339-340
HC-N3 helmet 339-340
HC-N4 helmet 339-340
HE Incendiary Tracer (HEI-T) 188
HE Shi-wen, Private 363
headlights 131-132
HEAT-T (*see* High-Explosive Anti-Tank – Tracer)
HEDP (*see* High Explosive Dual Purpose)
HEI (*see* High-Explosive-Incendiary)
HEI-T = HE Incendiary Tracer 188
Helicopter 248, 258-276, 301-305, 375-384, 386-394, 414-422
Helmet 15, 129, 182, 255-257, 339-346
Helmet illumination 15
Helmet-mounted display 255-257
Hengchun Base 433-446
HESH (*see* high explosive squash head)
Hex-Glinder 351-352
Hexagonal Ballistic Ceramics 340
Hexal P-30 71, 72
HF (*see* Hsiung Feng)
HH-R65A1/A2/A3 46-47
High Explosive (HE) 81, 85-87, 100, 182, 222-224, 241, 236-237, 241-243, 258-259
High Explosive Dual Purpose (HEDP) 236-237, 241
High Explosive Projectiles 242-243
High Explosive Squash Head (*HESH*) 222-224
High Explosive, Dual Purpose (HEDP) 81, 85-87
High Level Architecture *(HLA) Tactical Environment 301-304*
High Mobility Multipurpose Wheeled Vehicle (HMMWV; Hummer; Humvee) 96, 105-106, 283-288, 295-300
High Velocity – High Explosive Dual Purpose (HV-HEDP) 236-237
High Velocity/Self Destruct/Point Detonated 241
High-Explosive Anti-Tank – Tracer (*HEAT-T*) 100, 182, 222-224
Highway landing 270

HLA (*see* High Level Architecture)
HMX explosive 182
Hocheng Corporation 339-340
Honeycomb Structure 254, 306-311
HONG Rui-huang, Major 356
Howitzer 182, 364-368
Hsinchu 370-373, 386-394
Hsiung Feng 2 Anti-Ship Missile (Brave Wind) 375-384
HSU Hong-lin, Lt. Col. 357
HUANG Wei-zhe, Master Sgt. 361
Huatan County 270-271
Hukou Township 370-373
Hummer (*see* High Mobility Multipurpose Wheeled Vehicle)
Hunter 406
HV-HEDP (*see* High Velocity – High Explosive Dual Purpose)
hydro-pneumatic system 24

I

I/O Interface System 221, 301-304, 351-352
I2s3 221
Identification, Friend or Foe (IFF) 232, 235, 283-300,
IDF Unit Training Device (UTD) 247-248
iDPRA (*see* Intelligence Disaster Prevention and Relief Aiding System)
IEEE 802.11 Universal Standards 312-317 (IEEE = Institute of Electrical and Electronic Engineers)
IFF (*see* Identification, Friend or Foe)
Illumination Grenade 244-245
Imaging System 255-257
Immersive interaction (i2s3) Shooting Simulation System 221
IMON Adaptative Motion Cuing Control System 351-352
IMON Hex-Glinder 351-352
IMON Interactive Simulator 348-352
Incendiary 188
Indigenous Defense Fighter 247-248
Individual Field Tent 16, 150, 182
Industrial Technology Research Institute (ITRI) 361
Industrial, Scientific and Medical (*ISM*) 2.4 GHz band 312-317
Infantry 65, 102, 182, 364-368, 375-384
Information Operations (IO) 453-454
Information Security 353
Infrared 8, 20, 215-219, 253, 277-278, 289-294, 318-324, 348-350

Infrared thermal imaging technique 20
Injoy Motion Corp. 348-352
Input/Output (I/O) 221
Institute of Electrical and Electronic Engineers (*see* IEEE)
Integrated Communication System 281-282
Integrated Data I/O (input/output) 221
Integrated Satellite Phone 281-282
Intelligence 325-328
Intelligence Disaster Prevention and Relief Aiding System (iDPRA) 281-282
IR-on Indicator 217-219
Iron Fist (99th Marines) 433-446
Iron Sight 190-191
ISM (*see* Industrial, Scientific and Medical)
ISO14000 18
ISO14001 21
ISO9001 18, 21, 247-248
It's the Only Game In Town 461

J

jacketed soft point (JSP) *bullet* 148-149
Jamming 312-317
JIAN Ding-hua, Director 362
JSP (*see* jacketed soft point)

K

Kaohsiung 356, 364-368, 423-446, 452
KDB Cannon 74, 75
Kestrel Shoulder Launched Rocket 222-224
Kevlar Fibers 129
Kinematic Control System 10
Kinmen Island 414-422
Knife (see anti-cut clothing) 341-346

L

L62 74-75
L70 T92 Air Defense Gun 108-113, 289-294

LAADFCR (*see* Low-Altitude Air Defense Fire Control Radar)
Lambert Brightness 301-304
Landing Craft – Utility (LCU) 375-384
Laos 453-454
LARP (*see* Live Action Role Play)
Laser 14, 35-37, 103-104, 190-191
Laser indicator 103-104
Laser range finder 14, 102
Law Enforcement 184-186, 337, 462-467
Lead Semiwadcutter (SWC) 148-149
Leary, William 453-454
Li-Co Battery 17
Li-Iron Battery 17
LIAO Jun-wu Sgt. 357
Liar's Dice 461
Lien Yung (United Brave) Drill 357
Linear actuator platform 351-352
LINUX 306-311
Liquid Crystal Display (LCD) Monitors 179-180, 318-324
Live Action Role Play (LARP) 462-484
Lockheed Martin 452
Long Range System (LRS) 325-328
Longtan 453-454
Low-Altitude Air Defense Fire Control Radar (LAADFCR) 289-300
Low-Altitude Multi-Target Firepower Allocation System 283-288
Luminescence Map 10

M

M-9/DF-15 Missile 460-461
M109 Self-Propelled Howitzer 76-79, 364-368, 375-384
M110 8 inch (203 mm) self-propelled howitzer 76-77, 79, 375-384
M113 Armored Vehicle 375-384
M119A1/A5 76
M14 Rifle 46-47
M16 Rifle 76-77
M188A1/A2 Recovery Vehicle 78-79
M193 187
M203 grenade launcher 77, 83-84
M205 Powder Bag 91
M24 Sniper Rifle 58-59
M243 157-158

M257 157-158
M259 157-158
M260/M261 seven-tube 2.75-inch Folding-Fin Aerial Vehicle (FFAR) 258-259, 276
M2E2 (*see* T90)
M30 for HEAT-T M456A1   79, 182, 241
M32 74
M321787 159-160
M374A3 High Explosive (HE) Projectile 91-93
M39 77, 79
M41D Walker "Bulldog" Tank 74-75, 369
M430/M30A1 High Explosive, Dual Purpose (HEDP) 40mm Grenade 81, 241
M456A1 182
M48A3 tank 77-79, 100, 248, 306-311, 364-384
M48H Tank Simulator 248
M490 Projectile 182
M4A2/M1 76-77
M53A1 107
M56A3 High-Explosive-Incendiary (HEI) Cartridge 240
M60 Tank 100, 248, 306-311, 364-373
M60A3 Fire Control System 369
M60A3 Tank Simulator 248
M67 propellants for howitzer 182
M68 barrel 100
M69 Cannon 182
M776 Mechanical Time Fuze 244-245
M79 83-84
M80 bullets 125-126
M8A1 T63 Mortar 182
M9 Beretta 33-40
M918 Target Practice (TP) 82
M92FS Beretta 33-37
MA Ying-jeou, President 459-460
Madou District 271
Maintenance 386-394
Malibu Pub 452
Man-to-man training drill 184-186
Manportable 224
Map 10-11, 18
March Madness 461
Marine Corp (Taiwan) 223-224, 347, 433-446
Marine Corp Special Service Company (SSC)   433-446
Marine Corp Training Base 433-446
Materiel Production Center (MPC) 8-219
Matrix Control Station 318-324

Matsu 414-422
MC-1B Maneuverable *Parachute 138-139,* 395-413
McCallus, Michael 447-451
Mean Time Between Critical Failure (MTBCF) 283-288
Mean Time Between Failure (*MTBF*) 232, 234
Mean Time to Repair (MTTR) 283-288
Mechanical impact fuze 237
Mechanical Trigger 222-224
Mechanized Infantry 357, 259-360, 364-368
Melt-Castable Explosives 182
MEMS (Micro Electro-Mechanical System) Integrated Data I/O 221
Metal Fiber 253
Micro Electro-Mechanical System (MEMS) Integrated Data I/O 221
Microphone 179-180
Microwave 312-317
MIL-H-44099A 148-149
MIL-P-46593 Ballistic Standard 128
MIL-STD-1316 237, 244-245
MIL-STD-331C 242-245
MIL-STD-461C 312-317
MIL-STD-662 131-132, 148-149
MIL-STD-8100 318-324
MIL-STD-810D 318-324
MIL-STD-810E 312-317
Military Police 337, 363, 442-447
Military Police Motorcycle Motorcade 363
Ministry of National Defense (MND) 354-363
Minsyong Township 271
Mirage 2000 Fighter 375-384
MIS-S-1059A 187
Missile 231, 232, 234, 283-300, 460-461
Mission Control System (MCS) 227-231
MK-1 Multiple Grenade Launcher (MGL) 87
MK12- 1 77
MK19 MOD-III C208 fiberboard plate 81, 241, 468-484
MK52 MOD0 97-98
MK54 MOD2 98
MK83 MOD0 98
MLRS (*see* Multiple Launch Rocket System)
MMIC (*see* Monolithic Microwave Integrated Circuit)
Mobile device 281-282
Mobile Mortar System 13
Mobile Phased Array Radar 232-234
Mobile Point Defense Phased Array Radar System 283-288

Modem 283-288
Modular ballistic helmet (MBH) 131-132
Modular ballistic vest 127
Molybdenum 15
Monocular night-vision telescope 131-132
Monolithic Microwave Integrated Circuit (MMIC) 312-317
Morrison, James 453-454
Morse Code 301-304
Mortar 13, 23-24, 90-98, 103-106, 182, 242-245, 375-384
Motorcycle Motorcade 363
MT-1X Parafoil/HALO Parachute 395-413
MTBCF (see Mean Time Between Critical Failure)
MTBF (*see* Mean Time Between Failure)
MTTR (*see* Mean Time to Repair)
Multi-Functional Shovel 147
Multi-Functional Tent 182
Multi-Launch Rocket System 375-384
Multi-option fuze 246
Multi-Utilization Special Rifle 9
Multiple Launch Rocket System (MLRS) 277-278
Munitions 69-80
My Place Pub 452

N

National Defense Report 354-363
National Institute of Justice (*see* NIJ)
National Security Bureau 459-460
National Standardization Award 21
National Television Standards Committee (*see* NTSC)
NATO 54-57, 163-164, 339-340
Natural Disaster 11, 179-180, 281-282, 325-328
Navy 225-226
NBC (*see* Nuclear, Biological, Chemical)
Night Vision 14, 114-115, 131-132, 179-180, 192-203, 206-207, 213-219, 222-224, 248, 301-304, 341-346, 468-484
Night Vision Goggle (NVG) 14, 222-224, 301-304, 341-346
Night Vision Goggles (NVG) System 301-304
Night Vision Training System (NVTS) 248
NIJ III, IIIA 129, 339-346
NIJ IV 339-346
NIJ-STD-0101.04 118-124, 127

NIJ-STD-0106.01 148-149
NIJ-STD-0108.01 148-149
NIJ-STD-0801.01 127
No. 1 National Freeway Sun Yat-sen 270-271
Noise Suppressor (*see* Suppressor)
Non-Government Organization 462-484
Non-lethal weapons 178
Northrop 452
NTSC Color Video (NTSC = National Television Standards Committee)
Nuclear, Biological, Chemical Warfare (NBC) 151-152, 171-174
NUMA 306-311
nVision Simulated Night Vision Goggle 301-304

O

Occupational Health and Safety Assessment Series (*see* OHSAS)
Octol 71, 72
Oerlikon 74, 75, 188
Off-axis optical system 213-214
Off-Road Vehicle (OPV) 468-484
Offset carrier 312-317
OH-58D Kiowa Helicopter 258-259, 271, 273-276, 375-384, 386-394
OHSAS18000 18
OHSAS18001 21
One In-Cell 22" 348-350
Operational Detachment Alpha (ODA/Green Berets) 453-454
OPS-S-1365 187
OPS-S-1401A 187
Optic-Electro 235
Optical scope 114-115, 222-224
Optoelectronics Business 190-191, 204-207, 213-216
Order of Battle 486-490

P

P-230 Parachute 395-413
P-252 Parachute 395-413
Pager 312-317
Paladin 364-368
Panoramic Vehicle Imaging System 255-257

Parachute 11, 134-146, 182, 363, 395-413
Parachute Protective Boots 182
Parachutist Protective Boots 135
PASGT (*see* Personnel Armor System for Ground Troops)
passive camera system 318-324
Patrol Boats 318-324
Patrol Craft – Large (PLC) 375-384
PCI 306-311
PCMCIA (*see* Personal Computer Memory Card International Association)
PCS 312-317
PDSD (*see* Point-Detonating (PD), Self-Destroying (SD))
pencil beam 283-288
Penghu 414-422
People's Liberation Army (PLA) 363
Personal Computer Memory Card International Association (PCMCIA) 312-317
Personnel Armor System for Ground Troops (PASGT) of the US Army 129
Phase-Lock-Loop (PLL) 312-317
Phased Array Radar 283-288
Pingtung 356, 359, 395-413, 433-446
Plastic Bonded Explosives 182
Pneumatic Rope Thrower 175-177
Pneumatic System 306-311
Pocock, Chris 453-454
Point Defense Radar 283-288
Point Detonating (PD) Fuze 242-243, 246
Point Detonating Delay 246
Point-Detonating (PD), Self-Destroying (SD)/PDSD 240
Point-Initiating/Self-Destruct (PIBD/SD) Grenade Fuze for High Velocity – High Explosive Dual Purpose Cartridge (HV-HEDP) 236-237
point-to-point data link 312-317
Polaris Off-Road MRZR-4 468-484
police (*see* Law Enforcement) 184-186
Political Warfare Academy 453-454
Political Warfare Special Task Force (PWSTF) 453-454
polyester 253
Polymer Matrix Composite Material 254
Portable Pneumatic Rope Thrower 11
Post Video Processing (PVP) 301-304
PRC-37A (PRC = Portable, Radio, Communication) 289-294, 312-317
Prefabricated High Explosive (PFHE) 111
Primary Image Fabrino 306-311
Propellant 258-259
Prostitution 452
protective clothing 121-122

protective gloves 341-346
Psychological Warfare (PW) 453-454
PT2 Platform 318-324
pubs/bars 452
Pulse Doppler Radar 235

Q

Quick Reaction Company (MP) 363

R

radar 112-113, 230, 232, 235, 253, 279-280, 283-400
Radar Scattering Camouflage Net 253
Radar Section Electronic System Division 289-294
Radiated Scan 283-288
radio 281-294, 312-317
radome 254
rail mounting 206-209
raincoat 150
range finder 277-278
Ranger 570 468-484
Ray Ting 2000 MLRS 375-384
RDX 71, 72, 182
reconnaissance 226, 258-259, 271, 273-276, 318-328, 453-454
red dot reflex rifle scope 190-191, 341-346
REDICE 306-311
Reflective Memory Bus 301-304
Reflex Collimator Sight LED (Light Emitting Diode) 191
relay transmission system 325-328
Remington M24 Snipe Rifle (SWS) 58-59
Remington Rifles 54-59
Remote Controlled Grenade Launcher System 12
Remote Weapons Stations 114-115
Remote. Interface. Unit (*see* RIU)
Rende District 271
rifle barrel 60-61
Rifle Bolt Sensor 348-350
Rifle scope 341-346
Rifle Shooting Simulator System 45, 182

RIU system Structure 306-311
Robotic Vehicles 26
rocket 182, 222-224, 258-259, 276-278
rocket propellant 182
rope thrower 11, 175-177
RS-170 Monitor 318-324
RTAI 306-311

S

sabot 77
Saigon 453-454
salary 2-4
satellite phone 281-282
SC/PRC-37A 312-317
School of the Americas 453-454
Science and Technology Exhibition Center 362
Sea Dragon Frogmen 414-422
Seabob Dive Jet 347
SEAL Teams 414-422
Search Radar 295-300
Security Force Assistance 453-454
Selected Ion Monitoring (SIM) 230
Self-Destruct (SD) Grenade Fuze for High Velocity – High Explosive Dual Purpose Cartridge (HV-HEDP) 236-237
Self-destruct mechanism 237, 240, 241
Semiwadcutter (SWC) 148-149
Servo Motors 225-226
SG1 0300 Workstation Computer 301-304
SHI Jian-zhi, specialist 361
shoes 16
Shooting Simulation System 221
short take-off and landing (*STOL) 325-328*
short-range automated defense weapon system 225-226
Shoushan Camp 433-446
Shovel 147
Shower 151-152, 171-173
Shrink Small-Outline Package (*see* SSOP)
sight and bore – axis of a mortar 103-104
sighting-axis 103-104
signal flare 11, 153-156
signal processing 234, 294-300, 348-350

signal transmission distance 279-280
silencer (*see* Suppressor)
silicon carbide 341-346
silver 15
SIM Mode 230
simulation 221, 248-250, 281-282, 301-304, 306-311, 338, 348-352
Six-Axis 351-352
Skulls 442-447
Skydiving Demonstration Team 144-146
sleeping bag 182
Smart Phone 281-282
Smoke grenade 244-245
Smokeless Powder (SPDN) 97-98
SMP 306-311
SMT (*see* Surface Mount Technology)
Snepp, Frank 453-454
Sniper 8, 9, 32, 58-59, 60-67, 182, 359
Sniper School 65
SOIC-8 312-317 (SOIC = Surface Mount Integrated Circuit)
Spatial Disorientation Simulator 248
SPDF (*see* Flashless Powder) 97-98
SPDN (*see* Smokeless Powder) 97-98
Special Coating Materials 182
Special Combat and Assault Vehicle (SC-09A) 468-484
Special Forces 61, 63-65, 131-132, 138-146, 260-270, 305, 318-328, 363, 375-384, 395-422, 433-447, 453-454, 462-484 (*see also* Army Aviation and Special Forces Command)
Special Reconnaissance 453-454
Special Service Company 433-446
Special Task Force 455-457
Special Weapons and Tactics (SWAT) 462-467
Sports shoes 16
Spring Memorial Services 363
SQ (delayed action of 0.027 to 0.063 s) Fuze 242-243
SS-102 Leather Shoes 182
SS-95 formal shoes 16
SS-99 Leather Shoes 182
SSOP-28 (SSOP = Shrink Small-Outline Package) 312-317
Standard Missile (SM) 283-288
Star-Shaped Center Hole Extruded Double Base Propellant Grain 258-259
Stealth 253-254
Stewart hydraulic pneumatic 351-352
Stinger missile 359
SU Ren-bao, Col. 361

Sub-Calibre Training System for Mortars 13
Submarine mast 279-280
Suppressor 61, 62
Surface Mount Integrated Circuit (*see* SOIC)
Surface Mount Technology (SMT) 312-317
surface proximity 246
Surveillance 201-203, 226, 283-288, 318-328, 363
Surveying Equipment 20
SWAT (*see* Special Weapons and Tactics)
SWC (*see* Semiwadcutter)
System and Weapon Control Console 295-300

T

T10B Round Parachute 140-141, 395-413
T10R Reserve Round Parachute 142-143, 395-413
T3-75 Gas Mask 163-166
T3-92/94 Gas Mask 163-164, 167-168, 169-170
T39 Submachine Gun 29, 31
T3K1 Sniper Rifle 8
T4-102 Decontaminated Solution 17
T4-103 Light Duty Decontamination System 173-174
T4-86 NBC Shower Equipment 171-172
T4-98 Field Movable Shower 151-152
T41 Machine Gun 29, 31
T51 Pistol 0.45 (Model 1911A1) 29, 41
T57 Machine Gun 29, 31
T63 Mortar Shell (D-1-D) 94, 182
T65K1/K2 Rifle (AR-15) 29, 31, 46-47, 182, 187
T74/75 Platoon Machine Gun (FN MAG General Purpose Machine Gun/FN Minimi) 29, 31, 82, 114-115, 182, 187, 195-197, 240, 225-226, 395-413, 468-484
T75 Pistol (Beretta 92) 29, 31, 33-34, 182, 183
T77 Submachine Gun 29, 31, 182-183
T85 Assault Rifle/Grenade Launcher 29, 31, 52-53, 83-84, 157-158, 182, 187
T86 Assault Rifle 29, 31, 46-47, 94, 187
T90 Heavy Machine Gun (Browning M2E2) 29, 32, 77, 79, 182
T91 Grenade Launcher 9, 29, 32, 46-47, 52-57, 88-89, 182, 187, 348-350, 395-413, 468-484,
T91 Shooting Training Simulator 348-350
T92 Air Defense Gun 75, 108-113
T93 Sniper Rifle 29, 32, 58-59, 74, 79, 182
T97 Pistol (Beretta) 38-40

tactical rail 9, 131-132
Tactical Reconnaissance Group 325-238
Taichung 386-394
Tainan 271
Taipei 433-446
Taipei After Dark 452
Taiping 325-328
Taiwan Defense Industry Development Association 353
Taiwan Marine Corp 223-224
Taiwan Military – Law Enforcement Tactical Research and Development Association (TTRDA) 462-467
Taiwan Shooting Sports Association/Ding Fwu 462-467
Taiwan Strait 270-271, 460-461
Taiwan Strait Crisis 460-461
Tally (*see* Target In Sight)
Tank 77, 98, 100, 248, 255-257, 306-311, 351-352, 364-373, 375-384
Taoyuan 453-454
Target Acquisition System (TAS) 227-231
Target In Sight (Tally) 265-270
Target Infrared Flare 8
Target Smoke 8
target surveillance 283-288
targeting systems 318-324
TC-1 (*Tien Chien*) Missile 227-231, 289-300
TC100 Sub-Calibre Mortar Training Simulation System 95-96
TC103 81mm HE Mortar Round Extended Range (ER) 90-91
TC63 Mortar 107
TC69 Mortar 107
TC74 187
TC75 Bullet 183
TC79 187
TC86 HE 40mm Grenade 84
TC88 78, 79, 85, 188
TC88 HE Dual Purpose (HEDP) 85
TC88 HE Incendiary Tracer (HEI-T) 188
TC89 Marking Cartridge 184-186
TC90 HE Dual Purpose (HEDP) grenade 86
TC91 Hand Signal Flare 155-156
TC91 HE Dual Purpose (HEDP) 87
TC92 76, 77
TC94 7.62mm 8
TC96 Smoke Canister 157-158
TC97 97-98
TCXO (*see* Thermally Compensated Crystal Oscillator)

telescope 131-132, 201-203
tent 150-152, 182
Terrain Communication Obstacles 279-280
TFT LCD (*see* Thin Film Transistor Liquid Crystal Display)
TH-67 Creek Helicopter 248, 301-305
thermal camera 102, 114-115, 253, 277-280
Thermally Compensated Crystal Oscillator (TCXO) 312-317
Thin Film Transistor Liquid Crystal Display (TFT LCD) Monitors 179-180
Three Dimensional (*see* 3D)
Thunder 2000 MLRS 375-384
Time Fuze 244-245
TNT 71, 72, 94, 182, 395-413
torture 453-454
Tourison, Sedgwick 453-454
Towed Targets 295-300
TP-T M490 projectiles for 105mm M69 Cannon 182
tracer 188
track radar 289-300
Track While Scan (*TWS*) 235
Training 13, 45, 65, 82, 95-96, 221, 247-248, 301-304, 306-311, 337-338, 348-350, 395-413, 433-446
Training and Simulator System Research Division 301-304
Trammel Launcher 178
transmit/receiver (T/R) module 312-317
tripod 103-104
TS-103 Mortar Rectification Instrument (MRI) 103-104
TS-67 (M53A1) 107
TS-83A1 Night Vision Goggle 192-194
TS-84 Night Vision Scope 195-203
TS-91B Rifle Sight 8, 204-205
TS-93 Night Vision Weapon 8, 206-207
TS-95 Sniper Sight 9, 208-209
TS-96 Night Vision 14, 210-219
Tsoying 423-432
Tsoying Naval Base 414-422
TTRDA (*see* Taiwan Military – Law Enforcement Tactical Research and Development Association)
tungsten balls 100
Turnkey 301-304
TW-DIDA 353
TWS (*see* Track While Scan)
Type 18650 battery 103-104
Type 97 15
Type Oerlikon KGB Cannon 74, 75

typhoons (*see* natural disasters)

U

UH-1H Huey Helicopter 375-384, 386-394
UH-60M Black Hawk 386-394
UHMWPE (*see* ultra-high-molecular-weight polyethylene)
ultra-high-molecular-weight polyethylene (UHMWPE, UHMW) 125-126
Ultra-Low Insertion Loss 254
Unconventional Warfare 453-454
Unit Training Device (UTD) 247-248
Unmanned Aerial Vehicle (Drone) 136-137, 232, 234, 325-335, 359, 375-384
Unrestricted Warfare 363
Up/down converter chips 312-317
Upper Receiver T-91 54-57
Urban Warfare 363
US 6,911,413 B2 161-162
US 7,130,439 B2 169-170
US Army 129, 447-451
US Central Intelligence Agency (CIA) 453-454
US Congress 453-454
US defense attaché 447-451
US defense contractors 452
US Embassy (*see also* American Institute in Taiwan) 453-454
US Marine Corp 447-451, 468-484
US Military Assistant Advisory Group (MAAG) 447-451
US National Institute of Justice (*see* NIJ)
US Navy 414-422, 447-451, 460-461
US Navy SEAL 414-422
US Special Operations Command 468-484
US-Taiwan Business Council 459-460
US-Taiwan Defense Industry Conference (2016) 353, 459-460
USS Bunker Hill 460-461
USS Independence 460-461

V

V50 ballistic test 129, 131-132, 148-149, 339-340
Vehicle-mounted Mobile Air/Surface Fire Control Radar System 235
Veterans of Foreign Wars (VFW) 452

Veterans of the Army 101 ARB Association 414-422
Video Cassette Recorder 318-324
Video Life Detector 179-180
Vietnam War 453-454, 468-484
Virtual Reality 337-338
Virtual Training System 45
VR Military/Police Battle Field Training 337
VTF MK 117 NSD 97-98

W

Walkie-Talkie 281-282
Waterbag 123-124
Waterproof 16, 133
Weightlifter 363
Wireless communication 279-280, 283-288, 312-324
Wireless communication transmit/receiver (T/R) module 312-317
Wireless LAN 312-317
Workstation Computer 301-304
World War II 447-451

X

X-band (36 Channel) 283-288
X-band Pulse Doppler Radar and Gun Control 235
X98 Semi-Auto Sniper Rifle 8
XT104 Submachine Gun 29, 32, 41-42
XT105 Multi-Utilization Special Rifle 29, 32, 43-44
XTC100 Illumination Flare – Parachute 11
XTP-101 (T-75) single-barrel for Navy Platform or Land Mobile Platform 225-226
XTR-102 (T-75) dual gun for Fixed Station 225-226
XTS-102 Mortar Rectification Instrument 13-14

Y

YAN Pei-wen, Major 363
Yangmei 325-328

Z

Made in the USA
Las Vegas, NV
29 May 2022